An Introduction to Text-to-Speech Synthesis

Text, Speech and Language Technology

VOLUME 3

The titles published in this series are listed at the end of this volume.

An Introduction to Text-to-Speech Synthesis

by

Thierry Dutoit
Faculté Polytechnique de Mons,
Mons, Belgium

KLUWER ACADEMIC PUBLISHERS
DORDRECHT / BOSTON / LONDON

A C.I.P. Catalogue record for this book is available from the Library of Congress.

ISBN 0-7923-4498-7

Published by Kluwer Academic Publishers,
P.O. Box, 3300 AA Dordrecht, The Netherlands.

Kluwer Academic Publishers incorporates
the publishing programmes of
D. Reidel, Martinus Nijhoff, Dr. W. Junk and MTP Press.

Sold and distributed in the U.S.A. and Canada
by Kluwer Academic Publishers,
101 Philip Drive, Norwell, MA 02061, U.S.A.

In all other countries, sold and distributed
by Kluwer Academic Publishers,
P.O. Box 322, 3300 AH Dordrecht, The Netherlands.

Printed on acid-free paper

02-0998-200 ts

Printed in the Netherlands.

à Alice et Catherine

Contents

List of Figures ... xiii
Foreword ... xix
Preface ... xxi
Acknowledgments ... xxv

Chapter One: Introduction ... 1
1.1. What is speech made of? ... 1
1.1.1. The acoustic level ... 2
1.1.2. The phonetic level ... 5
1.1.2.1. Vocal fold vibration : physiology and acoustics ... 5
1.1.2.2. The international phonetic alphabet ... 6
1.1.2.3. Articulatory phonetics ... 7
1.1.3. The phonological level ... 8
1.1.4. The morphological level ... 10
1.1.5. The syntactic level ... 11
1.1.6. The semantic level ... 12
1.1.7. The pragmatic (or discourse) level ... 13
1.2. What is a TTS system? ... 13
1.3. How do *we* read? ... 14
1.3.1. The reading process ... 15
1.3.2. Seeing ... 16
1.3.3 Thinking ... 20
1.3.4. Saying ... 23
1.3.5. Hearing ... 24
1.4. Yet another speech synthesizer? ... 26
1.5. Automatic reading: what for? ... 30
References ... 32

PART ONE
FROM TEXT TO ITS NARROW PHONETIC TRANSCRIPTION

Chapter Two: Grammars, Inference, Parsing and Transduction ... 37
2.1. Basic concepts and terminology ... 38
2.2. Regular grammars (Chomsky type 3) ... 40
2.2.1. Definition ... 40
2.2.2. Use ... 42
2.2.3. Regular inference ... 44
2.2.4. Regular parsing ... 46
2.3. Context-free grammars (Chomsky type 2) ... 47
2.3.1. Definition ... 47
2.3.2. Use ... 48
2.3.3. Context-free inference ... 49
2.3.4. Context-free parsing ... 49
2.4. Extensions of context-free grammars ... 50

2.5. Lexicons, feature structures, and the PATR notation 52
2.6. Summary 55
References 55

Chapter Three: NLP Architectures for TTS Synthesis 57
3.1. Data formalisms 58
3.2. Rule formalisms 62
3.2.1. MLDSs and compiled multilevel rewriting rules 64
3.2.2. FSs and bottom-up chart-parsed DCGs 66
3.3. Summary 69
References 69

Chapter Four: Morpho-Syntactic Analysis 71
4.1. Preprocessing 73
4.1.1. Text segmentation 73
4.1.2. Sentence end detection 74
4.1.3. Dealing with abbreviations 75
4.1.4. Recognizing acronyms 75
4.1.5. Processing numbers 76
4.1.6. Dealing with idioms 77
4.2. Morphological analysis 77
4.2.1. Function words 77
4.2.2. Content words 78
4.2.2.1. Inflection 80
4.2.2.2. Compounding 82
4.2.3. Computational aspects 84
4.2.3.1. Organizing lexicons into efficient data structures. 84
4.2.3.2. Indexing lexeme and suffix groups 85
4.2.3.3. Unconstraining analysis 86
4.3. Contextual analysis 87
4.3.1. *N*-grams 89
4.3.2. Neural networks as taggers 93
4.3.3. Local nonstochastic grammars 95
4.4. Syntactic-prosodic parsing 100
4.5. Summary 101
References 102

Chapter Five: Automatic Phonetization 105
5.1. From text to phonemes: a long way 105
5.2. Two basic strategies 111
5.3. The morphophonemic module 113
5.4. The LTS transducer 115
5.4.1. Pronunciation treatises 117
5.4.2. Expert rule-based systems 118
5.4.3. Trained rule-based systems 120
5.4.4. Neural networks 123

5.5. Phonetic postprocessing 123
5.6. Proper names 125
5.7. Summary 126
References 127

Chapter Six: Automatic Prosody Generation 129
6.1. What is prosody? 129
6.2. Levels of representation of prosodic phenomena 130
6.3. Major components of prosody 131
6.4. The meanings of prosody 132
6.5. Intonation models 133
6.5.1. Acoustic models of intonation 134
6.5.1.1. Fujisaki's model 135
6.5.1.2. Acoustic stylization methods 136
6.5.2. Perceptual models of intonation 137
6.5.2.1. Automatic perceptual stylization 137
6.5.2.2. The IPO model of intonation 138
6.5.3. Linguistic models of intonation 139
6.5.3.1. Pitch contour theory 140
6.5.3.2. Tone sequence theory 142
6.6. Relationships between prosody and other aspects of speech 145
6.6.1. Lexicon and prosody 145
6.6.2. Syntax and prosody 145
6.6.3. Semantics, pragmatics and prosody 146
6.7. Syntactic-prosodic parsing 147
6.7.1. Hand-derived heuristics 149
6.7.2. Grammar-based systems 152
6.7.3. Automatic, corpus-based methods 155
6.8. Sentential stress assignment 160
6.9. From symbolic to acoustic representation of prosody 162
6.9.1. Generating timing 162
6.9.1.1. Duration "units" 163
6.9.1.2. Duration models and parameter estimation 164
6.9.2. Generating fundamental frequency 165
6.9.2.1. Generating *F0* with Fujisaki's model 166
6.9.2.2. Generating *F0* as sequences of stylized contours 166
6.9.2.3. Generating *F0* through sequences of tones 168
6.10. Summary 169
References 170

PART TWO
FROM NARROW PHONETIC TRANSCRIPTION TO SPEECH

Chapter Seven: Synthesis Strategies ... 177
7.1. Rule-based synthesizers ... 178
7.2. Concatenation-based synthesizers ... 180
7.2.1. Database preparation ... 180
7.2.2. Speech synthesis ... 183
7.2.3. Segmental quality ... 186
7.2.3.1. The choice of segments ... 187
7.2.3.2. The corpus ... 190
7.2.3.3. Segmentation ... 191
7.2.3.4. The model ... 192
7.2.3.5. The parametric speech coder ... 194
7.2.3.6. Prosody matching ... 194
7.2.3.7. Concatenation ... 195
7.3. Quality assessment ... 195
7.4. Summary ... 198
References ... 198

Chapter Eight: Linear Prediction Synthesis ... 201
8.1. The autoregressive (AR) model ... 201
8.2. A mathematical framework for linear prediction analysis ... 203
8.2.1. The linear prediction problem ... 203
8.2.2. The Yule-Walker equations ... 204
8.2.3. Covariance versus autocorrelation ... 205
8.2.4. A criterion for the choice of σ ... 206
8.2.5. The covariance method, with the Grahm-Schmidt algorithm ... 207
8.2.6. The autocorrelation method, with the lattice algorithm ... 210
8.2.7. The autocorrelation method, with the Levinson, Schur, and Split algorithms ... 212
8.2.8. Line spectrum pairs ... 213
8.3. Database compression ... 213
8.4. Prosody matching with the AR model ... 214
8.5. Segment concatenation ... 214
8.6. Speech synthesis ... 217
8.7. Segmental quality ... 217
8.8. Advanced production models ... 220
8.9. Glottal inverse filtering ... 221
8.9.1. The glottal autoregressive (GAR) model ... 221
8.9.2. Analysis algorithms ... 224
8.9.3. Segmental quality : further research issues ... 225
8.10. Conclusions ... 226
References ... 226

Chapter Nine: Hybrid Harmonic / Stochastic Synthesis 229
9.1. Hybrid models 230
9.2. Hybrid analysis 233
    9.2.1. Spectral analysis of speech 233
    9.2.2. Approximation criteria 235
9.3. Database compression 237
9.4. Prosody matching with the H/S model 238
9.5. Segment concatenation 239
9.6. Hybrid synthesis 241
9.7. Segmental quality 246
9.8. Summary 248
References 248

Chapter Ten: Time-Domain Algorithms 251
10.1. The TD-PSOLA "model" 253
10.2. Database compression 253
10.3. Prosody matching 254
10.4. Speech synthesis 256
10.5. Segmental quality with TD-PSOLA 257
    10.5.1 Phase mismatch 257
    10.5.2 Pitch mismatch 259
    10.5.3 Spectral envelope mismatch 259
10.6. Resynthesizing the segment database 261
    10.6.1. The resynthesis process 261
    10.6.2. Modified PSOLA synthesis 265
    10.6.3. Compression of the new segment database 266
    10.6.4. Segmental quality 266
10.7. Combining PSOLA with a parametric synthesizer 267
10.8 Conclusions 268
References 269

Chapter Eleven: Conclusions and Perspectives 271
11.1. Synopsis 271
    11.1.1. Natural language processing and TTS synthesis 271
    11.1.2. Digital signal processing and TTS synthesis 274
11.2. Prospects 277
References 279

Index 281

List of Figures

1.1. Block diagram of a typical speech recording system 2

1.2. Time waveforms and short-term Fourier transforms. 3

1.3. Narrow-band (top) and wide-band (bottom) spectrograms and time waveform of the utterance *Alice's adventures,* sampled at 16 kHz.. 4

1.4. Cross-section of the vocal apparatus (R. Boite, M. Kunt, *Traitement de la Parole*, Complément au traité d'électricité, Fig 1.1, p. 3, reproduced by permission of Presses Polytechniques et Universitaires Romandes; copyright © 1987, Presses Polytechniques et Universitaires Romandes) 6

1.5. Larynx cross-section as viewed from the top (R. Boite, M. Kunt, *Traitement de la Parole*, Complément au traité d'électricité, Fig 1.2.a, p. 3, reproduced by permission of Presses Polytechniques et Universitaires Romandes; copyright © 1987, Presses Polytechniques et Universitaires Romandes) 6

1.6. An example of voicing assimilation.. 9

1.7. A simple but general functional diagram of a TTS system. 14

1.8. A schematic data flow diagram of the oral reading process. 16

1.9. Cross-sectional view of a human eye (E.K. Kandel, J.H. Schwartz, *Principles of Neural Science*, 2nd edition, Fig. 28.2, reproduced by permission of Appleton & Lange, copyright © 1985, Appleton & Lange) 16

1.10. Each half of the visual field is processed by a separate hemisphere of the brain (E.K. Kandel, J.H. Schwartz, *Principles of Neural Science*, 2nd edition, Fig. 28.10, reproduced by permission of Appleton & Lange, copyright © 1985, Appleton & Lange) 17

1.11. A simplified view of the parallel distributed processing approach of letter perception: the neighbors of the letter *T* in the first position of a word (Rumelhart, McClelland, and the PDP research group (eds.), "Parallel Distrubuted Processing", vol. 1, pp. 3-40, reproduced by permission of MIT Press, Cambridge, MA; copyright © 1988, MIT Press) 19

1.12. A bilateral cooperative view of word recognition (M.M. Taylor, "Convenient Viewing and Normal Reading", in *Working Models of Human Perception*, B.A.G. Elsendoorn, H. Bouma, eds., Fig. 5, p. 303, reproduced by permission of Academic Press Limited; copyright ©1989 Academic Press Limited) 20

1.13. The Wernicke-Geschwind model for reading aloud (E.K. Kandel, J.H. Schwartz, *Principles of Neural Science*, 2nd edition, Fig. 52.1, reproduced by permission of Appleton & Lange, copyright © 1985, Appleton & Lange) 21

1.14. The auditory system (R. Boite, M. Kunt, *Traitement de la Parole*, Complément au traité d'électricité, Fig 1.7, p. 8, reproduced by permission of Presses Polytechniques et Universitaires Romandes, Lausanne; copyright ©1987, Presses Polytechniques et Universitaires Romandes) 24

1.15. Left - Isosonic curves in open field. Right - Auditory masking by a narrow band noise. 25

1.16. Wolfgang von Kempelen's talking machine (R. Lingaard, *Electronic Synthesis of Speech*, Fig. 1.1, p. 6; reproduced by permission of Cambridge University Press; copyright ©1985, Cambridge University Press) 28

1.17. Block-diagram of Dudley's voder. 29

2.1. A simple finite-state transition network that describes integer and decimal numbers in their full form and the related regular rules. 40

2.2. A simple Markov chain, that accounts for what the author of this material would modestly like the opinion of his readers about himself to be. 41

2.3. Some part of a finite-state transducer, which partially accounts for the phonological transcription of grapheme *c* in French.. 43

2.4. Left: A multilayer perceptron, with a single hidden layer. Right: The *i*th neuron of the *l*th layer. 44

2.5. Left: The maximal canonical grammar corresponding to the training database {aaa, aab, aba}. Right: A simpler grammar, obtained after having efficiently merged some states in the previous grammar. 45

2.6. Two laughing machines that obviously account for the same giggles (Gazdar, G., and C. Mellish, *Natural Language Processing in PROLOG*, Fig. 2.1 and 2.2, pp. 23 & 26, reprinted by permission of Addison Wesley Longman Ltd, copyright © G. Gazdar, C. Mellish 1989) 47

2.7. A simple recursive transition network which accounts for French sentences like *the pretty little black cat of my grand-mother drinks milk in the kitchen* 48

2.8. This tree cannot be described in terms of context-free rewriting rules. 49

2.9. An example of DAG and attribute value matrix representations of the feature structure for the word *eats*. 53

2.10. The graph unification process 54

3.1. The exponential complexity of the parsing problem: the more you know, the harder it is to know more 58

3.2. An old natural language processing strategy for TTS synthesis: linear exchange structures between sequentially organized processing modules 59

3.3. An example of a feature structure (depicted as a tree in this case, i.e. a DAG with no symbolic sharing of values) 60

3.4. An example of a multi-level data structure (MLDS) (H.C. van Leeuwen, and E. te Lindert, "Speech Maker : a General Framework for Text-to-Speech Synthesis, and its Application to Dutch", *Computer, Speech and Language*, vol. 7, n°2, 1993, Fig. 2, p.153, reproduced by permission of Academic Press Limited; copyright ©1993 Academic Press Limited) 61

3.5. MLDSs and FSs theoretically allow serial, hierarchical, or heterarchical scheduling. ... 62

3.6. The NLP module of a typical text-to-speech conversion system ... 63

4.1. A typical morpho-syntactic analyzer. ... 72

4.2. Describing acronyms with finite state automata. ... 76

4.3. The morphological structure of German words ... 82

4.4. Bigram for the sentence *Dogs like to bark* (J. Kupiec, "Robust Part-of-Speech Tagging Using a Hidden Markov Model", *Computer, Speech and Language*, vol. 6, n°3, 1992, fig. 1, p. 228, reproduced by permission of Academic Press Limited; copyright ©1993 Academic Press Limited) ... 92

4.5. A three-layer perceptron for part-of-speech disambiguation ... 94

5.1. Dictionary-based (top) versus rule-based (bottom) phonetization ... 112

5.2. Left: the phoneme HMM used in Van Coile (1991). Right: the simpler HMM model of Van Coile (1993). ... 120

5.3. Retrieval of the pronunciation of *a* in *behave* by trie search ... 122

6.1. Different kinds of information provided by intonation ... 132

6.2. A low-level acoustic description of the prosody of the French utterance: *Les techniques de traitement numérique de la parole...* ... 134

6.3. Fujisaki's production model of intonation ... 135

6.4. Declination lines obtained from an acoustic analysis ... 136

6.5. A straight-line acoustic stylization of the example of Fig. 6.2. ... 137

6.6. Delattre's ten fundamental intonations, embedded in a short dialogue ... 141

6.7. Finite-state grammar for H/L tone sequences ... 142

6.8. Automatic analysis of the intonation of the Dutch utterance *als je goed je best doet, zul je vast wel slagen* in terms of a tone sequence ... 144

6.9. A FSA for the simple but efficient chinks 'n chunks algorithm ... 150

6.10. Deriving prosodic phrases from syntactic ones, with more or less success ... 154

6.11. Transitions allowed between two states of a simple Markov chain accounting for the presence/absence of prosodic boundaries within a sentence ... 156

6.12. A yes/no decision tree for predicting prosodic boundaries in texts, using text-based information alone (Wang, M.Q., and J. Hirschberg, "Predicting intonational boudaries automatically from text : The ATIS domain", Proc. Speech and Natural Language Workshop, 1991, pp. 378-383 : Fig. 2, reproduced by permission of the authors) ... 158

7.1 A typical rule-based synthesizer ... 178

7.2 A general concatenation-based synthesizer ... 181

7.3. Average instantaneous power per phoneme in a non-equalized segment database for French 183

7.4. Example of a linear time alignment function for a segment with two sub-segments (e.g., a diphone). 185

7.5. Inside the segment concatenation block. 185

7.6. Parametric linear smoothing at the border of successive segments 186

7.7. An example of the ML-COC cluster splitting process for occurrences of the phoneme /e/ 189

7.8. Coarticulation affects the realization of the [ʃ] in [iʃe] and [uʃö] 191

7.9. Copy synthesis. 193

7.10. A general vocoder. 194

7.11. Comparing the discriminative power of intelligibility tests 197

8.1 The classical autoregressive model of speech production 202

8.2. Prediction vectors. 204

8.3. The forward prediction error vector $\mathbf{f_p}$ is the orthogonal component of $\mathbf{s_p}$ on the prediction subspace (a: general order p; b: N=3, p=2). 205

8.4. Forward prediction error vectors $\mathbf{f_m}$. 207

8.5. Backward prediction error vectors $\mathbf{g_m}$. 208

8.6. Building $\mathbf{f_2}$ from $\mathbf{f_1}$ and $\mathbf{g_1}$. 209

8.7. Prediction and PARCOR coefficients are different expressions of the same decomposition process. 210

8.8. The lattice inverse filter. 212

8.9. Linear interpolation of filter parameters 215

8.10. Roots of A(z) for diphones /ɔv/ (o) and /vɛ/ (∗)., for p=18 216

8.11. The lattice synthesis filter. 217

8.12. Extrinsic modeling errors with the covariance method 218

8.13. Wrong detection of formant frequencies and bandwidths when analyzing consonant [n] 219

8.14. The glottal autoregressive (GAR) speech production model 222

8.15. The glottal volume velocity waveform of Fujisaki and Ljungqvist (H. Fujisaki, M. Ljungkvist, "Proposal and Evaluation of Models for the Glottal Source Waveform", Proceedings of ICASSP 86, Tokyo, pp. 1605-1608; Fig. 2; reproduced by permission of the IEEE; copyright ©1986, IEEE.) 223

8.16. A (very schematic) geometrical interpretation of the GIF problem 223

8.17. Classical resolution of the GIF equations 224

9.1. Amplitude spectrum of a realization of the vowel [z] 231

9.2. The hybrid harmonic/stochastic model. 232

9.3. The effect of the analysis window length on the STFT of a periodic signal 234

9.4. Linear smoothing of harmonic amplitudes with the hybrid H/S model 240

9.5. Overlapping synthesis frames in the OLA approach. 244

10.1. The TD-PSOLA reharmonization process 252

10.2. Amplitude spectra of OLA frames extracted from the French vowel [a], for several values of F_R 254

10.3. Pitch and timing modifications with TD-PSOLA 255

10.4. Phase mismatch. 258

10.5. Pitch marks in diphone [an] are supposed to fall on the first negative peak of each period 259

10.6. Pitch mismatch 260

10.7. Spectral envelope mismatch 261

10.8. The MBE resynthesis operation of MBR-PSOLA. 262

10.9. The time-domain linear smoothing process 264

10.10. A spectrograph showing the effect of the time-domain linear smoothing process 265

Foreword

The field of speech synthesis has seen a large increase in commercial applications in the last ten years. As recently as 1986, there were only a few companies in the synthesis market, all exploiting one of two basic technologies—either formant-based phonemic synthesis or LPC-based diphone synthesis. While these approaches still form the basis of most text-to-speech products, new simpler waveform techniques have recently been developed, and improvements have been made in the older techniques.

Recent progress has been largely motivated by three factors: (1) the rapid increase of the ability of computers to perform tasks more rapidly, with lower cost, and more cheaply, (2) a large increase in the number of widely available text and speech databases, and (3) improvements in speech recognition and synthesis technology. For the first, the current ubiquity of speech in personal computers was difficult to foresee a decade ago. For both recognition and synthesis, faster and cheaper computers have been a major factor in the growth of speech applications.

Secondly, it has only been very recently that standard databases on CD-ROM have become widely available. As in automatic speech recognition, technological progress comes more rapidly when many research and development groups have simultaneous access to the same pertinent information. It has been difficult to model well the natural human processes of speech production and perception. Earlier synthesis researchers often relied on their own intuition and personal knowledge to develop so-called 'expert-system' (artificial intelligence) techniques, to simulate natural speech production. With the advent of relevant databases (both of speech - to better model the acoustics of the vocal tract - and text - to better understand the extraction of relevant information from text for speech synthesis), it has been convenient to examine much more speech (and more varied pronunciations from different speakers) than just a few years ago.

Synthesis applications have been significantly accelerated by the recent availability of practical speech recognizers. Many applications for speech require both synthesis and recognition, e.g., dialogues over the telephone, data entry. Thus the recent capability to control machines via voice (both accurately and inexpensively) has led to more use of synthesis as well.

Last and not least, there have been significant improvements in ways to do speech synthesis. These have led to considerably more natural-sounding computer speech. Earlier models yielded intelligible speech, whose quality was clearly inferior to that of human speech. Many commercial products still use these basic methods, and will gradually adapt their systems to take advantage of the ideas presented in this excellent book.

In the area of speech synthesis, there are no other current books that I can recommend for a good and comprehensive overview of the field. This text should remain a standard in the field of speech synthesis for years. The author is an expert in the field, as evidenced by his doctoral studies in speech synthesis and by the wide range of relevant topics he covers. The technical problems of speech synthesis are handled well in appropriate technical and mathematical detail. From the point of view of a speech researcher, this is exactly what he is looking for, to get a good understanding of speech synthesis. This book will serve well for technical people working in the field of speech processing, as well as for managerial people supervising the production of products and services related to speech synthesis.

Dutoit has succeeded well in his objective, to provide a comprehensive introduction to TTS (text-to-speech) for both engineers and linguists developing TTS or trying to grasp the wide range of details needed for TTS. It is impressive and unusual to find an author who can write well for an audience of both linguists and engineers. Far more often, linguists (even computational linguists) shy away from much mathematical and quantitative analysis, while engineers often ignore anything that cannot be expressed in mathematics or in programs. The technical literature is filled with treatises from either the linguistic OR engineering perspective, but rarely from someone who treats both domains equally and well. This book covers many of the improvements that have occurred in linguistic and speech processing since the development of MITalk in the late 1970s.

Finally, the European author brings a fresh and general view to a field that is often dominated by Americans (with our too restrictive view as English as the sole useful language). While the text concentrates on English, it often includes interesting details for synthesis of other languages, which reinforce the author's ideas.

Douglas O'Shaughnessy

Preface

The model of the engineering world that I learned in school was a simple one. An engineer is working at his desk when suddenly a bolt of pure inspiration strikes. "Eureka!" he cries, grabbing a pen to begin writing a seminal paper in the new field. The patent is granted by return mail, and a special issue of the Proceedings of the IEEE *has his picture on the cover. Companies race to produce the widget described in his paper, and within months he is being interviewed on "Lifestyles of the Rich and Famous." He lives happily ever after.* (...) *To say that some great invention coalesced out of primordial soup through a random instanciation of chaos theory is unappealing. I wouldn't say that myself in public, but sometimes late at night I wonder. This primordial soup is really pretty powerful stuff.*

R.W. Lucky, *IEEE Spectrum,* November 1994, p. 15

Audience

The aim of this textbook is to give a comprehensive introduction to text-to-speech (TTS) synthesis for those, engineers or linguists, who are trying to develop a complete TTS system or simply for people trying to understand what TTS synthesis is about. As a matter of fact, since very few people associate a good knowledge of signal processing with a comprehensive insight into language processing, synthesis mostly remains unclear. Both areas are investigated here in a progressive way, guiding the reader through the many possible solutions they provide to TTS synthesis problems, and trying to answer the questions he/she might have asked him/herself. Theoretical and practical issues are developed and compared, so as to highlight the constraints that have to be kept in mind when designing TTS systems. The book is not really self-contained. Some understanding of natural language processing and, more importantly, of digital signal processing will help. I have tried to provide readers with a top view of the problem, leaving some algorithmic details as black boxes to be uncovered in further readings. As such, the book also provides a structured presentation of the many papers published these last years in many areas related to TTS synthesis, which might also be useful to more experienced researchers.

Contents

Each and every synthesizer is a subtle, and more or less successful, combination of digital signal (DSP) and natural language processing (NLP). The particular DSP algorithms and NLP formalisms that each one exploits lead to typical synthetic speech features, addressing its segmental and suprasegmental quality. In a word: its intelligibility and naturalness. The book is therefore divided into two major parts: Part One is devoted to the NLP problems involved, while Part Two focuses on the DSP techniques that can deliver the expected high-quality synthetic speech, with a special emphasis on the so-called concatenative approach (as opposed to the rule-based approach, a very complete description of which is given in [J. ALLEN, S. HUNNICUT, D. KLATT, *From Text To Speech, The MITALK System*, Cambridge University Press, 1987]).

I had the feeling that quite a long introduction was necessary, given my personal experience of the frequent misunderstandings that occur when speech synthesis is under the limelight. **Chapter One** answers five fundamental questions related to speech analysis and synthesis, from linguistic, physiological, historical, and economical points of view. It provides a functional description of the human reading mechanism that progressively introduces the underlying complexity of TTS synthesis itself.

Part One begins with an introduction to formal language theory, automata, and NLP formalisms, in **Chapter Two**. Starting from Chomsky's hierarchy, regular and context-free grammars are shown to have initiated many helpful extensions, from stochastic grammars to unification-based formalisms. Coverage, parsing, and inference are examined in each case. **Chapter Three** examines the overall organization of NLP modules in a TTS synthesizer. It covers their internal data structure, whether linear, multilevel, or based on feature structures, and points out the related most relevant rule formalisms. **Chapter Four** describes the morpho-syntactic analyzer of most recent TTS systems as composed of a pre-processor, a morphological analyzer, a contextual analyzer, and a syntactic-prosodic parser. Each component is functionally described and available solutions are browsed. **Chapter Five** focuses on grapheme-to-phoneme conversion. After some examples showing why it does not functionally reduce to lexical database search, rule- and dictionary-based strategies are exposed. Finally, prosody generation, which currently remains the most intricate human faculty to imitate, is focused on in **Chapter Six.** The problem is set and shown to originate in shortcomings in the low-level (phonetic), medium-level (phonological), and higher-level (syntactic, semantic and pragmatic) modeling of rhythm and intonation. Existing tradeoffs are reviewed.

In **Part Two**, **Chapter Seven** introduces the digital signal processing module of a TTS synthesizer, in the form a general block-diagram description resulting from a series of choices that originate in technological and human constraints. Rule-based and concatenation-based synthesis strategies are exposed. Their capability to produce high-quality (HQ) speech is extensively debated. Subsequent chapters instanciate the functional blocks previously introduced in Chapter Seven, to some candidate algorithms for HQ synthesis. More specifically, the pros and cons of the autoregressive, hybrid harmonic/stochastic, TD-PSOLA,[1] LPC-PSOLA, MBR-PSOLA models in TTS synthesis are extensively discussed. Qualitative and quantitative results are presented on the basis of a real implementation of all models in a TTS system. A unified approach has been adopted, that reinforces differences and should help the reader in his comparison task. **Chapter Eight** reviews the very classical LPC synthesizer, often taken as the base quality for TTS systems using concatenation. The linear prediction framework is simply summarized through its geometrical interpretation, and its efficiency in a TTS system is discussed. It is concluded by an introduction to the first HQ candidate: the glottal inverse filtering algorithm. Its equations are geometrically interpreted and related to the LPC ones, and computation methods are derived. **Chapter Nine** focuses on the highly accurate but computationally intensive hybrid harmonic/stochastic models, under the banner of the well-known MBE one. Efficient segment concatenation, prosody matching,

---

[1] PSOLA/TD® is a Registered Trademark of France Telecom.

and synthesis algorithms are derived, and the ability of hybrid models to produce natural sounding speech is investigated. **Chapter Ten** switches to the currently most widely used time-domain techniques. After a comprehensive analysis of the pros and cons of TD-PSOLA, possible extensions or modifications are exposed, as the MBR-PSOLA (multiband resynthesis) or LPC-PSOLA methods, which combine the computational efficiency of the original algorithm with the flexibility of the MBE and LPC models, respectively.

The textbook is concluded by a necessary synopsis of the main ideas exposed in the previous chapters. Research perspectives are finally derived.

Languages

Although the problems addressed in the first part of this textbook, which is devoted to natural language processing, are typically language dependent, we have given most examples for English or French and some for German. As far as processing strategies and algorithms are concerned, however, we believe that most of our conclusions can be drawn, *mutatis mutandis,* from many other languages (at least for European ones). Part II deals with signal processing issues and is therefore much less sensitive to language peculiarities.

Acknowledgments

I would first like to thank the Faculté Polytechnique de Mons (FPMs) for its financial support throughout the writing of this book.

Many individuals have also contributed to its release. Jean Véronis and Nancy Ide (and indirectly Daniel Hirst) have partly initiated this work and supported it at Kluwer Academic Publishers. Piet Mertens provided critical advice for the first part of this book and participated in the elaboration of its plan. He also contributed to Chapter Six, which owes much to his experience on prosody modeling. Many thanks to Christophe d'Alessandro, Douglas O'Shaughnessy, and to my collegues at AT&T research labs, Jont Allen, Mark Beutnagel, Alistair Conkie, Juergen Schroeter, and Yannis Stylianou, for having critically reviewed the book and suggested many improvements (special thanks to Douglas for his foreword). Many thanks to Vincent Pagel, too, for his intensive programming and testing of my synthesizer and for all sorts of fruitful discussions (not to forget Celine Egea, for her incredible pitch). I am indebted to Veronique Aubergé, Paul Bagshaw, Gerard Bailly, Frederic Beaugendre, Olivier Boeffard, Hervé Bourlard, Françoise Emerard, S. Frenkenberger, Kjell Gustavson, Julia Hirschberg, Richard Horne, Volker Kraft, Mats Ljungqvist, and David Yarowski for having kindly provided some helpful information on their work and on related issues, and to Yves Laprie, whose Snorri software has made it possible to produce high-quality spectrograms. To a larger extent, I am also greatly indebted to the primordial soup of R.W. Lucky.

Without René Boite, Henri Leich, and Joel Hancq, this work would simply not be. I dare say they committed me as a scientist. I also do not forget all the members of our Circuit Theory and Signal Processing Laboratory, whose good humor and agreeable natures contributes greatly to the tranquillity of our working place.

Although modern writers are armed with increasingly powerful tools, such as automatic spelling checkers, thesauruses, online dictionaries, grammar correctors, and even translators, Nancy Dutoit, Julian Beever, Béatrice Pothier, and the anonymous copy-editor contracted by Kluwer have definitely convinced me that one cannot reasonably spare human proofreaders. I am more grateful to their several hundred billion neurons than my non-native English can express.

I cannot conclude these acknowledgments without thanking my wife, Catherine, who could have lots to complain about this book. After all, it was not always pleasant competing with a speech synthesizer! She knows how I appreciate her understanding. Many thanks to my mother, sisters, and grandparents, too, for having always supported me.

Thierry Dutoit

Chapter One

INTRODUCTION

Alice was beginning to get very tired of sitting by her sister on the bank, and of having nothing to do: once or twice she had peeped into the book her sister was reading, but it had no pictures or conversations in it, "and what is the use of a book," thought Alice "without pictures or conversation?"

Lewis Carroll, *Alice's Adventures in Wonderland*

Language is the ability to express one's thoughts by means of a set of signs, whether graphical (as with the Roman, Cyrillic, or Arabic alphabets, or the Chinese ideograms), gestual (like the sign language for the deaf-mute), acoustic (as with speech; one often refers to *acoustic gestures*), or even musical (like the whistle language used by some Pyrenean shepherds). It is a distinctive feature of human beings, who are the only creatures to use such a structured system. *Speech* is one of its main components. It is by far the oldest means of communication between people and it is also the most widely used. No wonder, then, that people have extensively studied it and often tried to build machines to handle it in an automatic way. In this introductory chapter, we shall try to answer briefly some of the many questions that arise when we first confront the problem of synthesizing this rather complex signal.

1.1. What is speech made of?[1]

The information conveyed by speech can be analyzed in many ways. Speech scientists generally distinguish several nonmutually exclusive levels of description—namely, the *acoustic, phonetic, phonological, morphological, syntactic, semantic,* and *pragmatic* ones. As we shall see, they are related to the transformations that are needed to establish a bridge between the two surface forms that occupy a central position in this book: speech and text.

---

[1]Of course, it is impossible, in an introduction to speech synthesis, to give a complete answer to this question. The following pages are merely aimed at presenting partial but self-contained material for a background explanation. Many points will be further developed in subsequent chapters.

1.1.1. The acoustic level

Speech physically appears as a variation of the air pressure, caused and emitted by the articulatory system. Acoustic phoneticians[2] study our speech by transforming it into an electrical signal with the adequate transducer: a microphone. In modern recording systems, the resulting electrical signal is then digitized : it is successively low-pass filtered,[3] sampled, and quantized (Fig. 1.1). It may then be submitted to various digital signal processing operations, so as to highlight its acoustic traits: *fundamental frequency* (often denoted as *F0*), *intensity*, and *spectral energy distribution*. Each acoustic trait is itself related to a perceptual quantity: *pitch*, *loudness*, and *timbre*.

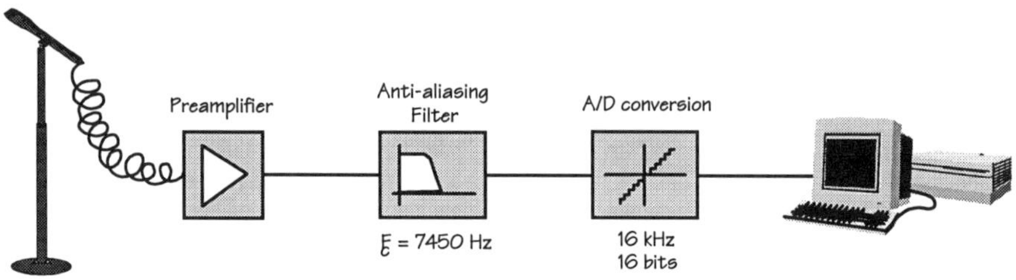

Fig. 1.1. Block-diagram of a typical speech recording system (sampling frequency = 16kHz in this case).

Figures 1.2 and 1.3 show several views of a given speech signal, each of which uncovers in its own way some of these acoustic characteristics.

The time waveform (bottom of Fig 1.3) reveals an alternation of silences together with fairly periodic segments or noisy ones, respectively denoted as *voiced* or *unvoiced* segments. Inside voiced portions, the fundamental frequency evolves slowly with time. It most often ranges from 70 to 200 Hz for men, from 150 to 400 Hz for women, and from 200 to 600 Hz for children. In contrast, the intensity of the signal may vary abruptly, even within voiced portions.

A simple computation of the *short-term Fourier transform* (STFT) uncovers its spectral contents. It is typically implemented as the Discrete Fourier Transform (DFT) of the digitized signal *x(n)* weighted with a finite length window *w(n)* (typically 30 ms long, over which one can reasonably assume that the signal remains stationary given the inertia of the articulatory system):

---

[2]In the sequel, we have presented the acoustic and phonetic levels as if they were independent, although, strictly speaking, the acoustic aspects of speech are also within the competence of a particular component of phonetics, called *acoustic phonetics* (the other ones being *articulatory* and *auditory* phonetics). Articulatory phonetics will be introduced in Section 1.1.2.

[3]Following the *sampling theorem*, signals should be sampled with a sampling frequency chosen to be at least twice their highest frequency component (see, for instance, Oppenheim & Schafer, 1975). They should be low-pass filtered before sampling otherwise. This is the case for speech synthesis, which uses sampling frequencies from 8 to 20 kHz while speech sounds still have important spectral components over 10 kHz.

$$x(n) \leftrightarrow X(n,k) = \sum_{i=0}^{N-1} x(i)\,w(n-i)\,e^{-jki\frac{2\pi}{N}}, \quad \text{for } k = 0,\ldots,N\text{-}1 \tag{1.1}$$

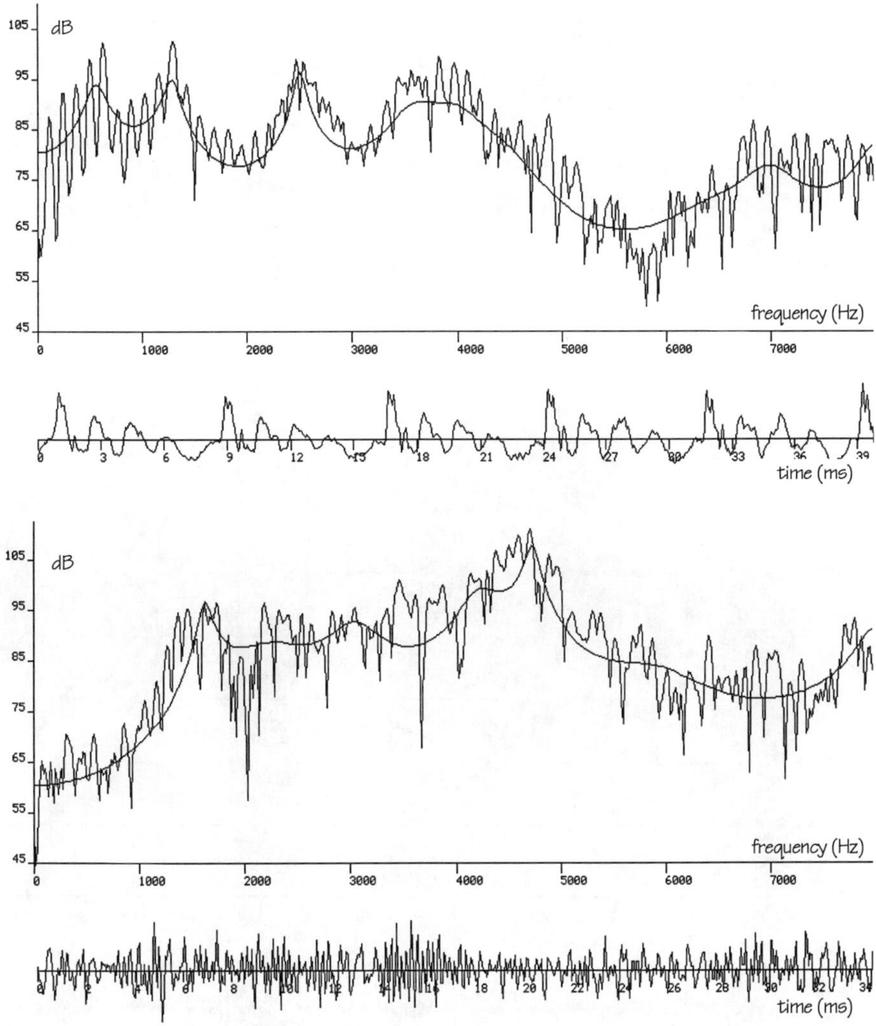

Fig. 1.2. Time waveforms and short-Term Fourier transforms (together with spectral envelopes) of the **[a]** (top) and of the **[ʃ]** (bottom) of the French word *baluchon*, using a 30 ms long Hamming weighting window.

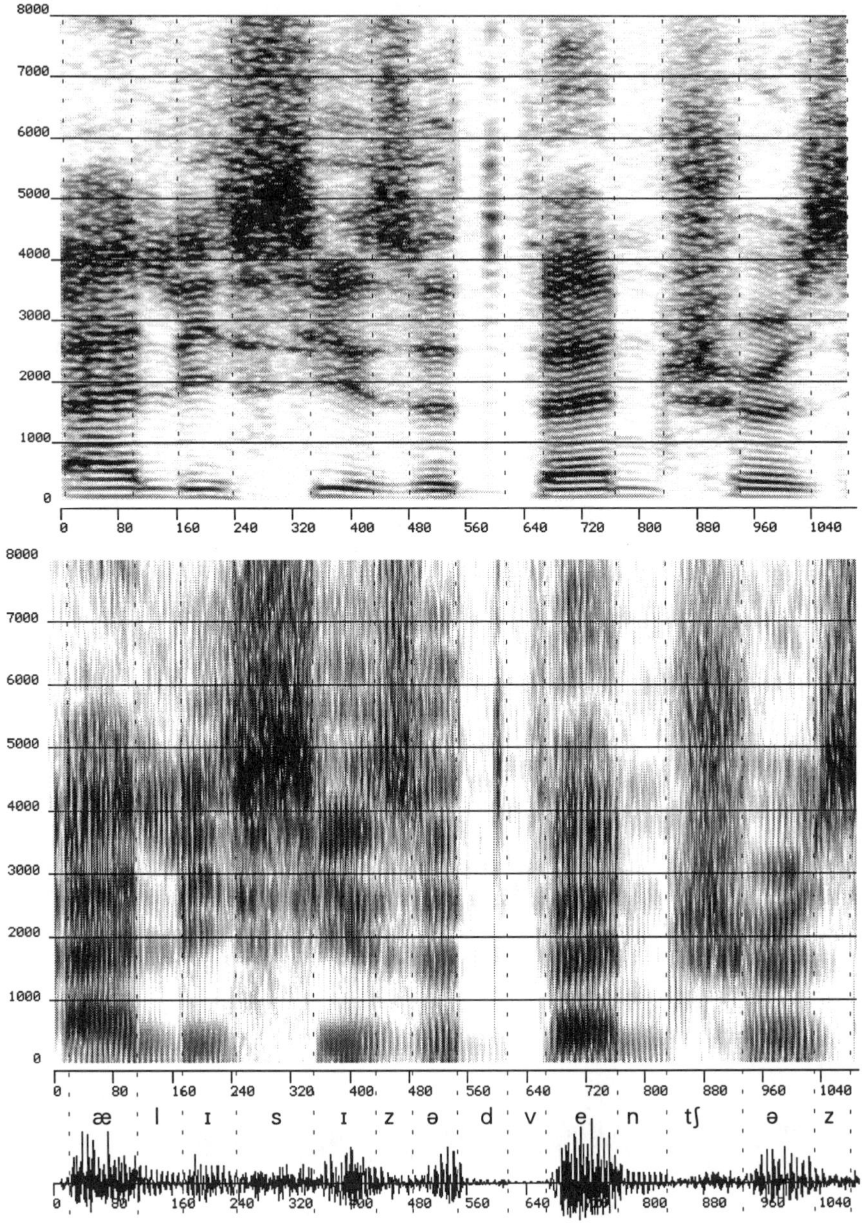

Fig. 1.3. Narrow-band (top) and wide-band (bottom) spectrograms and time waveform of the utterance *Alice's adventures*, sampled at 16 kHz. Hamming weighting windows have been used—respectively, 10 and 30 ms long.

Voiced parts appear as narrow spectral peaks, the center frequencies of which share a harmonicity relationship with the fundamental frequency, while the very nature of unvoiced spectra is stochastic. The overall spectral shape of both types of sounds, however, called the *spectral envelope*, exhibits broad peaks and valleys, called *formants* and *antiformants*. The temporal evolution of their center frequencies and bandwidths determine the *timbre* of the related sound. It turns out that most voiced sounds have a low-pass spectrum, with about one formant per kilohertz of bandwidth, of which only the first three or four have a real phonetic value. In contrast, unvoiced sounds generally appear as high-pass spectra.

Finally, the *spectrogram* gives a time-frequency representation of the signal by showing the temporal evolution of its short-term spectral amplitude as grey levels in a two-dimensional plot. Spectrograms are termed as narrow- or wide-band, depending the duration of the weighting window $w(n)$. Wide-band spectrograms are obtained with short windows (typically 10 ms). Voiced periods appear as sequences of dark vertical strips (which originate in the fact that the energy of the signal covered by the weighting window is a function of its position within the pitch period). Wide-band spectrograms highlight the spectral envelope of the signal and therefore make it possible to sketch the temporal evolution of formants. They are therefore an invaluable tool for acousticians and phoneticians. Experts can even *read* spectrograms—that is, recover the utterance from its time-frequency representation.[4] Narrow-band spectrograms are less frequently used. They rather put the fine spectral structure in evidence: harmonics in voiced sections appear as horizontal striations.

1.1.2. The phonetic level

Phoneticians also have a continued interest in studying the way speech signals are produced by the articulatory apparatus, presented in Fig. 1.4.

1.1.2.1. Vocal fold vibration: physiology and acoustics

Indeed, speech can be described as the result of the coordinated action of a number of muscles. The respiratory organs provide the energy needed to produce speech sounds, by forcing an air flow in the trachea and through the *vocal cords* (or *folds*). These are actually composed of two contiguous membranes, the tension of which is controlled by neighboring muscles (Fig. 1.5). They provide an aperture in the larynx, called the *glottis*. The air flows unimpeded through it during breathing, whispered speech, and when producing voiceless consonants. Voiced sounds on the contrary originate in the total obstruction of the larynx, which increases the air pressure up-stream and forces the vocal cords to open so as to release the air. The Bernouilli effect then causes a pressure drop, which suffices to close the glottis abruptly, and a cycle is completed, the duration of which depends on the tension of the vocal cords. It ensues that a glottal waveform composed of a sequence of pulses is fed through the vocal cavities—namely, the pharyngeal and oral cavities for most sounds. When the *uvula,* which is the tip of a soft tissue structure called the *velum*, is lowered, the nasal cavity is shunted with the oral one.

---

[4] For an impressive collection of wide-band spectrograms of American English sounds, see Olive *et al* (1993).

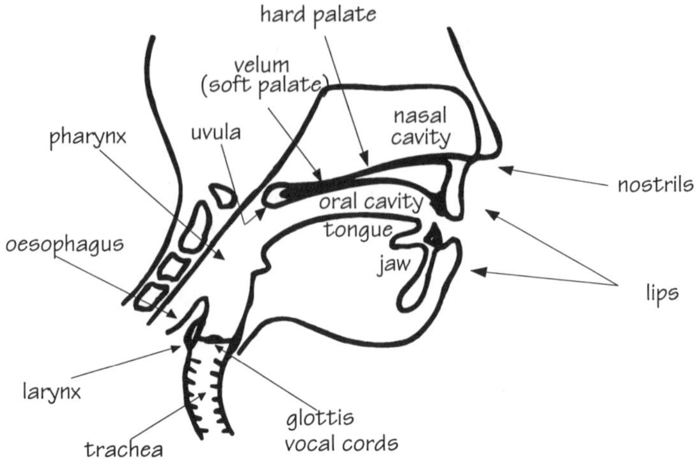

Fig. 1.4. Cross-section of the vocal apparatus (after Boite and Kunt, 1987).

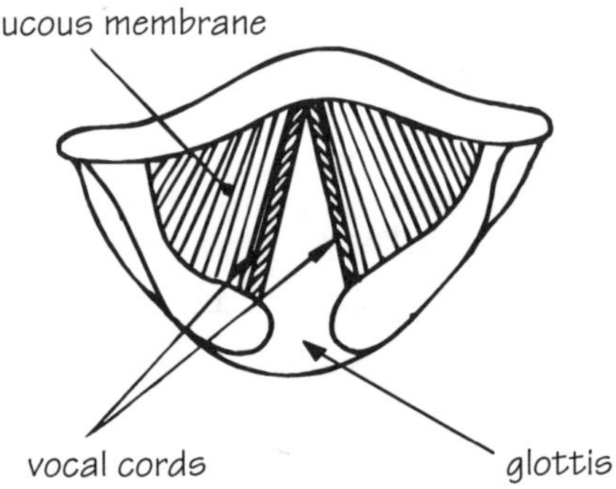

Fig. 1.5. Larynx cross-section as viewed from the top (after Boite and Kunt, 1987).

The acoustic traits of a speech sound are naturally related to its production. Its intensity is related to the air pressure above the larynx. Its pitch, which is simply the frequency of the opening/closure cycle of the vocal folds, is additionally determined by the tension of the glottal muscles. Its timbre originates in the dynamic filtering of the glottal pulses through successive acoustic cavities.

1.1.2.2. The International Phonetic Alphabet

The International Phonetic Alphabet (IPA) associates phonetic symbols to sounds, so that pronunciations can be written in a compact and universal way (see Table 1.1 for English; for a deeper insight, see IPA, 1949).

IPA	EXAMPLES	IPA	EXAMPLES
ɪ	*bit* [bɪt]	p	*pat* [pæt]
i	*bee* [biː]	t	*top* [tɒp]
ɛ	*red* [red]	k	*cut* [kʌt]
æ	*cat* [kæt]	b	*but* [bʌt]
ɑ	*art* [ɑːt]	d	*dig* [dɪg]
ɒ	*lot* [lɒt]	g	*go* [gəʊ]
ɔ	*all* [ɔːl]	θ	*thief* [θiːf]
ʊ	*put* [pʊt]	f	*laugh* [lɑːf]
u	*shoe* [ʃuː]	s	*list* [lɪst]
ʌ	*cut* [kʌt]	ʃ	*dish* [dɪʃ]
ə	*obey* [əbeɪ]	h	*hit* [hɪt]
ɜ	*learn* [lɜːn]	ð	*this* [ðɪs]
ɪə	*beer* [bɪə]	v	*give* [gɪv]
eɪ	*date* [deɪt]	z	*zoo* [zuː]
eə	*bear* [beə]	ʒ	*vision* [vɪʒən]
aɪ	*life* [laɪf]	l	*lot* [lɒt]
aʊ	*house* [haʊs]	r	*rat* [ræt]
ɔɪ	*boil* [bɔɪl]	j	*few* [fjuː]
oʊ	*soap* [soʊp]	w	*wet* [wet]
ʊə	*poor* [pʊə]	tʃ	*check* [tʃek]
m	*map* [mæp]	dʒ	*jam* [dʒæm]
n	*ton* [tɒn]		
ŋ	*ring* [rɪŋ]		

Table 1.1. The IPA symbols used in English.

1.1.2.3. Articulatory phonetics

It is convenient to group speech sounds into *broad phonetic classes*, related to their *manner* of articulation.[5] In English, one generally distinguishes seven groups—namely, *vowels, fricatives, stops, nasals, glides, liquids, diphthongs,* and *affricates.*[6]

Vowels [ɪ, i, e, æ, ɑ, ɒ, ɔ, ʊ, u, ʌ, ə, ɜ][7] differ from consonants in the degree of aperture of the vocal tract (and not, as one might think at first, in the level of activity of the vocal folds, already mentioned above and termed as *voicing*). If the vocal tract is open enough for the air pulsed by the lungs to flow without meeting any obstacle, a vowel is produced. The task of the mouth is then reduced to modifying the vocal timbre. If, on the other hand, the path ever narrows or even temporarily closes, the air flow gives birth to a noise: a consonant is produced. The mouth then fully becomes a speech production organ.

---

[5]It is also possible to group sounds as a function of their *place* of articulation: *labial* [m], *dental* [t], *alveolar* [s], *palatal* [ʃ], *velar* [k], *uvular* [ʀ] (in French), *pharyngeal* [ħ] (in Arabic), or *glottal* [h].

[6]In fact, this classification relates only to the normal phonation type, termed as *modal register* voicing. Other types include *whispering*, *breathy* phonation (the vocal folds are kept partially open), *creaky voice (*the vocal folds firmly tighten, so that speech sounds harsh), and *vocal fry* (the vocal cords are relaxed, which results in a very irregular and low-pitched speech, usually at the end of an utterance).

[7]These are only English vowels, of course. The examples that follow are given for English, too.

In the case of *nasals* [m, n, ŋ], the uvula is lowered, and the nasal tract provides the main transmission channel.

Fricatives originate in a constriction of the vocal tract, either at the glottis, at the hard palate, at the teeth, or at the lips. Voiceless fricatives [θ, f, s, ʃ, h] are produced by a turbulent noise from the glottis; voiced ones [ð, v, z, ʒ] combine periodic and aperiodic components: the vocal folds vibrate but they never really close.

Stops or *plosives* are the most dynamic sounds. They are articulated in three steps: the vocal tract first closes at some point, which results in a build-up of pressure, the release of which produces a transient burst. Voiced plosives [b, d, g] are distinguished from voiceless ones [p, t, k] according to the presence or absence of vibration of the focal folds during the build-up.

Glides and *liquids* combine features from both vowels and consonants. Like vowels, the central position of *glides* [j, w] (or *semi-vowels*, or *semi-consonants)* is quite open at times, but the sudden transition toward and/or away from this position produces a friction that is typical of consonants. *Liquids* [l, r] are vowel-like articulations but use nonvowel tongue shapes to effect partial closure of the oral tract.

Finally, *diphthongs* and *affricates* are combinations of vowels and consonants, respectively. *Diphthongs* [ɪə, eɪ, eə, aɪ, aʊ, ɔɪ, oʊ, ʊə] can be seen as slow and smooth changes between static vowels (as opposed to glides, which evolve more quickly). *Affricates* [tʃ, dʒ] are composed of a stop followed by a fricative, but the duration of their stop is smaller than in a simple concatenation.[8]

1.1.3. The phonological level

Phonology (sometimes called *functional phonetics*) is the necessary interface between phonetics and higher-level linguistics.

In the previous sections, we have described speech on an acoustic or physiological point of view as if it were not to convey any meaning. Speech sounds have therefore been presented independently. Phonology introduces abstract *linguistic* units (as opposed to *speech* units) called *phonemes.* A phoneme is the smallest meaningful contrastive unit in a language. It is thus neither defined on an acoustic nor physiological basis but on a *functional* one. Phonemes have no independent existence: they constitute a structured set of units in which each element is intentionally different from all the other ones. Besides, a list of the phonemes of most European languages has been set up as early as the end of the nineteenth century on the basis of the study of so-called *minimal pairs*, composed of pairs of words sharing all but one sound, which suffices to change their meaning (as for *put/pet*).

Phonetics could thus be better defined as the *study of the articulation of phonemes*. Phonemes can indeed appear in practical speech through a multitude of articulatory forms, called *phones*. These can result from voluntary changes in the articulation of a base sound, as often happens in regional pronunciations. Such variations do not give birth

[8]See Brosnahan and Malmberg (1970) for a more rigorous introduction to phonetics.

to new phonemes, since no semantic information is conveyed. They are called *free allophones*.

Independently of these free modifications, all phonemes are subject to important phonetic variations due to the influence of neighboring sounds on the dynamics of the vocal tract. This leads to *context-dependent allophones.* Articulatory movements can indeed be altered in a given context either to minimize the effort to produce for realizing an articulatory position given the previous one or in contrast to anticipate the next articulatory gesture. These effects are known as *reduction* (or *undershooting*), *assimilation*, and *coarticulation.* Coarticulatory phenomena are due to the fact that each articulator moves continuously from the realization of one phoneme to the next. They appear even in the most careful speech. In contrast, reduction and assimilation phenomena originate in physiological constraints and are sensitive to speaking rate. Assimilation is due to the overlapping of articulatory gestures and may result in the complete modification of a phonetic trait, hence its name (Fig 1.6 is a typical example). Reduction is rather due to the fact that articulatory targets are less likely to be reached in rapid speech than in slow speech. Some phonemes, such as glides, liquids, and stops, are particularly critical to these types of segmental variability.

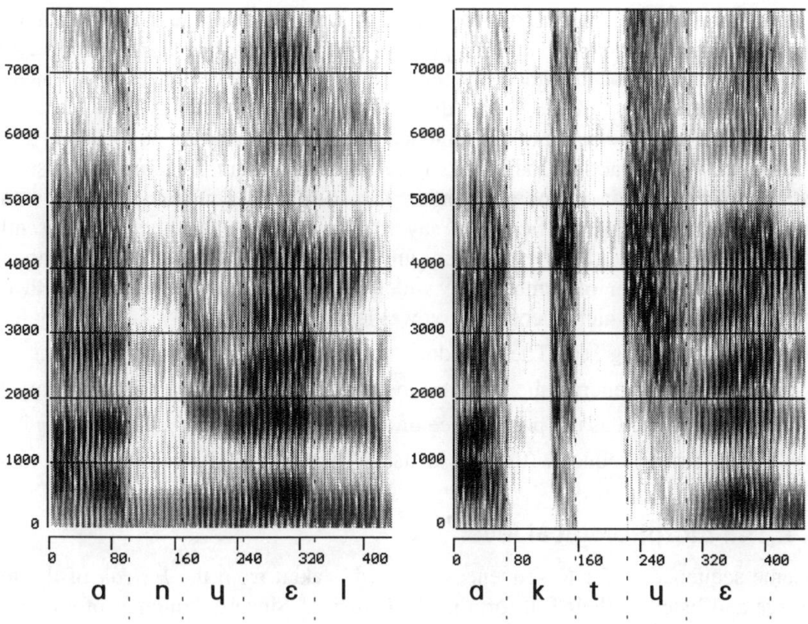

Fig. 1.6 An example of voicing assimilation. Left: the beginning of the French word *annuellement* [anɥɛlmɑ̃], where the French semivowel [ɥ] appears in a voiced context; right: the beginning of *actuellement* [aktɥɛlmɑ̃]: [ɥ] is completely devoiced by the preceding voiceless stop.

Phonemic representations also make use of the IPA, but symbols are surrounded by slanting lines instead of brackets. The distinction between phonetics and phonology clearly appears when trying to align phonemic strings with their phonetic counterparts: allophonic variations and coarticulations are accounted for only in the latter. For example, the French phoneme /ʀ/ can be voiced [ʀ] or not [ʀ̥] (*parent* - /paʀɑ̃/ - [paʀɑ̃], but *pitre* - /pitʀ/ - [pitʀ̥]). In English, /ŋ/ is sometimes pronounced [n], mainly in syllable-final position (*something* - /sʌmθɪŋ/ - [sʌmθɪn]). It is important to understand that phonetic symbols are aimed at transcribing what has actually been said, whether meaningful or not. If a nuance can be perceived, there should be a means to signal it in the phonetic transcription, but it should not necessarily appear in phonological descriptions. Besides, phoneticians generally distinguish two phonetic transcription systems, respectively referred to as *broad* and *narrow* systems. The former incorporates only as much phonetic information as is necessary to distinguish the functioning sounds in a language (therefore approaching phonemics very closely), and the latter incorporates as much more phonetic information as the phonetician desires or as he can distinguish. Narrow phonetic transcriptions separate allophones by adding diacritical marks to phonetic symbols to account for variations of manner and place of articulation (such as the breath mark [̥], the voice mark [̬], or the dental mark [̪]).

Eventually, a phonological description cannot claim to be complete if it does not account for the duration, the intensity, and the pitch of phonemes, insofar as they do bring some information that is not provided by the IPA symbols. These three components are collectively termed as *prosody*. The duration of phones within syllables and that of silences determine the *rhythm* of the sentence; their pitch constitute its *melody*. However, the definition of abstract prosodic units (one could call them *prosodemes*) raises many questions. Since it is a major problem in text-to-speech synthesis, it will be further debated in Chapter Six. Suffice it to say for the moment that there is presently no International Prosodic Alphabet, nor any universal prosodic transcription methodology. Only a limited number of symbols are widely used. Duration is indicated with length marks ([ː] and [ˑ]), which account for vowel lengthening (yet only as a ternary feature: short/half long [ˑ]/long [ː]). The IPA does not, however, provide a convenient way of transcribing pitch or energy movements. Instead, it accounts for *stress* (primary [ˈ] or secondary [ˌ]), defined as the prominence of a syllable, achieved by pronouncing it either louder, longer, or unreduced (that is, in its full form, as if it were not in a phonetic context).

1.1.4. The morphological level

Phoneme sequences refer to sequences of words, taken from the *lexicon* of the related language and listed in their full form in *dictionaries*.[9] Since the number of entries in an

[9]More specifically, not only phonemes can refer to words. In many languages known as *tone languages* (which roughly include Swedish, Chinese, and most African languages), tone (syllabic pitch contour) is lexically distinctive, in the sense that some words in the lexicon are distinguished by their tone alone. In order for dictionaries to account for them, the IPA includes some tone symbols (namely, the rise, fall, high tone, low tone, rise-fall, and fall-rise).

abridged dictionary of most European languages approaches 50,000, one can work out their lexical richness at several hundred thousand.

When studying a language, however, it is striking to notice that the words it is made of, although very numerous, often share some of their spelling, as if they were formed from other smaller words or parts thereof (for example, *image, images, imagine, imagination, imagery, image-maker,* and so on). *Morphology* is the part of linguistics that describes word forms as a function of a reduced set of meaningful units, called *morphemes,* and subsequently separated into *stems* and *affixes* (themselves separated into *prefixes, infixes,* and *suffixes*).

One generally distinguishes *inflectional* morphology, which accounts for morphological features such as gender, number, mode, tense, or person (*image, images*), *derivational* morphology, which studies the construction of words from various syntactic categories from a common stem (*image, imagine, imagination, imagery*) and *compounding* morphology, whose task is to explain how two or more stems can be combined to form a new one (*image+make=image-maker*).

Both the pronunciation, the part of speech category, and the meaning of surface words, can be explained in terms of these morphemic components. Hence the importance of morphology in the context of natural language processing.

1.1.5. The syntactic level

Not all sequences of words from the lexicon of a language result in a correct sentence. Indeed, the list of permissible sentences, although infinite in natural languages, is restricted by their *syntax.* Syntax is not to be confused with the rules that are used to describe it, which are organized into *grammars*. Most words lose their individuality when dealt with by grammatical rules: only their *part of speech category* is examined. These are in turn defined as lists of words that are interchangeable for a given grammar.

For example, the (oversimplified) grammar–

> *sentence = noun group + conjugated verb*
> *noun group = determiner + noun [+ preposition + noun group],*

in which braces denote optional components—disallows the sequence *determiner+conjugated verb* as in *my singed* or *the reads.*

In addition to setting up a border between well- and ill-formed sentences, grammars allow us to describe the hierarchical organization of sentences (by way of declarative rules). The operation of finding the syntactic structure of a sentence with respect to a given grammar is called *parsing*. The resulting information has an appreciable advantage over purely linear lexical descriptions: it highlights the possible internal structure(s) of a sentence. For instance, the sequence of words

> *Time flies like an arrow*

can be aligned with the following syntactic structures:

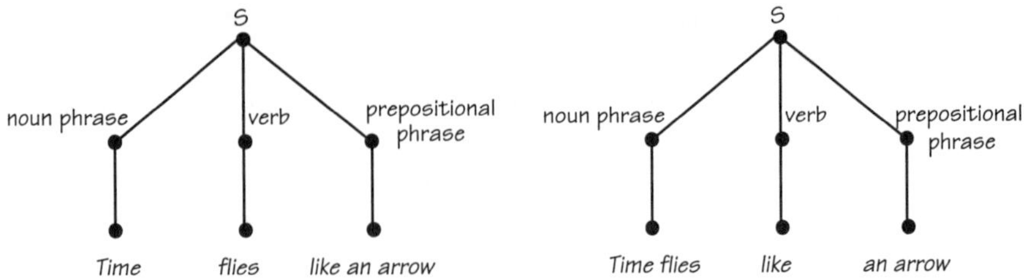

Let us finally recall that it is generally possible to account for the syntax of a language with many different grammars, depending on the part of speech categories, the rule definitions, and the *formalism* chosen for the rules. Traditional grammars are merely one of them. Yet they are not particularly well-suited for a computer implementation, as they assume a prior knowledge and use of the language. The Grevisse grammar, for example, which has been widely used as a reference in primary and secondary Belgian schools, defines the verb as *"the word of the sentence which expresses either the action performed or undergone by the subject, or the union of the attribute to the subject."* In contrast, so-called *formal* grammars have appeared in the early fifties in the context of a new science termed as *computational linguistics* in order to describe the syntax of natural languages in a form that can be directly exploited with computers. It turns out that this objective has never been met, at least not with strictly syntactic grammars.[10]

1.1.6. The semantic level

Although syntax drastically restricts the set of well-formed sentences, it does not constitute an exhaustive criterion for acceptability. Many more sentences should be ruled out, simply because they have no meaning at all (as for *the yellow politeness cries bread*). This basically originates in the confusion that is intentionally made between words belonging to the same part of speech category. The study of word meanings, how they are related to one another, and the bases for lexical choice is the subject of *lexical semantics*.

Among the important questions to be answered by semantic representations, for instance, problems of scope ambiguity loom large. Such a simple sentence as

John did not come to New York by plane.

can be understood as

Someone else did.
John departed from New York by plane.
John went elsewhere.
John came to New York with another means of conveyance.

depending on the scope of *not* (although all these meanings match the same syntax).

[10]Formal grammars, however, are still intensively used to describe computer languages. They have also been extended with semantic features, that make them powerful tools for natural language processing. Last but not least, formal grammars and their stochastic extensions have received a very strong mathematical attention these last twenty years, mainly in the context of the *automatic* processing of linguistic structures. We shall investigate these topics more formally in Chapter Two.

However, when *semantic features* (such as *abstract/concrete*, *animate/inanimate*, color, pattern, shape) are attached to words in the lexicon, the number of classes rapidly increases (and the number of elements in each class decreases) so that words tend to recover their individuality. As expected, the complexity of rules that account for the semantic relationship between words also increases, as does the size of the lexicon. In this regard, it should be noticed that the distinction between syntax and semantics is intrinsically blurred.[11] As a matter of fact, syntactic descriptions are often semantically bearing, as in the example of the previous section. To a larger extent, any syntactic analysis based on an important number of part of speech categories inevitably (although sometimes not explicitly) refers to semantic features.

Semantic grammars and parsers are still being studied by computer linguists. Only partial coverage (i.e. parsers for a restricted semantic area) has been achieved yet.

1.1.7. The pragmatic (or discourse) level

As opposed to semantic meaning, which is often termed as context independent, pragmatic meaning is defined as *context dependent*. Everything that refers to the context and the intentions of the speaker and addressee, has something to do with pragmatics.[12] Its area of interest covers subjects such as *presuppositions, conversational implicatures,* and *indirect speech acts*. It is, unfortunately, much less developed than semantics.

1.2. What is a TTS system?

Intuitively, the ultimate goal of a text-to-speech (TTS) synthesizer is to read *any* text, whether it was directly introduced in the computer by an operator or scanned and submitted to an optical character recognition (OCR) system. Reading should be intelligible and natural.

Let us try to be clear. There is a fundamental difference between the system we are about to discuss here and any other talking machine (as a cassette player, for example) in the sense that we are interested in the automatic production of *new* sentences.

This definition still needs some refinement. Indeed, specific talking machines termed as *voice response systems* produce artificial speech by simply concatenating isolated words or parts of sentences. They are, however, applicable only when a limited vocabulary is required (not more than 100 words, in general) and when the sentences to be pronounced share a very restricted structure, as is the case for the announcement of arrivals in train stations, for instance. In the context of TTS synthesis, it is impossible (and luckily useless) to record and store all the words of the focus language. It is thus more suitable to define text-to-speech as the *production of speech by machines, by way of the automatic phonetization of the sentences to utter.*

---

[11] Strictly speaking, the distinction appears more clearly when one refers to semantics as the analysis of truth conditions for logic propositions. This will not be dealt with here.

[12] To a larger extent, as Gazdar and Mellish (1989) soundly note, it is used "to sweep up all the difficult aspects of meaning that researchers wish to postpone considering". This adds to the difficulty of delineating the limits between semantics and pragmatics.

Figure 1.7 introduces the functional diagram of a general TTS synthesizer. As human reading, it comprises a natural language processing (NLP) module, capable of producing a phonetic transcription of the text to be read, together with the desired intonation and rhythm, and a digital signal processing (DSP) module, which transforms the symbolic information it receives into natural-sounding speech. The particular formalism chosen for the intermediate information varies from one synthesizer to another. In the following, we will term it as *narrow phonetic transcription* to emphasize that it is not composed of a mere sequence of phonemes but rather of a list of allophone names augmented with prosodic information (in a form that will be specified later). It is even more detailed than the standard narrow phonetic transcription mentioned in Section 1.1.3, which provides no normalized means of accounting for prosody.

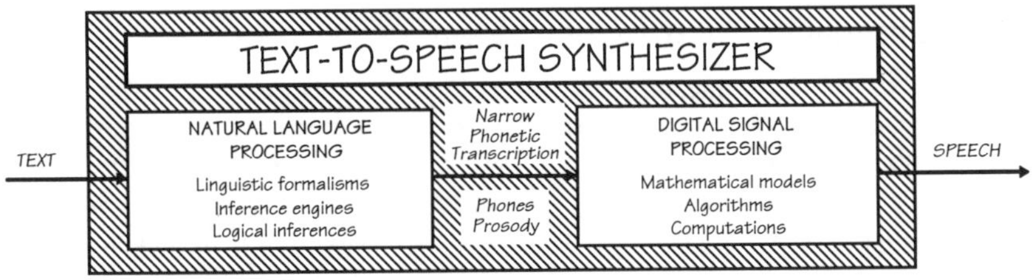

Figure 1.7. A simple but general functional diagram of a TTS system.

From this clear partition of synthesis tasks into symbolic and numerical ones, it is interesting to make a distinction between the segmental and suprasegmental features of a TTS system, which are closely related to it. We shall term as *segmental quality* the efficiency of the machine to produce natural-sounding speech sounds, assuming the NLP module actually delivers high-quality (humanlike) information. It obviously has a lot to do with the ability of the DSP module to account for coarticulation effects. On the other hand, we shall refer to *suprasegmental quality* as the richness of the prosodic contours it is capable to exploit. This is mainly the responsibility of the NLP module. *Naturalness* and *intelligibility*, leading properties in TTS synthesis, clearly depend on both.

1.3. How do *we* read?

To some extent, all of us are eminent linguists, the little but essential difference between us and them being self-consciousness. A quick examination of the many operations that are carried on when we read will now suffice to verify this profound insight.

After all, is not the human being potentially able to correctly pronounce an unknown sentence, even from childhood? We all have, mainly unconsciously, a good knowledge of the reading rules of our native tongue. They were transmitted to us, in a simplified form, in elementary school, and we improve them year after year. However, it would be a bold claim indeed to say that it is only a short step before the computer is likely to equal the human being in that respect. Despite the present state of our knowledge and techniques and the progress recently accomplished in the fields of signal processing and artificial intelligence, we would have to express some reservations. For, as will be seen, the reading process draws from the furthest depths, often unthought of, of the human intelligence.

Let us bow to the facts: there is still a long way to HAL, the brilliant talking computer of *2001, A Space Odyssey*.

1.3.1. The reading process

Physiologically speaking, modern science has a deep knowledge of the way the external organs that are exploited during the act of oral reading work. One could hardly maintain this assertion as far as the neurobiological mechanisms that allow us to read are concerned: the higher we go in the organization of our reading faculties, the less we know about them.

As a matter of fact, even though the functioning of isolated neurons has been extensively studied (a recapitulation of which is given in the impressive volume edited by Kandel and Schwartz, 1985), we have very little information on their overall organization: we are in the same situation as a scientist standing in front of a machine with hundreds of billions of well-known elementary processing units, connected in a very intricate way, and wondering how it effectively works and why it occasionally breaks down.[13] Yet the input and output signals of all units are basically identical: sequences of pulses everywhere.

As far as reading is concerned, our principal means of investigating the rules that govern the processing of information in our brain are general experiments. Specialists try to interpret them into facts about the *functional* way the sensory information provided by our eyes is treated to command the articulatory muscles. These range from purely logical deductions about every day facts, to interpretations of medical experiments on the brain, through considerations on the hypothetical origin of some well-known speech disorders.

A reasonable description of the reading process therefore unmistakably calls for functional aspects as well as physiological ones, as in Fig. 1.8, which introduces a simple data flow diagram of the act of reading. The image is seized by the sensory neurons of our eyes and transmitted in the form of electrical stimuli to our gray matter, where it is processed to command the motor neurons responsible for the correct activation of our lungs, vocal folds, and articulatory muscles. This results in the production of speech, permanently monitored by the brain, mainly through the organs of hearing, in order to adjust in real time the configuration of the vocal tract, together with the energy of the uttered signal and the intonation applied to it. In the following, we shall briefly examine, as a first introduction to the complexity of the reading process, the respective contributions of the organs of reading, from a physiological point of view, as well as the functional part played by the brain for each of its connections to the outer world.

---

[13]The extraordinary potential of our brain is revealed in a striking way by its own deficiencies, excesses, or dysfunctionings. Readers interested in an impressive description of such neurological anomalies should refer to Sacks (1987).

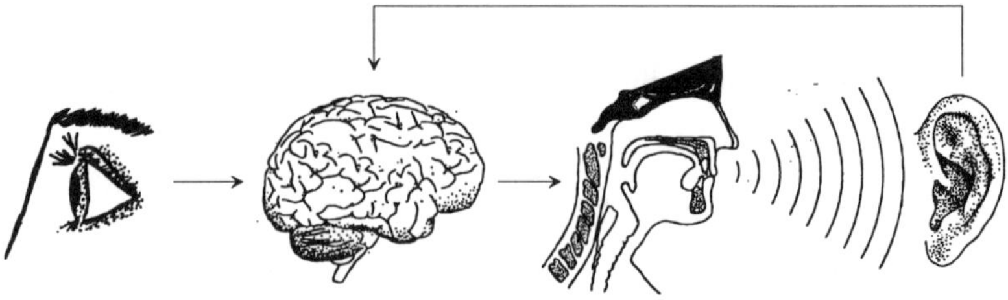

Fig. 1.8. A schematic data flow diagram of the oral reading process (adapted from Holmes, 1988).

1.3.2. Seeing

Figure 1.9 presents the schematic cross-section of an eye. The image is captured by the *pupil,* through the opening adjusted by the *iris*, inverted by the *lens*, and cast on the *retina*. The human retina contains two types of photoreceptors—namely *cones* and *rods*. Cones are the cells that detect form and color and are responsible for day vision; rods enable night vision. The maximum concentration of cones on the retina is encountered in a small area called the *fovea*, which therefore has the greatest spatial resolution. The sensory information collected by the cones is grouped in the *optic disc*, or *blind spot*, a region of the retina that has no photoreceptors (hence its name), and transmitted to the visual cortex through the *optic nerve*.

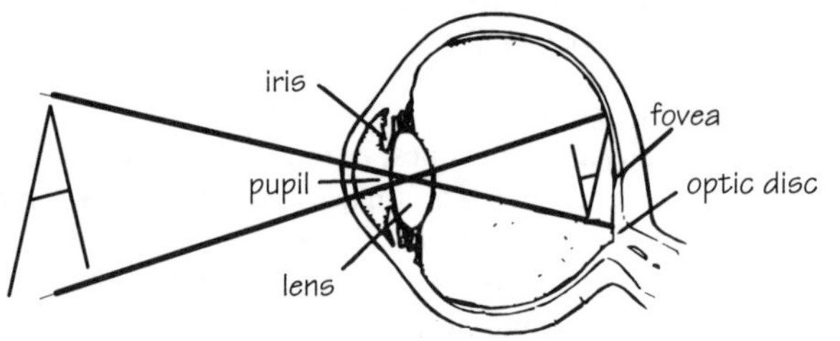

Fig. 1.9. Cross-sectional view of a human eye (after E.K. Kandel, J.H. Schwartz, *Principles of Neural Science*, 2nd edition, reproduced by permission of Appleton & Lange, copyright © 1985, Appleton & Lange).

It is often wrongly believed that the sensory information seized by each eye is processed by the contralateral hemisphere of the brain. What actually happens is that the neurons in each retina are separated in two groups, which respectively constitute the *temporal* and *nasal* hemiretinas. Both are projected to a different hemisphere, in a region called the *lateral geniculate nucleus*, in such a way that only the nasal pathways cross (Fig. 1.10). Thus, it is more adequate to say that each half of the brain is concerned with the contralateral *hemifield of vision*.

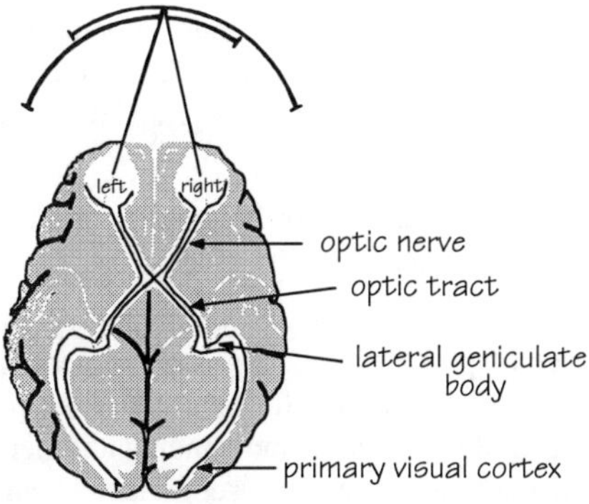

Fig. 1.10. Each half of the visual field is processed by a separate hemisphere of the brain (after E.K. Kandel, J.H. Schwartz, *Principles of Neural Science*, 2nd edition, reproduced by permission of Appleton & Lange, copyright © 1985, Appleton & Lange).

The lateral geniculate nucleus has been shown to enhance the antagonisms between the center and the surround of the vision field. It sends its results to the gray matter, in a region called the *primary visual cortex* in the occipital region of the brain, where they are further analyzed so as to decompose the visual world into short line segments of various orientations (to be used for the discrimination of form and movement) and combine the output from the two eyes (to be used for depth perception). Perception itself is elaborated only in the higher-order visual cortices, of which much less is known.

Information in each area of the visual cortex is clearly processed in a *distributed* way (as opposed to a *centralized* one). As a matter of fact, the cortex is composed of six layers organized into vertical columns, so that neighboring columns in a given area do the same process on data originating from neighboring regions in the previous area. The overall organization of processing has been shown to be a combination of *hierarchical* and *distributed* computing (which are basically antagonist designs). Hierarchy is highlighted when observing cell activation from area to area as a function of the visual signal presented to the eye. At each level, each cell sees more than do the cells at a lower level, and higher cells have a greater capacity for abstraction. Complex cells, for example, survey the activity of groups of simple cells, which in turn survey the activity of groups of geniculate cells, and so on. On the other hand, parallelism originates in the fact that different types of information are sent to different areas of the brain, where they are processed simultaneously. More than six maps of the visual field have been detected in the area that directly follows the primary visual cortex, for example.

This has led many researchers to conceptualize *perception*[14] itself as a distributed feature. Cells in many areas would code for a particular aspect of the stimulus: shape,

[14]This should not be confused with *sensation*. The latter has a clear physical mapping: sensory information is provided by corresponding *sensory neurons* (as the cells on the retina), while "perception neurons" cannot be adequately defined. Perception, indeed, is mostly an intuitive concept.

depth, movement, and color. Only the simultaneous activation of specific cells in these areas would result in a given perception. One often refers to the analogy with the individual silver halide grains of a photograph: the grains themselves do not represent the photograph of a face, but the ensemble does.

Similar conclusions can be obtained from functional observations and simulations, independently of any physiological data.

Among the useful observations in favor of the parallel processing hypothesis, there is clinical evidence of a series of partial losses of visual perception due to localized brain damage termed as *visual agnosias*. Patients, for example, lose *independently* their ability to perceive colors, or forms and patterns, or movements and depths. Hence the idea that different aspects of visual perception must be separately localized and simultaneously examined.

As early as the 1980s, the notion of *parallel distributed processing* received a large consent among the scientific community, after some impressive demonstrations of its powerful capabilities were given in many areas, and in the context of letter perception in particular, given the relative simplicity of the input data (two-dimensional stimuli, composed of binary coded pixels). McClelland and Rumelhart (1981), especially, have shown in a computer simulation that many experimental findings on the role of context on the perception of letters can be accounted for by a relatively simple model that associates visual features, letters, and words, to the nodes of a three-layer network (one layer for each type of information ; see Fig. 1.11). The nodes in each level are fully interconnected and partially connected to the nodes of the neighboring levels (there is actually no connection between nodes in nonneighboring levels). Each node is characterized by its time-varying *activity*, which is a function of its past activity (through a time-decay factor) and of the influence of the nodes connected to it. Connections can be *excitatory* or *inhibitory*, depending on their relative consistency (a typical example of inconsistency being the case of two different words, or letters, or features). Perception is assumed to result from the congruence of the activity of the related nodes.

Among other things, this model has highlighted the importance of distributed and interactive representations to explain the so-called *perceptual advantage of letters in words*—the fact that letters are easily recognized when they are embedded in a word. As a matter of fact, the activation of the features of a letter rapidly contributes to the activation of one or more words, which in turn feed back to the letter level and enhance its perception. Similarly, it has explained the *enhanced perception of letters in words under conditions of degraded input* (typically, when some part of the letter is blurred). It is then easy to imagine that the effect of lexical, syntactic and semantic contexts on word perception itself could be accounted for by a similar principle. When reading texts, indeed, it is common to ignore, unconsciously, some typographical errors (one character is missing or has been replaced by another one), even those that tend to modify the normal pronunciation of the word they affect. We scarcely decrypt words entirely. The estimation of some of their letters, together with a quick glance at their length, often suffices for a complete recognition: our linguistic knowledge enables us to *infer* the whole information from some part of it.

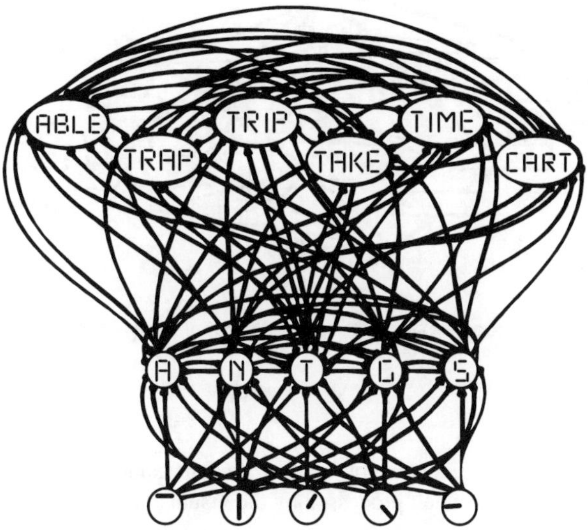

Fig. 1.11. A simplified view of the parallel distributed processing approach of letter perception: the neighbors of the letter *T* in the first position of a word (after McClelland and Rumelhart, 1981).

Notice that some hierarchy remains present in the aforementioned model, in the form of several layers focusing on increasingly complex percepts. Another form of hierarchical organization can be put in evidence in our everyday reading experience. It suffices to notice that our reading fluency is severely impeded whenever we encounter a complex word we have seldom read before (what about *proopiomelanocortin*?). It has been pointed out in many studies that two types of processes actually coexist: the slow and the fast ones (see Taylor,1988, for instance). The fast processes tend to be holistic, the slow to be analytic (hierarchically organized). In normal reading, the fast processes often produce enough perception to continue the reading without the need to wait for the results of the slow processes.[15] The information they are able to gain, however, is limited, so that when accuracy is demanded, or when the material is difficult, slow processes dominate. This motivates the introduction of the bilateral cooperative model of Fig. 1.12, in which the left and right branches, respectively, account for the slow and fast processes. Sensory data are processed by both branches simultaneously, in order to extract their phonetic, syntactic, and semantic structure. The output of each analytic process on the left can be used by the holistic process on the right.

[15]This distinction between *iconic* and *analytical* recognition clearly appears in *prosopagnosias*, due to some lesions of the right occipital lobe of the human brain. Patients fail to recognize faces at first sight but rather need some time to analyze them so as to extract their most typical features, which in turn induce the final recognition process.

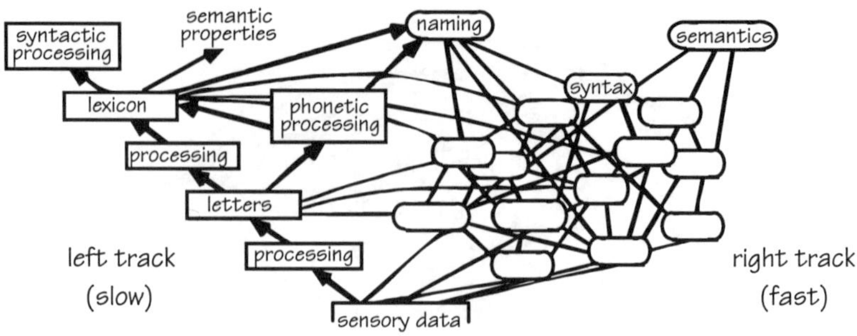

Fig. 1.12. A bilateral cooperative view of word recognition (after Taylor, 1988).

1.3.3. Thinking

If perception is quite an abstract concept, what is *thought* made of? What happens to the visual information once it has been processed by the visual cortex?

A partial answer to these questions can be found in the study of language and reading disorders, respectively termed as *aphasias* and *alexias,*[16] and of their neurobiological justification by the *Wernicke-Geschwind* model, presented in Fig. 1.13. According to this model, pronouncing a text involves transfer of the visual information from the primary visual cortex to higher-order visual cortex, and from there to a specific region of the association cortex termed as *angular gyrus* and supposed to be involved in the integration of many sensory channels (visual, auditory, and tactile). It is then projected to the so-called *Wernicke's area*, situated in the left hemisphere, where the perception of the words are formed and where the related memory images are stored. The images are conveyed, by means of the *arcuate fasciculus*, to *Broca's area,*[17] also in the left hemisphere, where the isolated perception of words is combined into a grammatical and semantic structure and where the memory for word articulation is stored. Finally, the sound pattern of the sentence is transmitted to the facial area motor cortex, which is in charge of the articulation.

The justification of the Wernicke-Geschwind model resides mainly in the number of predictions is allows about aphasias. First, it predicts the effect of a lesion in Wernicke's area (which results in the so-called *Wernicke's aphasia*): patients are incapable of understanding the text they read (some of them, aloud), while their spontaneous speech remains fluent (in rate, rhythm, and melody). They also fail to express the ideas they have in mind: their speech is fluent but empty! Second, it predicts *Broca's aphasia*, caused by a lesion in Broca's area. Comprehension is maintained, but both reading and spontaneous speech are strongly impaired. Patients typically use keywords only and

---

[16]More specifically, alexias are acquired reading difficulties, not to be confused with *dyslexias*, which are congenital.

[17]Wernicke and Broca's areas have received the names of their discoverers, two neurologists of the ninetieth century who have greatly contributed to our modern understanding of language disorders.

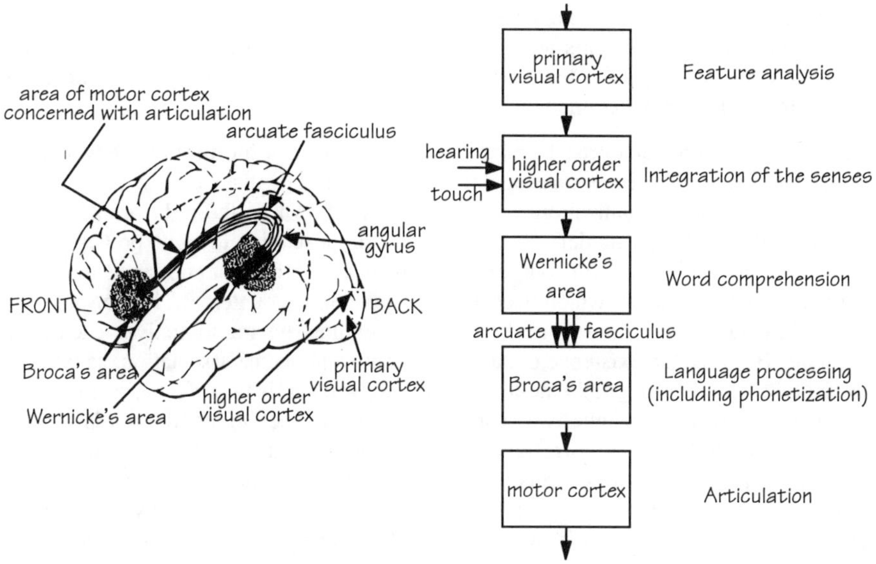

Figure 1.13. The Wernicke-Geschwind model for reading aloud (after E.K. Kandel, J.H. Schwartz, *Principles of Neural Science*, 2nd edition, reproduced by permission of Appleton & Lange, copyright ©1985, Appleton & Lange).

mainly word roots (singular nouns, infinitives, and no articles). Third, it foresees a lesion that could leave Broca's and Wernicke's areas intact but simply disconnected from each other: *conduction aphasia*. Comprehension is good and spontaneous verbal output remains fluent, but since Broca's area receives no input from Wernicke's one, reading aloud is strongly impaired.

The model also efficiently explains some typical cases of alexias, as the one that is reported in Kandel and Schwartz (1985), discovered by the end of the ninetieth century by the French neurologist Jules Dejerine. A patient had two localized lesions: one in the left visual cortex and one in the visual part of the *corpus callosum*, the main pathway between the left and right hemispheres. As a result, his left hemisphere, which, as we said in the previous section, processes the right vision field, could no longer see, while the left one had a perfect vision. And since both hemispheres were disconnected from each other, the information about the left vision field could not activate the language areas of the left hemisphere.

What precisely happens in Broca's and Wernicke's areas is still mostly unknown. What is quite certain, however, is that word perception and comprehension precedes deeper linguistic analyses, such as phonetic encoding or syntactic analysis. This leads us to think that, when reading, we first associate meanings to words and let these meanings provide us with the way they are uttered. From being purely phonetic (a child reads words as connected syllables, without understanding what he/she reads[18]), our lecture becomes more semantic with years. This actually makes all the difference between people who need an internal uttering of a text to understand it[19] and those for whom spelling and meaning are in close connection.

Such a concept-to-phoneme transformation is likely to happen for most of the words that we use currently, the ones we recognize in the twinkling of an eye, with a parallel distributed processing organization as the one that is shown on the right branch of Fig. 1.12 (parallel distributed models have also been proposed for thought processes; see Rumelhart *et al.*, 1986). Our (unconscious) linguistic enlightenment entitles us to quickly find the associated concept, which itself induces a phonetic counterpart, in the same way we instantaneously put a name on a familiar face without having to analyze each of its lineaments. Besides, the existence of such a concept-to-phoneme association, as opposed to an analytic, rule-based, grapheme-to-phoneme conversion, is in perfect agreement with the fact that, during a dialogue, we scarcely refer to the spelling of the words we use: concepts are directly put in correspondence with pronunciations. However, we do not completely get rid of the slower analytic approach, which is the one we were taught to use at the very beginning of our reading education. When unknown words are encountered, especially those with a peculiar spelling, it may still come into action.

Last but not least, the pronunciation of a sentence requires to associate it with a coherent melody, rhythm, and intensity—a prosody. Much remains to be done to know how this is being achieved. The only physiological data we have on this aspect of speech originates in recent discoveries on disturbances of the affective components of language: *aprosodias*. It has been found, indeed, that prosody (as well as emotional gesturing) depends on certain specialized areas in the right hemisphere. What is more, the anatomical organization of the right hemisphere seems to mirror the organization of its left counterpart in that respect: patients with lesions in the region that mirrors Broca's area are unable to organize prosody fluently, while lesions in the region corresponding to Wernicke's area impede its comprehension.

It is a wonder that we are able to start reading a sentence far before having seen its end. It must be that we unconsciously decompose each sentence into groups of words, to which we attach a consistent intonation. Or, to put it more precisely, it can be shown that the prosody we apply on sentences is closely related to their syntactic structure, even if semantic information is sometimes necessary.[20] We should nevertheless beware of thinking that we possess an integrated syntactic analyzer! On the contrary, it is much more likely that we implicitly perform a semantic analysis of the text we read, from which we can obtain, *a posteriori*, its syntactic analysis. This led the Belgian linguist

---

[18]Following the *syllabic* or *semisyllabic* methods, children now learn to read from basic elements, such as letters and syllables, to complete words. This was not the case in the golden age of the so-called *global* method, which considered words as entities to recognize in one trial. This technique provides faster readers, but it is also believed to contribute in the degeneration of their spelling ability.

[19]This phenomenon, called *subvocalization*, limits the reading speed to about 150 words per minute, that is a normal speaking rate, while skilled readers can read up to 900 words per minute.

[20]The relationship between syntax and prosody will be examined in Chapter Six.

Nicolas Ruwet to write that traditional grammars are rather "supposed to be helpful for an intelligent reader who, somehow or other, already dominates the syntax of a language" (Ruwet, 1968).

To cut a long story short: far beyond the simple grapheme-to-phoneme transcription of each isolated word that one might have expected to take place in our brain, it is all of our linguistic competence that is exploited at any level of the reading process. Besides, the operation is so complex that, if ever one tries to establish its rules and fix its parameters, the phenomenon looks nondeterministic: the same person might most often pronounce a given sentence in a variety of ways, depending on its (the sentence's) or his (the reader's) context, and the effect (emphasis, double meanings, and so on) that is intended.

1.3.4. Saying

Reading a text aloud does not only imply recognizing a sequence of characters and understanding the resulting words, word groups, and sentences. It eventually requires the many articulatory muscles of the articulatory apparatus already presented in Fig. 1.4 to be commanded in a coordinated way.

Many studies have shown that this task is carried out by some specialized cortical areas: the facial parts of the *premotor* and *motor cortices*, which exist in both cerebral hemispheres. Besides, these areas have been included in the aforementioned Wernicke-Geschwind model, which in turn can explain the anatomical basis for some typical disorders in the understanding or in the execution of learned movements (such as speaking), even though no paralysis can be found. Such deficits in the cognitive components of learned movements are termed as *ideomotor apraxias* and are often accompanied by aphasia to some extent. This is easily understood in the case of conduction aphasia. Patients with conduction aphasia, which originates from the disruption of the arcuate fasciculus of the left parietal lobe, are still able to understand what they read, but the related information can no longer be transmitted to Broca's area (hence the associated aphasia) or to the premotor cortices. As a result, they cannot articulate the sentences they read, while they can articulate spontaneous speech. Other types of apraxia may result from surgical destruction of some part of the corpus callosum (which connects the left hemisphere to the right one). Patients are unable to carry out articulatory commands with their right facial muscles, while their left ones perform perfectly well.

Obviously, these areas are the source of a classical problem in phonetics and speech research: that of the inextricable difference between physical and linguistic descriptions of speech—between what is actually said and what is really meant to be said. As mentioned by Lindblom (1989): *"in spite of several decades of phonetic research on many languages, we still encounter serious difficulties when it comes to specifying phonological units in such a way that their phonetic description will remain invariant across the large range of contexts that the communicatively succesful real-life acts present to us."* The question is: what type of target do the premotor cortices refer to when transforming symbolic phonemes into genuine articulatory gestures: articulatory, acoustic, or auditory ones?

It turns out to be none. The quest for invariance has led to the so-called target theories, which state that phonemes can be described in terms of articulatory goals that would not be attained given the sluggishness of the articulators. The resulting *undershoot* would be a function of the duration and context of articulatory gestures, while the goals would remain invariant. These theories have proved imperfect with years, although they have yielded considerable insight into the problem. It has also been suggested that speech signals themselves could contain absolute physical invariants in the form of feature templates (in a given parametric space) corresponding to phonetic segments. The many difficulties that still arise when trying to automatically divide continuous speech into phones provide some indication on the roughness of such an assumption. Eventually, *auditory gestures* have been thought to exhibit invariance: although articulatory targets could change, their perception would remain constant. Again, experimental findings have shown up the weakness of these ideas. In practice, we are even perfectly incapable of uttering twice a sentence exactly the same way, whether from an articulatory, acoustic, or auditory point of view. This suggests that linguistic invariance has to be defined only at the level of listener comprehension—that is, on a *cognitive* ground (Lindblom, 1989): the variability that speech signals exhibit is merely a function of informational constraints that arise in speaker-listener interactions.

1.3.5. Hearing

Last but not least, the sense of hearing plays an important part, although often ignored, in the act of reading aloud. As soundly noted by Harvey Fletcher (1953): "We speak with our ears"[21].

Physiologically speaking, speech sounds are collected by the auditory apparatus,

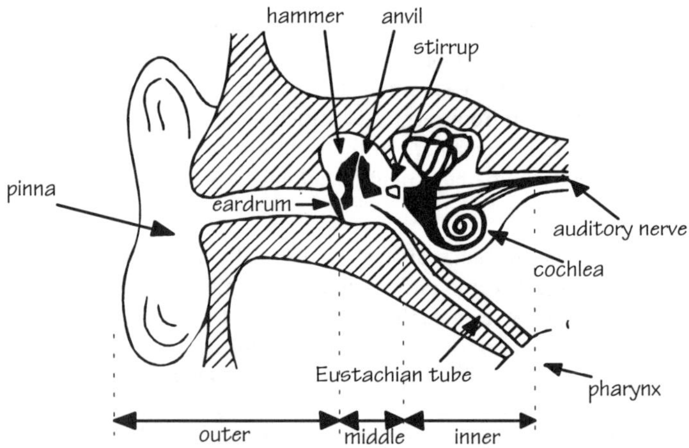

Fig. 1.14. The auditory system (after Boite and Kunt, 1987).

[21] Quoted in Allen (1996).

subdivided into the *outer ear*, the *middle ear,* and the *inner ear* (Fig. 1.14). The vibration of the *eardrum* (the *tympanic membrane*) is transmitted through the *ossicles* (*hammer - anvil - stirrup*) to the fluid-filled *cochlea* (the inner ear), where the mechanical-to-neural transduction takes place. The cochlea is coiled as a helicoidal spiral. It contains the *organ of Corti*, which rests on the *basilar membrane*. This membrane is partially covered with so-called *hair-cells.* The movement of cilia at the top of these cells triggers electrical firings transmitted to the brain by the *auditory nerve*. Each sinusoidal sound initiates a traveling wave along the basilar membrane, the peak amplitude of which occurs at a different position as a function of the frequency of the stimulus (Von Bekesy, 1960). Given their distribution along the cochlea, each hair cell is most sensitive to a different frequency band, so that one might broadly approximate the processing performed by the inner ear as spectrum analysis.

The organization of acoustic data transmission and processing in the brain is amazingly similar to its visual counterpart. Contributions of both ears are sent to both halves of the brain, whose parallel[22] hierarchical processing results in a distributed perception of sounds.

What is actually perceived is quite another matter. Its functional study has evolved into a specific science called *psychoacoustics.* Without going into details about the considerable contribution brought by psychoacousticians to the study of speech, it is interesting to be aware of some basic facts.

Fundamentally, the ear does not respond equally to all frequencies. Fig. 1.15 displays the so-called *isosonic curves*—the curves of equal perceptual intensity, or *loudness* (measured in sones), as a function of frequency. They show that the maximum sensitivity is achieved in the band [500 Hz, 10 kHz].

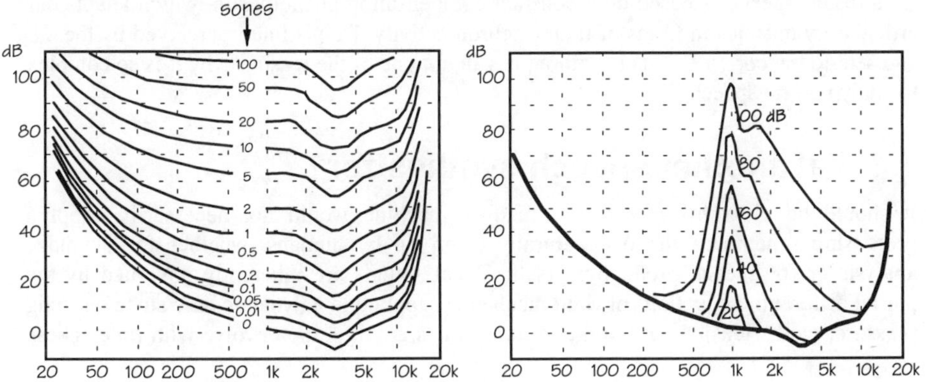

Fig. 1.15. Left : Isosonic curves in open field. Right : Auditory masking by a narrow-band noise: first limen of audibility as a function of the frequency and of the power of the masking noise (after Zwicker and Feldtkeller, 1981; see also Wegel and Lane, 1924).

---

[22]Many maps of the frequency spectrum have been found in the *primary auditory cortex.*

Moreover, sounds can be *masked* by oneanother. This psychoacoustical property of the human ear is examplified in Fig. 1.15, which shows how a narrow-band noise centered on 1 kHz locally elevates the first limen of audibility (the minimum acoustic power for a sine wave to be just-noticeable, as a function of frequency). Not only does any sound affect one's ability to hear, but its effect is related to its power in a nonlinear way (as shown again in the left panel of Fig. 1.15).

Yet we have not examined what is ultimately *understood*! Linguistic knowledge naturally interferes with psychoacoustical properties when listening to speech, exactly as it does when reading texts. Lexical, syntactic, and semantic knowledge is extensively employed for this task. Phones, indeed, are never uttered alone. The phonetic context in which they are embedded is obviously used by the brain. Some phonemes therefore convey more information than others, insofar as they are less likely to appear in a word for a given language, so that their accurate recognition helpfully restricts the search space for their plausible neighbors[23]. Similarly, phoneme sequences are organized into words, which have to obey a syntax and constitute a meaningful sentence. Hence, the deep knowledge of a language (of *symbolic* data) brings some tremendous help in recognizing speech sounds (*physical* data). Besides, the study of these interactions is covered by a specific area of linguistics and psychology, called *psycholinguistics.*

In a word, the complexity and the amount of data processing performed by the brain to organize the sense of hearing and the capability of understanding prevents the effect of the feedback control loop of Fig. 1.8 from being easy to investigate. The part played by this loop during oral reading, however, is hardly negligible (Perkell, 1981).[24] Its effect is mainly felt during the acquisition of speech capabilities (deaf children learn to speak with some difficulty) or the loss of hearing ones (sooner or later, most hearing impaired persons end up experiencing difficulties in controlling their intonation and voice energy). As a result, speech is based on a constant compensation phenomenon, which adapts our articulatory muscles in terms of motor neuron activity, from stimuli perceived by the ear and sent to the cortex.[25] This is without any doubt one of the reasons why it is so complex to study natural speech.

1.4. Yet another speech synthesizer?

It should be clear by now that a reading machine would not necessarily adopt a processing scheme as the one naturally taken up by humans, whether for language analysis or for speech production itself. Vocal sounds are inherently governed by the partial differential equations of fluid mechanics, applied in a dynamic case since our lung pressure, glottis tension, and oral and nasal tracts configuration evolve with time. These

---

[23] In contrast, there is experimental evidence that *nonsense* phone sequences are heard as independent sounds: to correctly identify the sequence, each phone must be correctly identified (Fletcher, 1953; see Allen, 1994 for a discussion)

[24] To what extent would you be able to speak correctly without hearing yourself?

[25] Incidentally, it is interesting to know that it is the delay inherent in this feedback loop, and variable from one person to another, coupled with the sensitivity of the associated transducer (in this case, the frequency sensitivity of one's ear) that conditions one's ability to sing in tune! (see Kawahara *et al.*, 1996, for instance)

are controlled by our cortex, which takes advantage of the power of its parallel structure to extract the essence of the text read: its meaning. What is more, the control dynamics are continuously adjusted by the aforementioned perception feedback loop. Even though, in the current state of the engineering art, building a text-to-speech synthesizer on such intricate models is almost scientifically conceivable,[26] it would result anyway in a machine with a very high degree of (possibly avoidable) complexity, which is not always compatible with economical criteria. As Lindblom (1989) denotes: "after all, planes do not flap their wings"!

Results are more important than means. No wonder, then, that the inner functioning of TTS systems developed up to now has often deviated from its human counterpart. We shall now briefly examine how. However, even though the history of speech synthesis is an interesting subject *per se*, we do not mean to investigate here the many steps involved in it throughout the centuries: we could, indeed, hardly do it better than Calliope (1989), Lingaard (1985), Liénard (1977), and Flanagan (1972), to which we invite the interested reader to refer. Not to mention Klatt (1987), who presents the TTS family in the form of amazing genealogical trees. The paragraphs that follow merely offer a selective synopsis of the prominent events and individuals.

Each and every synthesizer is the result of a particular and original imitation of the human reading capability, submitted to technological and imaginative constraints that are characteristic of its creation time. From the (misleadingly) talking heads of the ancient Greeks to the end of the eighteenth century, the ability to speak has stimulated considerable interest among our ancestors, although no synthesizer worthy of the name has emerged. There is no lack of passages in our literature to confirm it, as in this quotation from the famous mathematician and engineer Leonhard Euler (1761):

> "It would be a considerable invention indeed, that of a machine able to mimic speech, with its sounds and articulations. I think it is not impossible."

Euler was all but a dreamer: not more than eight years later, Wolfgang von Kempelen began to study the feasibility of a genuine talking machine, which he would spend twenty years to refine. In 1791, he disclosed its functioning in *The Mechanism of Speech and the Construction of a Speaking Machine*. It actually constituted a mechanical analogue of the articulatory system, composed of a main bellows, reed and india-rubber cup to simulate the lungs, vocal cords and mouth, respectively (Fig. 1.16). Small pipes acted as nostrils, additional levers enabled the trained user to produce fricative noise by letting the air flow through auxiliary pipes, and secondary bellows simulated the expansion of the vocal tract after mouth closure when uttering stop consonants. On the whole, his machine could mimic about twenty speech sounds.

Such mechanical analogues have further been investigated until the beginning of the twentieth century, mostly under the impulse of von Kempelen 's pioneering work, in concert with progress in the understanding of vowel quality. Faber, Willis, Wheatstone, Bell (!), Lloyd, Paget, and Riesz were some major contributors.

---

[26]Intensive research on articulatory synthesis, neural networks, semantic analysis, and speech recognition give evidence of it.

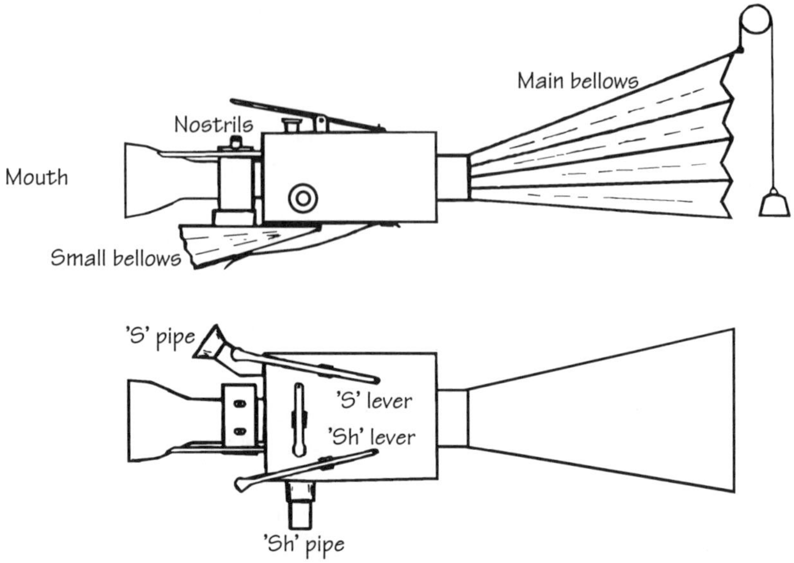

Fig. 1.16. Wolfgang von Kempelen's talking machine (after Lingaard, 1985).

It was not until the early 1930s that scientists began to introduce electrical models of the phonation mechanism and analysis systems capable of extracting their parameters, in order to perform automatic speech synthesis and recognition. The first electrical synthesizer aimed at producing connected speech, Dudley's *voder,* came out of the Bell Labs in 1939. As illustrated on Fig. 1.17, it implemented a spectrum synthesis of speech by feeding a series of ten band-pass filters connected in parallel with a common excitation signal: either a periodic signal, the pitch of which was adjusted with a pedal control, to produce voiced sounds or some noise to simulate unvoiced ones. The voiced/unvoiced decision was commanded by a wrist bar. All outputs were amplified independently, each amplification factor being commanded by a potentiometer assigned to a separate key, and summed. Three additional keys were used to introduce transients so as to reproduce stop-consonants. It succeeded in producing intelligible speech. As for von Kempelen's machine, the voder was to be played by an experienced operator, trained to perform the required articulatory-acoustics transformation in real time. Cooper, Liberman, and Borst have later automated it in an arrangement called the *voback* (now referred to as the *pattern playback machine*) by replacing the keyboard by a series of contiguous photocells illuminated proportionally to the amplitude along the frequency axis of a spectrographic display (Cooper *et al.,* 1952).

In fifty years, synthesis techniques and technologies have considerably changed. The advent of digital computers has made it possible to simulate synthesizers before (or instead of) realizing them in hardware. On the other hand, VLSI circuits now make it possible to produce, on a single application-specific integrated circuit, a complete speech synthesizer.

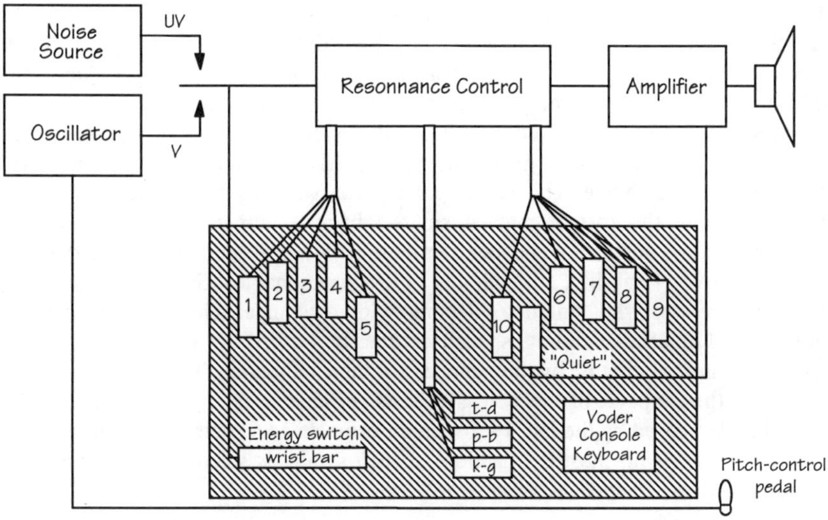

Fig. 1.17. Block-diagram of Dudley's voder (after Dudley *et al.*, 1939).

The first step in this direction has been taken in 1960, when speech analysis and synthesis techniques split into two paradigms referred to by Linggard (1985) as the *system* and *signal* approaches. In the first one, also termed as *articulatory synthesis,* the speech production mechanism itself is modeled with more or less physiological details. In the other, sometimes denoted as *terminal-analogue synthesis*, it is the speech signal that is modeled, using whatever means are convenient. Both approaches have evolved independently, with quicker results from signal modeling, given its intrinsic simplicity. As opposed to the system approach, indeed, it does not imply to *explain* coarticulatory effects in terms of articulatory gestures but simply to *describe* the related acoustic waveforms. While articulatory synthesis is still an open problem, two major terminal-analogue synthesizers—namely, *formant* synthesizers and *linear predictive* ones—have emerged and progressively been enhanced in the 1970s, under the impulse of the pioneering work of Fant, Flanagan, Holmes, Itakura, Rabiner, Stevens, and many others.

In parallel with the development of speech models and, accordingly, of speech analysis and resynthesis techniques, there has been, as early as the year 1965, a considerable interest in the use of computers as *text-to-speech* synthesis systems, as a step forward from the mere speech duplication task they had previously been assigned to. Allen, Holmes, Klatt, Mattingly, Shearme, and Umeda were some of the main characters of those days. They lay down the bases of our modern speech synthesis systems, by carefully and successfully examining the problems involved, both at the signal and the language levels. As we have seen in this introductory chapter, indeed, being able to produce speech from text requires a considerable amount of knowledge, from phonetics to pragmatics, and raises important questions on the best way of producing natural sounding speech from a symbolic, phonetic input.

The concept of *high quality TTS synthesis* eventually appeared in the mid-1980s as a result of important developments in speech and natural language processing techniques, mostly due to the emergence of new technologies (digital signal and logical inference processors). High quality is now a must for the expansion of the speech products family. Most of the commercially developed systems, indeed, suffer from a lack of voice quality and naturalness that has hampered their broad diffusion. As Levinson *et al.* (1993) note: "Acceptance of a new technology by the mass market is almost always a function of utility, usability, and choice. This is particularly true when using a technology to supply information where the former mechanism has been a human." Utility refers to the availability of new, easier, or cheaper services. It is not a controversial matter, as we shall see below. Usability, though, is still questionable. State-of-the-art speech synthesizers are not well suited to dealing with discourse. Their ability to produce intelligible and natural-sounding speech is clearly limited by the poorness of their emotional dynamics—that is, by the plainness of their intonation. This is undoubtedly the reason why the appeal of speech synthesis products among the general public considerably decreased by the late 1980s, an unfortunate consequence of which being the uncomprehending, if not hostile, reactions speech synthesis scientists still have to face: "Yet another robot-like speech synthesizer?"

Hopefully, there seems to be a revival now, as a result of recent progress in the areas of signal and language processing, mostly due to the availability of powerful computers at very low cost. Synthesizers are about to be part of the household equipment. We do hope that this book will contribute in some way to this resurrection.

1.5. Automatic Reading: what for?

The potential applications of high-quality TTS systems are numerous. Here are some examples:

Telecommunications services. TTS systems make it possible to access textual information over the telephone. Knowing that about 70 percent of the telephone calls actually require very little interactivity, such a prospect is worth being considered. Texts might range from simple messages, such as local cultural events (cinemas, theaters), to huge databases that can hardly be read and stored as digitized speech. Queries to such information-retrieval systems could be put through the user's voice (with the help of a speech recognizer) or through the telephone keyboard (with DTMF systems). One could even imagine that our (artificially) intelligent machines could speed up the query when needed, by providing lists of keywords, or even summaries. In this connection, AT&T has recently organized a series of consumer tests for some promising telephone services (Levinson *et al.*, 1993). They include: Who's Calling (get the spoken name of your caller before being connected and hang up to avoid the call), Integrated Messaging (have your electronic mail or facsimiles being read automatically over the telephone), Telephone Relay Service (have a telephone conversation with speech or hearing impaired persons using *ad hoc* text-to-voice and voice-to-text conversion), and Automated Caller Name and Address (a computerized version of the "reverse directory"). These applications have proved acceptable, and even popular, provided the intelligibility of the synthetic utterances was high enough. Naturalness was not a major issue in most cases.

Language education. High-quality TTS synthesis can be coupled with a computer aided learning system and provide a helpful tool to learn a new language. To our knowledge, this has not been done yet, given the relatively poor quality available with commercial systems, as opposed to the critical requirements of such tasks.

Aid to persons with disabilities. Voice handicaps originate in mental or motor/sensation disorders. Machines can be an invaluable support in the latter case: with the help of an especially designed keyboard and a fast sentence-assembling program, synthetic speech can be produced in a few seconds to remedy these impediments. Astro-physician Stephen Hawking gives all his lectures this way. The aforementioned telephone relay service is another example. Blind people also widely benefit from TTS systems, when coupled with optical character recognition systems (OCR), which give them access to written information. The market for speech synthesis for blind users of personal computers will soon be invaded by mass-market synthesizers bundled with sound cards. DECtalk™ is already available with the best SoundBlaster™ cards now, although not yet in a form useful for blind people.

Talking books and toys. The toy market has already been touched by speech synthesis. Many speaking toys have appeared, under the impulse of the innovative Speak'n Spell™ from Texas Instruments. The poor quality available inevitably interferes with the educational ambition of such products. High quality synthesis at affordable prices might well change this.

Vocal monitoring. In some cases, oral information is more efficient than written messages. The appeal is stronger, while the attention may still focus on other visual sources of information. Hence the idea of incorporating speech synthesizers in measurement or control systems, as in airplane cockpits to prevent pilots from being overwhelmed with visual information.

Multimedia, man-machine communication. In the long run, the development of high-quality TTS systems is a necessary step (as is the enhancement of speech recognizers) towards more complete means of communication between people and computers. Multimedia is a first but promising move in this direction.

Fundamental and applied research. TTS synthesizers possess a very peculiar feature that makes them wonderful laboratory tools for linguists: they are completely under control, so that repeated experiments provide identical results (as is hardly the case with human beings). Consequently, they make it possible to investigate the efficiency of prosodic models. A particular type of TTS systems, which are based on a description of the vocal tract through its resonant frequencies (its *formants*) and denoted as *formant synthesizers*, has also been extensively used by phoneticians to study speech in terms of acoustical rules. In this manner, for instance, articulatory constraints have been enlightened and formally described.

To a wider extent, research on speech systems has provided a great stimulus to many other areas of science and engineering, including linguistics in general, information theory, system modeling, signal processing, computer science, and technology. As Allen (1985) notes: "These speech systems provide excellent examples for the study of complex systems, since they raise fundamental issues in system partitioning, choice of descriptive units, representational techniques, levels of abstraction, formalisms for knowledge representation, the expression of interacting constraints, techniques of modularity and hierarchy, techniques for characterizing the degree of belief in evidence, subjective techniques for the measurement of stimulus quality, naturalness and preference, the automatic determination of equivalence classes, adaptive model parameterization, tradeoffs between declarative and procedural representations, system architectures, and the exploitation of contemporary technology to produce real-time performance with acceptable cost." Many of these topics will be addressed in the remaining chapters of this book.

References

ALLEN, J., (1985), "A Perspective on Man-Machine Communication by Speech", *Proceedings of the IEEE*, vol. 73, n°11, pp. 1541-1550.

ALLEN, J.B., (1994), "How do Humans Process and Recognize Speech", *IEEE Transactions on Speech and Audio Processing*, vol. 2, n°4, pp. 567-577.

ALLEN, J.B., (1996), "Harvey Fletcher's Role in the Creation of Communication Acoustics", *Journal of the Acousical Society of America*, 99, pp. 1825-1839.

BOITE, R., and M. KUNT, (1987), *Traitement de la parole*, Presses polytechniques romandes, Lausanne.

BROSNAHAN, L.F., and B. MALMBERG, (1970), *Introduction to Phonetics*, Hetter & Sons, Cambridge.

CALLIOPE, (1989), *La Parole et son Traitement Automatique,* CNET-ENST, Collection technique et scientifique des télécommunications, Paris, Masson, pp 410-414.

COOPER, F.S., P.C. DELATTRE, A.M. LIBERMAN, J.M. BORST, and L.J. GERSTMAN, (1952), "Some Experiments on the Perception of Synthetic Speech Sounds", *Journal of the Acousical Society of America*, 24, pp. 597-606.

DUDLEY, H. , R.R. RIESZ, and S.A. WATKINS, (1939), "A Synthetic Speaker", *Journal of the Franklin Institute*, 227, pp. 739-764.

FLANAGAN, J.L., (1972), *Speech Analysis, Synthesis, and Perception*, Springer Verlag, Berlin, pp. 204-210.

FLETCHER, H., (1953), *Speech and Hearing in Communication*, Krieger, Huntington, NY.

GAZDAR, G., and C. MELLISH, (1989), *Natural Language Processing in Prolog: an Introduction to Computational Linguistics*, Addison-Wesley, Reading, MA.

GREVISSE, M. , (1980), *Le Bon Usage*, Duculot, Gembloux, 11th edition, p. 668.

HOLMES, J.N. , (1988), *Speech Synthesis and Recognition*, 1988, Van Nostrand Reinhold, London.

IPA, (1949), *The Principles of the International Phonetic Association : a Description of the International Phonetic Alphabet and the Manner of Using it, Illustrated by Texts in 51 Languages*, International Phonetic Association, Dept of Phonetics, University College of London.

KANDEL, E.K., and J. H. SCHWARTZ, (1985), *Principles of Neural Science*, parts I,II,III, Elsevier, pp. 1-283.

KAWARA, H., KATO, H., and J.C. WILLIAMS, (1996), "Effects of Auditory Feedback on F0 Trajectory Generation", *Proceedings of the International Conference on Speech and Language Processing*, Philadelphia, vol. 1 pp. 287-290.

KLATT, D. H., (1987), "Review of Text-to-Speech Conversion for English", *Journal Acoustical Society of America,* 82, 3, pp. 737-793.

LEVINSON, S.E., J.P. OLIVE, and J.S. TSCHIRGI, (1993), "Speech Synthesis in Telecommunications", *IEEE Communications Magazine*, pp. 46-53.

LIENARD, J.S., (1977), *Les Processus de la Communication Parlée - Introduction à l'Analyse et à la Synthèse de la Parole*, Masson, Paris.

LINDBLOM, B.E.F., (1989), "Phonetic Invariance and the Adaptive Nature of Speech", in B.A.G. Elsendoorn and H. Bouma eds., *Working Models of Human Perception*, Academic Press, New-York, pp. 139-173.

LINGAARD, R., (1985), *Electronic Synthesis of Speech*, Cambridge University Press, Cambridge, pp 1-17.

McCLELLAND, J.L., and D.E. RUMELHART, (1981), "An Interactive Activation Model of Context Effects in Letter Perception: Part I. An Account of Basic Findings", *Psychological Review*, vol. 88, n°5, pp. 375-407.

OLIVE, J.O., A. GREENWOOD, and J. COLEMAN, (1993), *Acoustics of American English : A Dynamic Approach*, Springer-Verlag, Berlin.

OPPENHEIM, A.V., and R.W. SCHAFER, (1975), *Digital Signal Processing*, Prentice-Hall, Englewood Cliffs, NJ.

PERKELL, J.D.S., (1981), On the Use of Feedback in Speech Production, in T.Myers, J.Lavers & J.Anderson, eds., *The Cognitive Representation of Speech*, Amersterdam.

RUMELHART, D.E., P. SMOLENSKY, J.L. McCLELLAND and G.E. HINTON, (1986), "Schemata and Sequential Thought Processes in PDP Models", in *Parallel Distributed Processing*, MIT Press, Cambridge, MA, vol. 2, pp. 7-57.

RUWET, N., (1968), *Introduction à la Grammaire Générative*, Plon, Paris.

SACKS, O., (1987), *The Man Who Mistook his Wife for a Hat*, HarperCollins Publishers, NY. French edition: *L'homme qui Prenait sa Femme pour un Chapeau*, Seuil, Paris.

TAYLOR, M.M., (1988), "Convenient Viewing and Normal Reading", in B.A.G. Elsendoorn & H. Bouma, eds., *Working Models of Human Perception*, Academic Press, pp. 293-315.

VON BEKESY, G., (1960), *Experiments in Hearing,*, McGraw-Hill, New-York.

WEGEL, R., and LANE, C., (1924), "The Auditory Masking of One Pure Tone by Another and its Probable Relation to the Dynamics of the Inner Ear", Physics Review, 23, pp. 266-285.

ZWICKER, E., and R. FELDTKELLER, (1981), *Psychoacoustique*, CNET-ENST, Collection technique et scientifique des télécommunications, Masson, Paris.

Part One

FROM TEXT TO ITS NARROW PHONETIC TRANSCRIPTION

And I'll tell you a secret—I can read words of one letter! Isn't THAT grand! However, don't be discouraged. You'll come to it in time.

Lewis Carroll, *Through the Looking Glass*

We introduced in Fig. 1.7 the diagram of a general TTS synthesis system. Going one level deeper in our description is almost straightforward as far as the NLP module is concerned. Indeed, since the synthesizer needs to know the narrow phonetic phonetic transcription of the words to utter—that is, the sequence of phones to utter as well as the related prosody, which are obviously not explicitly specified in the text to be read—we can easily imagine that our NLP module will contain two dedicated sub-blocks: a letter-to-sound module (LTS), and a prosody generator (PG), as shown in the figure.

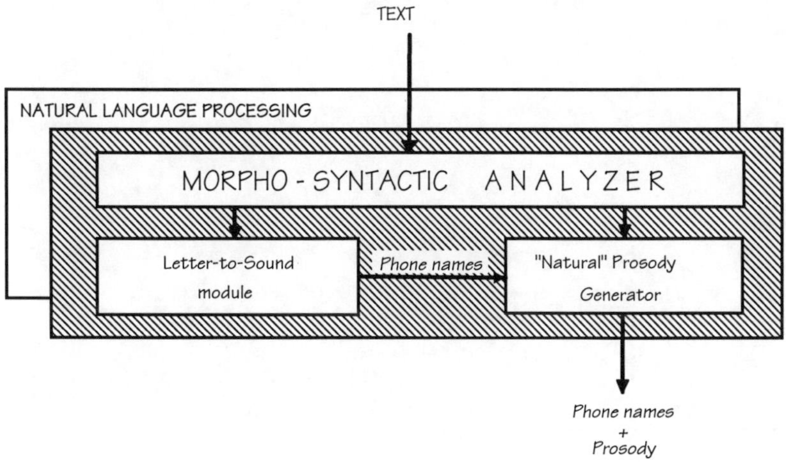

The functional skeleton of the NLP module.

What is less obvious at first sight, is the need for a powerful morpho-syntactic analyzer (MSA). Indeed, when questioned on their reading ability, most non-specialists believe that the sounds they utter share a one-to-one relationship with the letters they originate from and that the intonations they produce have no other origin than the punctuation of the sentence they read. This, however, is in strong contradiction with the intricate mental processes that we have highlighted in Chapter One. In practice, neither the LTS module nor the PG one have a real chance of success if they are not adequately supplied with high-level information, such as the part of speech of isolated words and the dependency relationship between connected words for the LTS block, and some form of a hierarchical description of sentences in the case of the PG one.

As a result, TTS synthesis is now increasingly involved in computational linguistics. Therefore, before entering into details about the problems addressed by MSA, LTS, and PG modules, in Chapters Four, Five and Six, respectively, it is worth introducing natural language processing itself, by saying a word on the tools available to the practical computational linguist. This is done in Chapter Two.

Notice, finally, that the figure above does not claim to be architecturally bearing. All it says is that there is no *functional* reason to introduce a feedback data flow from LTS or PG to MSA, while a data flow from LTS to PG is needed, since intonative and rhythmic decisions partly depend on the particular sequence of phonemes to be pronounced, via syllables.[1] *Architectural* issues on data and rule formalisms (how are data flows organized?) and control strategies (how are block firings scheduled?) will rather be debated in Chapter Three.

[1]In all generality, rhythmic constraints (such as the speech articulation rate, which influences the realization of some phonetic assimilations) also have to be taken into account in LTS. Practically, these can be seen as global parameters in TTS systems, so that no data flow is really needed from PG to LTS.

Chapter Two

GRAMMARS, INFERENCE, PARSING AND TRANSDUCTION

"Speak English!" said the Eaglet. "I don't know the meaning of half those long words, and, what's more, I don't believe you do either!"

Lewis Carroll, *Alice's Adventures in Wonderland*

At the very beginning of my own work on TTS synthesis, I was looking for a short but accurate introduction to grammars, parsers, and lexicons; an introduction in which the related terminology would be comprehensively explained, with not too many details, but not being too cursory either; an introduction in which the goals and, better still, the coverage limits of grammar models and the efficiency limits of parsers could be perceived.[1] I now know that this research was bound to fail, for even though grammars are often based on simple facts, their implications are particularly hard to appreciate. What is more, linguistic theories often insist on their peculiarities, rather than on their common properties. I thus grasped pieces of information throughout my research and painfully tried to form a coherent view. It was not, however, until I read the inescapable (Gazdar and Mellish, 1989), the didactic (Charniak, 1993) and (Winograd, 1983), the very complete (Sabah, 1989), the rigorous (Miclet, 1984) and the brilliant (and concise) (Gibbon, 1991), and after I had attended the comprehensive *introduction to natural language processing* courses of Prof. Martin Kay at the ELSNET summer school on prosody (University College of London, 1993) that I began to have deeper insights in this intricate area. The following pages are merely a concise presentation of some basic facts extracted from these references. It does not claim to be complete nor truly accurate, for it inevitably oversimplifies some aspects of grammars. It should be seen as a first introduction to computational linguistics, a map on which essential topics are positioned and shown to constitute if not a coherent set then at least a collection of interrelated approaches. We hope that it can nevertheless be of some help to readers who, like I did, are trying to find their way in what appears at first sight to be a *linguistic maze*.

[1]Following my own conviction that the real expert is one who knows what things *cannot* do and why not.

2.1. Basic concepts and terminology

The words *syntax* and *grammar* are often wrongly interchanged. Syntax (*sun-tacho* in Greek: to weave together) is the set of hidden dependency relationships among the words of a language, considered through their part of speech only, which makes all sequences of words not equally acceptable. Grammars are a formal way to express these constraints. One refers to *the* syntax of a language, but *many* grammars can describe it. Furthermore, grammars are not restricted to syntactic descriptions. They should be seen as tools to highlight the hidden hierarchical organization of symbolic data strings, whatever their linguistic bearing: sequences of phonemes, morphemes, words, meanings, intonation patterns. All respect a particular syntax.

Grammars really were first formalized by Chomsky [1957]. He described so-called *phrase structure grammars* as sets of *rewrite* (or *production*) *rules* of the form:

$$\alpha \rightarrow \beta \qquad (2.1)$$

in which α and β are any sequences of *symbols*[2] chosen from the *vocabulary* (the set of admissible symbols) **V**, composed itself of two disjoint subsets: $\mathbf{V_T}$, the *terminal* vocabulary (i.e., the set of all terminal symbols) and $\mathbf{V_N}$, the *non-terminal* vocabulary. Nonterminal symbols differ from terminal ones in that they never appear in surface forms but participate in the description of sentences. We shall write single elements of $\mathbf{V_T}$ as lowercase roman characters (*a, b, c, . . .*), single elements of $\mathbf{V_N}$ as upper case roman characters (*A, B, C, ...*), and sequences of elements of **V** (*phrases*) as Greek characters (α , β, γ, . . .). Rule (2.1) simply states that α can be replaced by β.

By introducing restrictions on α and β, Chomsky defined a complexity scale ranging from 3 (the most restrictive grammars) to 0 (the most including ones). Even though phrase structure grammars are less used today, at least in their initial form, this classification is still applied, for it is strongly connected to the algorithmic complexity and storage requirements of the corresponding parsers. Phrase structure grammars also have an important didactic function, in that they allow progressive introduction of the grammar terminology.

We shall systematically distinguish three complementary problems related to language processing, respectively termed as *inference, parsing,* and *transduction*:

1. Assuming a given complexity for the language to parse (that is, choosing the right set of restrictions on α and β), one has to find the production rules that account for a finite subset of its surface forms, termed as the *training database*, from an initial nonterminal symbol *S* (an abstraction for *sentence*). This problem is known as *grammar inference.* It does not, in general, admit a single solution. In order to restrain the number of plausible grammars for a given training database, it is generally assumed that the database is *complete* for the grammar to be found—that is,

    - All the terminal symbols are included, at least once, in the database;

[2]Symbols are understood here in a very wide sense: they may be characters, phonemes, morphemes, words, and so on.

- All the rules of the grammars have to be used, at least once, to generate the database.

Yet the number of solutions is generally so high that it is hardly possible to enumerate them. Furthermore, completeness can never be guaranteed for natural language databases. As a result, two distinct grammars estimated from the same training data set most often generate different languages, with the training data as a common subset.

2. Once the rules have been established, it is necessary to design an efficient *parsing algorithm*, which is able to analyze, in a finite time and with limited memory, any sentence of the language examined, in terms of the allowed grammatical rules. In other words, a parser (sometimes also called an *acceptor*) has to "prove" that it is possible to obtain a given sentence by rewriting *S* with the rules of a grammar. It is therefore necessary to develop efficient tree search methods, as parsing can be seen as the search for a path in a tree from *S* to the sentence to parse. Each branch in the tree corresponds to a rule to apply.

   Parsing is a non-trivial task, due to some major difficulty: *nondeterminism*, which arises when, at a given analysis point, several rules are equally applicable at first sight. The parser then has to make an arbitrary choice and try to justify it by finding a way to end up with an accepted parse of the current sentence. If it ever fails to do so, it has to *backtrack* to its erroneous decision (termed as a *backtracking point*) and make another choice. This implies, of course, that all backtracking points have to be memorized. It may also happen that several choices lead to accepted parses (*I see / the man with a telescope, I see the man / with a telescope*). Such unsolved nondeterminism is called *ambiguity*.

3. Most of the time, one wants not only to know if a sentence belongs to the language covered by a given grammar (that is, if it can be parsed with that grammar) but also what rules should be used to prove it. This, indeed, is not strictly within the competence of the parser. It can easily be achieved, however, by associating, with each branch in the aforementioned search tree, the corresponding rule. More generally, one can associate, with each such branch, a production rule of the grammar corresponding to the language to analyze, and a production rule of another language. Such a machine is called a *transducer.* It parses an input language with the first production rules associated to the branches of the related parse tree and uses the second rules to produce phrases in another language. Transducers are intrinsically reversible—that is, the "input" language can be considered as the output one and conversely.

We shall now review the first two classes of Chomsky's hierarchy—namely, *regular* and *context-free* grammars, which are widely used in computational linguistics—and try to present the set of grammar formalisms that are proved, or believed, to be equivalent to the related phrase structure grammars. We shall then briefly touch upon useful extensions of context-free grammars, and end this chapter with an introduction to *unification grammars* which depart from Chomsky's formalism.

2.2. Regular grammars (Chomsky type 3)

2.2.1. Definition

Regular grammars, also called *finite-state grammars*, are the far simplest formal descriptions one could imagine. They restrain rewrite rules to the form :[3]

$$A \rightarrow bC \tag{2.2}$$

$$A \rightarrow b \tag{2.3}$$

It is straightforward to see that such rules can equivalently be expressed by *finite-state automata* (FSA) and conventionally illustrated by a *finite-state transition network* (FSTN), which describe rules as possible transitions (depicted as *arcs*) between a finite number of states (denoted by nonterminals in the rules), each transition producing one terminal symbol in the surface string (Fig. 2.1). The languages FSA produce are said to be *regular.*

If an FSA is deterministic—that is, if it does not allow several transitions from a given state to be associated with the same symbol—then it is possible to determine the current state at any point of the analysis on the basis of the already parsed symbols. If it is not, some lookahead is needed, unless the FSA is ambiguous, in which case several paths can be found in the network for a given string.

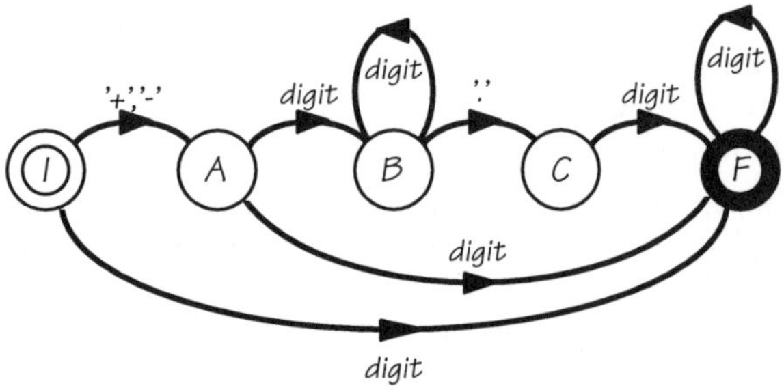

I -> '+' A
I -> '-' A
A -> digit B
A -> digit F
B -> digit B
B -> '.' C
C -> digit F
F -> digit F

Fig. 2.1. A simple finite-state transition network that describes integer and decimal numbers in their full form and the related regular rules. The initial and final states (*I* and *F*) are respectively depicted as white and black rings. Internal ones (*A*, *B*, and C) are shown as circles.

Probabilistic FSAs, in which transitions are additionally associated with probabilities, are also of a great interest for computational linguists. They allow to describe the possible

[3]Rule (2.2) actually defines a right branching grammar, for each rule shifts the current position in the parsed string one character to the right. Left-branching regular grammars change (2.2) in to $A \rightarrow Cb$.

strings of a language as non equally probable, as in the simple and personal example of Figure 2.2. Notice that *transition probabilities* from any state sum to one. If a probabilistic FSA is non-ambiguous, then the probability of a given string is given by the product of the transition probabilities for the only path which accounts for it.

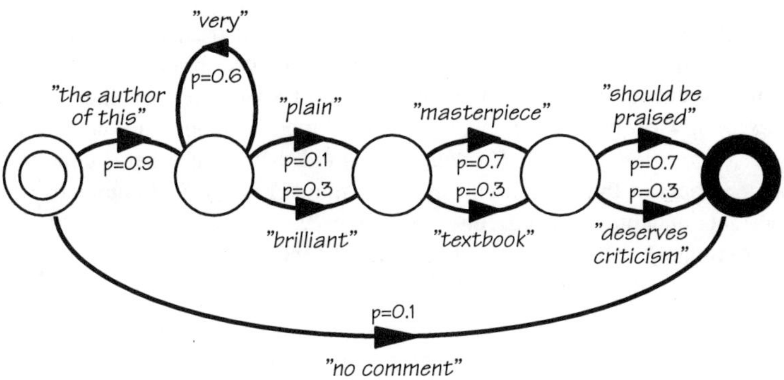

Fig. 2.2. A simple probabilistic FSA that accounts for what the author of this material would modestly like the opinion of his readers about himself to be. Several nonequally probable transitions are allowed from each state. When this model is used for parsing, however, it is possible to determine the current state at any point of the analysis on the basis of the already parsed words.

In the case of ambiguous probabilistic FSAs, the probability of a string is computed as the sum of its transition probabilities on all the related paths. *Hidden Markov models* (HMMs) fall in that category.[4] Furthermore, since transitions are no longer univocally associated to characters, HMMs often assume more generally that each state can *a priori* produce any character, independently of the next transition. Hence the distinction between *emission* and *transition probabilities* (Rabiner and Juang ; 1986). Emission probabilities are associated with states and account for the production of characters, while transition probabilities are associated with arcs. With this option in mind, the probability of a string $w_1, w_2, \ldots, w_N$ given a path $l = (q_{l,1}, q_{l,2}, \ldots, q_{l,N})$ is given by the product of all the related transition and emission probabilities:

$$P(w_1, w_2, \ldots, w_N | q_{1,l}, q_{2,l}, \ldots, q_{N,l}) = \prod_{j=1..N} P(w_j | q_{l,j}) P(q_{l,j} | q_{l,j-1}) \quad (2.4)$$

[4]More specifically, we refer here to *discrete* HMMs, which model sequences of *symbols* taken in a discrete set: the vocabulary. HMMs would be *continuous* otherwise.

2.2.2. Use

One can summarize the restriction imposed by regular rules by stating that, at any position in the parsed string, *the information brought by all previous symbols can be compacted into a* finite *number: the index of the current state*. Consequently, finite languages (languages with a finite set of admissible strings) are always regular.[5] In contrast, regular grammars are inherently incapable of realistically accounting for languages in which an unbounded number of imbricated bracketed expressions can theoretically be encountered. As a matter of fact no FSA can tell whether sentences like *the names of my brother, aunt, cousin,*.[add as many member of the family as you wish]... *are, respectively, John, Martha, Paul,* [add their names here] contain as many names as members of the family, except if this number is *a priori* bounded. Apart from this important inadequacy, even finite languages can considerably benefit from more complex grammars,[6] in the sense that higher-level generalizations may result in a reduced set of rules, with an increased coverage.[7]

FSAs can also be used to transform strings (to translate simple languages into one another) by simply associating symbol mappings with transitions, instead of characters alone (Fig. 2.3). In this case, they are preferably termed as *finite-state transducers* (FSTs). The corresponding rewriting rules are often expressed in the form

$$a \rightarrow b\,/\,l\,\_\,r \qquad (2.5)$$

which states that *a* is rewritten as *b* when surrounded by *l* and *r,* where *a, b, l,* and *r* are terminal symbols. It has been shown by Johnson (1972) that, provided some constraints are added on the application of such rules (like *no re-rewriting is allowed*), they define *regular relations* between their inputs and outputs, which can in turn be expressed by an FST. It is important to notice that, although rules like (2.5) are often termed as contextual, they define nothing else than *regular* transducers (and certainly not, as often wrongly stated, *contextual* ones, at least in Chomsky's terminology). As a matter of fact, the transitions from state to state are only determined by *terminals a, l,* and *r.*

FSTs are extensively used in phonology and morphology, as they have interesting features. They are, for example, inherently reversible. What is more, their mathematical properties have been deeply formalized. In particular, it has been shown by Johnson (1972) that regular relations are *closed under serial composition*, which means that a system of ordered rewriting rules such as (2.5) can always be merged into a single FST. Efficient compilers are now available (Kay and Kaplan, 1994; Mohri and Sproat, 1996).

---

[5]The converse, however, is not true (see again Fig. 2.1: sequences of digits can be *a priori*, of infinite length).

[6]A language of a given complexity can always be accounted for by a grammar of higher complexity (corresponding to a lower number in Chomsky's hierarchy). This operation, however, is not always profitable, for the algorithmic complexity of parsers increases accordingly.

[7]It should be quite obvious to engineers that the free responses of a digital filter with fixed-point implementation (when its input, output, and all its internal signals are restricted to be sequences of integer numbers coded with a finite number of bits, which implies that they are bounded) constitute a regular language! As a matter of fact, the initial values of the internal memories of the filter, which constitute a finite set, suffice to compute a free response. The filter recursion, however, which is obviously not a regular rule, suffices to describe it.

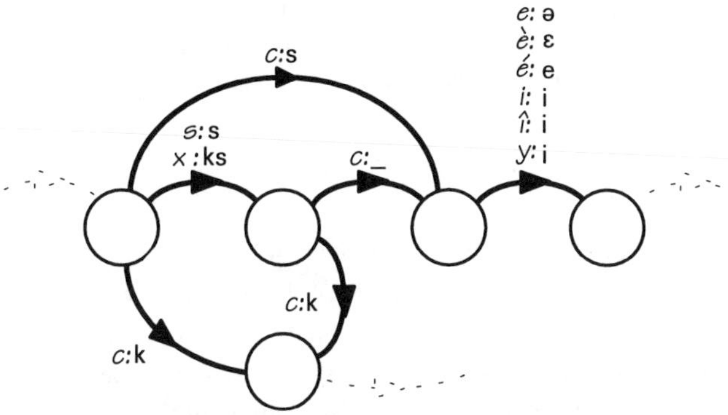

Fig. 2.3. Some part of a finite state transducer, which partially accounts for the phonological transcription of grapheme 'c' in French (as in '*excède* /ɛksɛ:d/, *disciple* /disipl/, *rescapé* /ʀɛskape/, *cidre* /si:dr/, *code* /kɔ:d/, *etc...*'). Labeled arcs of the form 'x : y' mean that x is transduced into y when transitions are made. Current positions in both strings are then shifted *n* and *m* characters to the right, respectively, where *n* and *m* are the numbers of characters in x and y.

In the context of probabilistic FSAs, the restriction introduced at the beginning of this section can be understood as: *the only information affecting the probability of an output symbol, or of the next state, is the current state.* This is the so-called *Markov assumption,* which plays a central role in the development of efficient inference and parsing algorithms. probabilistic FSAs play a very important part in language analysis in the context of speech synthesis and recognition, mostly in the form of so-called *n-grams,* which make the additional assumption that *the only information affecting the probability of a symbol is the* n *previous symbols.* They will be described in Section 4.3.1. Recent developments in computational linguistics have led to the concept of *weighted finite state automata* (WFSA), which associate weights to arcs and extend the notion of probabilistic FSA beyond HMMs (see Pereira and Riley, 1996).

While we are defining FSTs, and after having introduced the use of HMMs as stochastic FSTs, it is worth noticing that regular transduction rules can also be implemented in the form of so-called *multilayer perceptrons* (MLP), a particular type of *artificial neural networks.*[8] Artificial neural networks are based on simplistic models for biological neuron behavior and the interconnections between these neurons. They are basically computational architectures defined by a network topology, a node nonlinearity, and a learning (inference) algorithm, in the spirit of the parallel distributed processes introduced in Section 1.3.2. Most often, the nonlinear function is a limiter or a sigmoid function. The MLP is the far most widely used network. It is composed of an input layer

[8] Readers looking for a good introduction to neural networks may refer to the unescapable (Lippman, 1987). For a deeper understanding of MLPs, (Bourlard and Morgan, 1994) is a good reference.

and an output layer separated by one or more hidden layers of nodes, with each layer connected to the next layer, feeding its node values forward (Fig. 2.4). If there are L layers in the network, the ith output node of the lth layer has the value:

$$v_i^{(l)} = g\left(\sum_{j=1}^{M_{l-1}} w_{ji}^{(l)} v_j^{(l-1)} - w_{0i}^{(l)}\right) \quad , \qquad i = 1,\ldots, M_l, \quad l = 1,\ldots, L \quad , \tag{2.6}$$

where M_l is the number of nodes in the lth layer, $w_{ji}^{(l)}$ is the connection weight between the jth node of the l-1th layer and the ith node of the lth layer, $w_{0i}^{(l)}$ is the threshold of the ith node, and $g(.)$ is the nonlinearity. The M_0 inputs are written as $v_j^{(0)}$, for $j = 1, \ldots, M_0$. An input pattern corresponding to a, l, and r of (2.5) is fed at the input of the network, and the outputs of each neuron of the output layer are compared. The one that is closest to '1' wins, and the associated symbol, b, is deduced. It is straightforward to notice that such a network simply implements a FST!

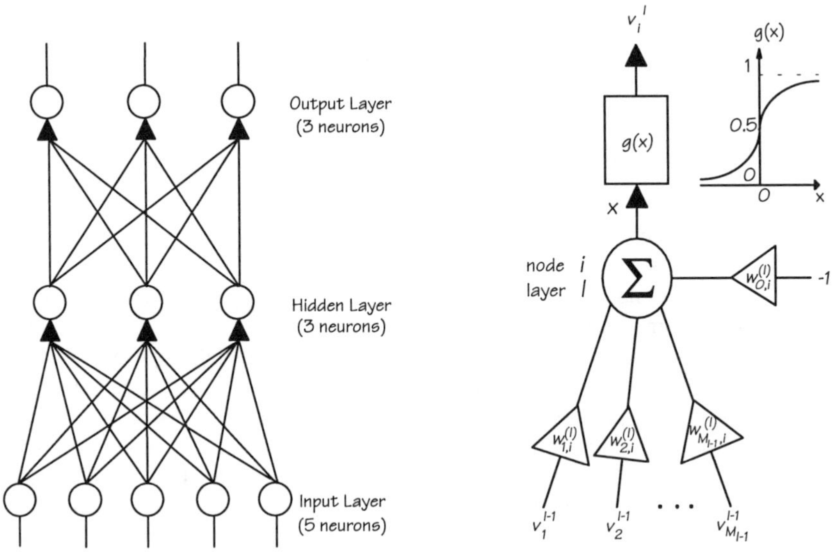

Fig. 2.4. Left: A multilayer perceptron, with a single hidden layer. Right: The ith neuron of the lth layer.

2.2.3. Regular inference

Finding a (nonprobabilistic) regular grammar that describes a given finite training database is a trivial problem,[9] since it suffices to build a separate FSA for each entry in the database and to merge their input state (Fig. 2.5). This defines the so-called *maximal canonical grammar*. A lot of other solutions can then be obtained by trying to further

[9]It is, by the way, part of the *sequential circuits synthesis* problem in electrical engineering.

merge other states (Miclet, 1979). Recently, however, some very efficient solutions have emerged from the mathenatical framework of weighted finite state automata (see the work of Mohri, 1994)

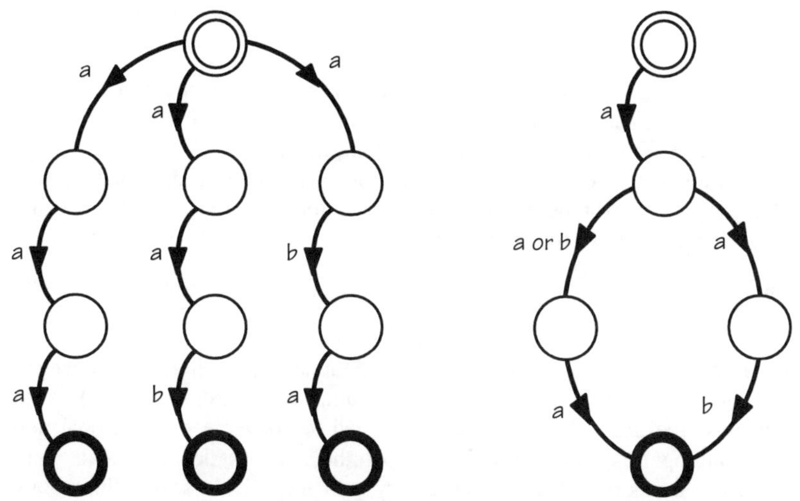

Fig. 2.5. Left: The maximal canonical grammar corresponding to the training database {aaa, aab, aba}. Right: A simpler grammar, obtained after having efficiently merged some states in the previous grammar.

In the context of probabilistic finite state grammars, an efficient inference algorithm exists for hidden Markov models: the Baum-Welch algorithm (also known as the *forward-backward* algorithm) (Rabiner and Juang, 1986). It provides an accurate but computationally expensive solution to the inference problem by iteratively refining the emission and transition probabilities that maximize the likelihood of all the training strings, taking all their admitted paths in the model into account:

$$\max\left(\sum_{\substack{\text{training}\\ \text{strings } k}}\left(\sum_{\text{paths } l}\left(\prod_{\text{words } j} P(w_{k,j}|q_{k,l,j})P(q_{k,l,j}|q_{k,l,j-1})\right)\right)\right), \tag{2.7}$$

in which $w_{k,j}$ is the jth word of the kth training string and $q_{k,l,j}$ is the state associated with word $w_{k,j}$ in the lth admitted path for the kth training string, so that $P(w_{k,j}|q_{k,l,j})$ and $P(q_{l,j}|q_{k,l,j-1})$ are the emission and transition probabilities to estimate. The well-known Viterbi (Forney, 1978) algorithm is often used as an efficient alternative. It maximizes the probability of the best path only, for all the strings in the training database:

$$\max\left(\sum_{\substack{\text{training} \\ \text{strings } k}} \left(\prod_{\substack{\text{words } j \\ \text{(bestpath)}}} P(w_{k,j}|q_{k,j})P(q_{k,j}|q_{k,j-1}) \right) \right), \tag{2.8}$$

in which $q_{k,j}$ is the state associated with word $w_{k,j}$ in the most likely path for the kth training string.

Eventually, training an MLP for a given regular transduction task implies optimizing its weights and thresholds so as to make it best account for the (input, output) pairs of a training database. The most widely used training method is the error back-propagation algorithm (Rumelhart *et al.*, 1986). It is a modified gradient procedure that minimizes the sum of the squared error between the desired outputs d_i, $i=1,\ldots,M_l$, and the actual outputs $v_i^{(L)}$, $i=1,\ldots,M_l$ for the training data:

$$\min \sum_{\substack{\text{(input,output)} \\ \text{pairs } p}} \varepsilon_p, \qquad \text{with} \quad \varepsilon_p = \sum_i (d_i - v_i^{(L)})^2 , \tag{2.9}$$

Starting from random initial values, weights and thresholds are iteratively changed in the direction of the negative gradient of (2.9), until a minimum of (2.9) is reached. Unfortunately, the network cannot be guaranteed to converge to the global minimum. Both the training time and the efficiency of the network depend on its number of neurons. Some tradeoff is therefore necessary, which results in severe limitations in the complexity of the transduction that state-of-the-art MLPs can achieve, even when training is performed on super-computers. The efficiency of the network is also strongly related to the format of the input data. The best results are obtained when the input information is arranged in such a way that similar values of *a, l,* and *r* in (2.5) correspond to similar input patterns.

2.2.4. Regular parsing

Even though parsing nondeterministic regular grammars reduces to trying to find a path from the initial to the final state that accounts for the input string, it implies the ability to make a choice between alternatives when they arise (see Fig. 2.6 for a simple example). Two basic strategies inherited from artificial intelligence can *a priori* be used—namely, *depth-first* and *breadth-first* search. The former accepts the first choice of each alternative until reaches a dead end. It then comes back to the last choice, makes another one, and so on. It therefore examines the search tree in its vertical dimension first, hence its name. On the opposite, the latter investigates states in the order of their depth in the tree.

It can be shown that a nondeterministic regular grammar can always be transformed into a deterministic one. Regular parsing then eventually confines itself to a trivial problem, in both the probabilistic and nonprobabilistic cases (hidden Markov models being excluded). It can be solved in linear time—that is, parse time increases linearly with both the number of symbols in the sentence to be parsed and the number of states in the model.

Fig. 2.6. Two laughing machines that obviously account for the same giggles (after Gazdar and Mellish, 1989). Only the left one is deterministic: whatever the current state, the next transition is unambiguously dictated by the input string. The right machine is non-deterministic : there are two possible transitions when an *a* is seen.

As far as HMMs are concerned, linear time can also be achieved, with the already mentioned Viterbi algorithm, which uses dynamic programming to provide the most likely path in a hidden Markov model, given the string to parse.

Eventually, MLPs also transduce an input string in linear time, since each symbol is transduced at a time. The tranduction time for a single character is obviously proportional to the square of the number of neurons in each layer of the network, since each layer is fully connected to its neighbors.

2.3. Context-free grammars (Chomsky type 2)

2.3.1. Definition

Context-free (or *algebraic*) grammars somewhat release the formal constraints imposed by rules (2.2) and (2.3), by allowing the right side of rules to include any sequence of terminal and/or nonterminal characters:

$$A \rightarrow \alpha \tag{2.10}$$

As in the case of regular grammars, context-free grammars can straightforwardly be depicted in the form of so-called *recursive transition networks* (RTNs), composed of a number of labeled finite-state automata, modified in such a way that state transitions can be associated to nonterminal labels (Fig. 2.7). Each transition network corresponds to all the rules that have its label as left part. Recursion arises when a transition network calls itself or when cross calls are encountered.

Stochastic context-free grammars have been developed as well. They can be hidden or not (that is, deterministic or not). As for regular Markov models, they associate probabilities to rules, so that whenever several parses arise for a given string (in the ambiguous case), a decision can be made on the basis of their probabilities (computed as before, as the product of the probabilities of the related rules).

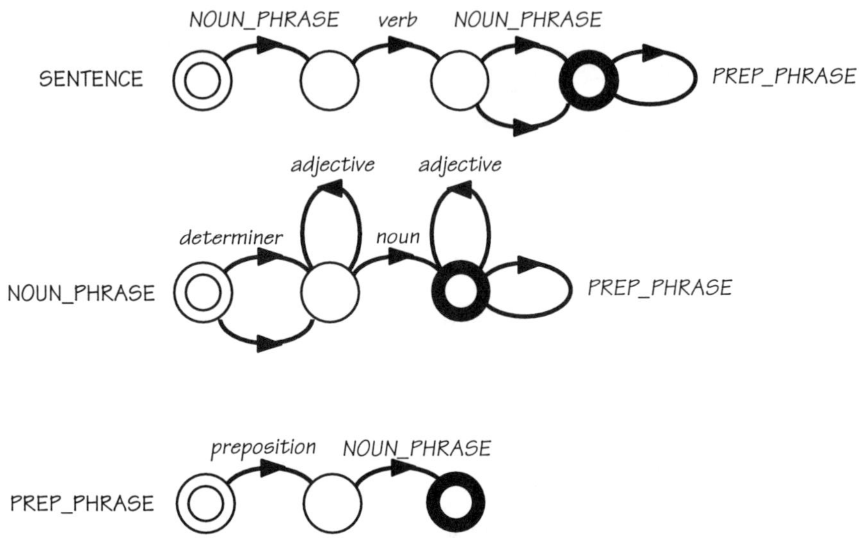

Fig. 2.7. A simple recursive transition network which accounts for French sentences like *Le joli petit chat noir de ma grand-mère boit du lait dans la cuisine* (*the pretty little black cat of my grandmother drinks milk in the kitchen*).

2.3.2. Use

As such, context-free phrase structure grammars (CF-PSG) allow us to generate an infinite number of imbricated bracketed expressions, with rules of the form

$$A \rightarrow bAc \quad (2.11)$$

Consequently, they can describe hierarchical trees with an *a priori* infinite number of stages between root and leaves. Their advantage lies in the ability to capture intrinsically embedded structures within strings.

CF-PSGs are extensively used in programming languages, for which they are often presented in their *Backus Naur form* (BNF). With this notation, terminals are written as such, nonterminals are surrounded by < and >, and ➔ is replaced by *::=*, like in

<Sentence> ::= <Noun_Phrase> verb <Prepositional_Phrase> (2.12)

Their descriptive power is limited by the fact that branches in the tree cannot overlap (Fig. 2.8). What is more, taking morphological facts into account (such as number or gender agreement) requires to split non-terminals into as many subtypes as morphological variations (NOUN, for instance, is split into FEM_SING_NOUN, FEM_PLUR_NOUN, MASC_SING_NOUN, MASC_PLUR_NOUN). This increases the number of rules exponentially.

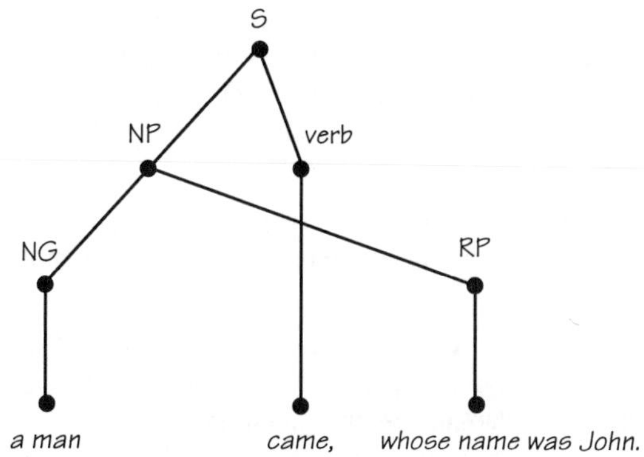

Fig. 2.8. This tree cannot be described in terms of context-free rewriting rules.

As for finite-state transducers, RTNs can be used as a means of converting an input sentence into an output sentence, by associating pairs of symbols (both terminals and non-terminals this time) to arcs, in which case they are termed as *pushdown transducers* (PT). Notice that although PTs can translate simple languages into each other (intrinsically, RTNs are nondirectional), they can do little more than word-by-word translation. This, however, does not mean that CF-PSGs could not be used for *parsing* natural languages efficiently.

2.3.3. Context-free inference

Since a given training database is always finite, and it has been said in Section 2.2.1.2. that finite languages can always be described by regular grammars, the number of context-free solutions to an inference problem is still greater than the number of regular solutions. Again, heuristics have to be used, based on hypotheses on the forms of rules. Moreover, human intervention is often useful.

In the case of stochastic CF grammars, an inference algorithm exists, known as the *inside-outside* algorithm (Lari and Young, 1990). It has some similarity with the *forward-backward* method implemented in the Baum-Welch algorithm, of which it can be seen as an extension. Its complexity, however, is cubic in both the length of the observation sequence and in the number of nonterminals, which severely limits its use.

2.3.4. Context-free parsing

As opposed to regular parsing, context-free parsing is not trivial. *Top-down* and *bottom-up* parsing strategies are available. The former starts from the *S* symbol and tries to rewrite it into the current surface form, while the latter does the opposite. Top-down context-free parsers implement a *pushdown automaton*: each time the parser enters a new network, it stores its current state on a stack, and each time a final state is reached, a new current state is un-stacked (except if the stack is empty, in which case parsing is finished). Top-down parsers are not particularly efficient for highly nondeterministic grammars, because a great number of rules can be erroneously applied before being invalidated by the surface form. This number even turns out to be infinite for grammars with recursive rules (which describe infinite loops). In contrast, bottom-up parsers are more *data-driven*: they always keep an eye on the surface string. This can mainly be done in three ways:[10]

1. With the so-called *shift-reduce* algorithm, which is an undirected bottom-up strategy. When nondeterministic rules are encountered, their (reverse) application is tested in a fixed order.
2. With the *left-corner* strategy, in which the choice for the rules to be tested first is determined by the current symbol in the rewritten string (either terminal or not). It is particularly efficient for languages in which groups are introduced by specific markers.
3. Finally, with the *head-driven* (or *head-corner*, also called *island-based*) method, which differs from the left-corner method in the sense that group heads are used in place of left markers. In the context of natural language processing, for example, verbs are first detected to orient the parsing process.

The left-corner and head-driven methods are extensively used in NLP. They have the advantage that, even if a wrong decision was made at a given backtracking point, well-formed structures can be subsequently memorized, and reused when other choices are tried, a strategy that is often referred to as *chart parsing* (see Charniak, 1993, for instance).

A classical context-free parsing algorithm has been given by Earley (1970). It parses nondeterministic type 2 languages in cubic time—that is, parse time increases proportionally to the cube of the length of the input string. Non-ambiguous grammars, however, can be parsed in square time (and even in linear time in most cases, provided the grammar formalism is slightly more constrained). Recent developments in computational linguistics have even provided methods for approximating context-free grammars with FSAs (Pereira & Roe 92).

2.4. Extensions of context-free grammars

Context-sensitive and *unrestricted* grammars—Chomsky's types 1 and 0—have scarcely been used in computational linguistics. They are not easily open to computer implementations. Inference and parsing are still open problems. In contrast, most developments in modern computational linguistics have focused on extending context-free grammars with a view to increasing their power to a type 1 or 0 one.

Transformational grammars have been proposed by Chomsky. They combine formal context-free rules with so-called *transformations*, which account for the fact that identical tree structures can appear under various surface forms (active/passive, and so on). They are, however, more generation than parsing oriented.

---

[10]Each of these strategies can be expressed in just a few lines of Prolog code (none for the top-down strategy, since Prolog uses a top-down inference engine).

Augmented transition networks (ATN) (Woods, 1970) are a first type of extension. They add *conditions* and *actions* to the transitions of RTNs. Conditions restrict the use of transitions. They can be based on properties of all previously parsed constituents, including the current one. Actions are used to memorize information on the syntactic structure of already parsed constituents (such as *gender* and *number* for noun phrases, and *mode* and *tense* for verbs, including any kind of information, like *agent, aim,* and so on., if needed) and to set flags that can be further checked by subsequent conditions. A formal ATN description language has been developed by Woods, and compilers exist that optimize parsing time (see Winston, 1979 and Harris, 1985 for LISP and Pascal implementations, respectively). On the opposite of FTNs and RTNs, ATNs are explicitly directional. Besides, they have to be implemented in a *procedural* way, as opposed to FTNs and RTNs which are essentially *declarative.*

Definite-clause grammars (DCG) (Pereira and Warren, 1980 ; see also Pereira and Shieber, 1987) basically differ from phrase structure grammars in the sense that they incorporate variables in both the left and the right parts of the rules. *Term unification* (as in Prolog) is then used to derive their values. A customized DCG rule formalism has also been defined, which is very similar to a context-free description, as in this (simplified) example:

sentence(NP,VP,Subject) - - >
nounphrase(NP,Number,Person),
verbphrase(VP,Number,Person),
{Subject=NP}. (2.13)

in which nonterminals are lower case words, and variables are indicated by an initial capital letter. Rule (2.13) simply states that a sentence is accepted as composed of sub-strings called *NP* and *VP* if one can find a noun phrase composed of *NP*, with number and person given by *Number* and *Person*, followed by a verb phrase composed of *VP* with the same number and person, in which case the subject of the sentence is *NP,* or, equivalently, that a sentence is composed of a noun phrase (its subject) and a verb phrase, which share number and person features.[11]

DCGs are extensively used in natural language processing, as their Prolog implementation is immediate. Most Prolog interpreters even directly accept the DCG rule notation. Their weakness originates in their closeness to Prolog: DCGs typically provide left-to-right, top-down, depth-first parsers, which are known to be quite inefficient, since they cannot remember already partially parsed material (as opposed to chart-parsers). Another disadvantage is that they imply a fixed term order within rules. In the example (2.13), for instance, querying the Prolog interpreter for *sentence (Subject, "I", "sleep")* would generally fail.[12] Similarly, DCGs do not provide an elegant way to handle underspecification. If, for instance, one wants to tell the parser that *the* can be singular or plural, and assuming determiners are listed as *determiner (Spelling, Number)*, then he has to add a rule of the form *determiner(the,_)'* in which _ is an unbounded variable.

---

[11]Notice that, although such a DCG does nothing else than *parsing* a sentence, it can be easily turned into a transducer, so as to *output* its structure as a result of the parsing task.

[12]This might seem at the very least desirable to an engineer; it is typically, and sensibly, repulsive to a computational linguist.

2.5. Lexicons, feature structures, and the PATR notation

In contrast with the phrase structure grammars introduced in the previous section, recent advances in computational linguistics have given birth to a new family of grammars, termed *unification grammars* (which should be understood as *'graph unification'*, rather than as term unification, which is typically the case of DCGs), as an evolution from the transformational formalism. These include three leading formalisms: *lexical functional grammars* (LFG) (Bresnan and Kaplan, 1982), *generalized phrase structure grammars* (GPSG) (Gazdar *et al.*, 1985), and the more recent and widely accepted *head driven phrase structure grammars* (HPSG) (Pollard and Sag, 1987, 1991). They have in common a lexical approach to syntax and semantics, in which linguistic rules are progressively incorporated in the lexicon in the form of *features,* so that the only rule that is finally maintained is the very general *unification* principle. In that respect, they can be seen as a lexical version of DCGs: variables no longer appear in rules but rather in lexical entries. Besides, the study of computational issues in this area is of a great importance for phrase structure grammars themselves. As a result, lexicons have taken a central position in today's computational linguistics.

Basically, a lexicon[13] is a list of words with additional syntactic (and possibly morphological and semantic) information, that is supposed to be useful to the parser. Among the many ways of describing lexical items, the *PATR* formalism (Schieber, 1986) is now widely accepted among computational linguists. It encodes each entry in the form of so-called *categories*—sets of *feature* (or *attribute*)-*value* pairs. PATR objects are therefore also termed as *feature structures* (FSs). They are typically denoted as

```
TypeOfObject NameOfObject :
      <Feature1>=Value1
      <Feature2>=Value2
      etc.,                                                    (2.14)
```

in which TypeOfObject, NameOfObject, Feature1, Feature2, etc. stand for any symbolic name. Each feature can itself take a complete category as its value, or a simple atom. In the case of category-valued features, feature names are added subcategory feature names, as in

```
TypeOfObject NameOfObject :
      <Feature1>=Value1
      <Feature2 SubFeature1>=Value2
      <Feature2 SubFeature2>=Value3
      etc.                                                     (2.15)
```

PATR categories, however, should not be confused with the hierarchical trees that we have introduced in the previous section. They fundamentally differ from trees in that features (or subcategory features) can symbolically *share* their values, as denoted by

---

[13]We will restrict the use of the term *dictionary* to the conventional printed object for human use and let the *lexicon* be a component of linguistic theory or of a NLP system.

TypeOfObject NameOfObject :
    <Feature1>=<Feature2> , (2.16)

which states that Feature1 has the same value as Feature2, *whatever the value is* (atom or category) *at any time*. Sharing is therefore stronger then a simple copy of values.[14]

Additionally, features can be *unspecified*, that is any feature that is not explicitly listed in a category description has *a priori* any value. They can even be *underspecified*: referencing a feature name without subfeature names when they exist (in the case of a category valued feature) automatically covers the complete feature value (the related category).

PATR categories (both lexical entries and rules) can actually be depicted by so-called *directed acyclic graphs* (DAGs), of which hierarchical trees are only a particular case.[15] Features are associated to arcs in the DAG, but their order is not important. An alternative to DAGs is to display feature structures in the form of *attribute-value matrices* (AVMs) in which shared categories are indexed by numbers (Fig. 2.9), which is helpful when the number of sharings in a DAG becomes important.

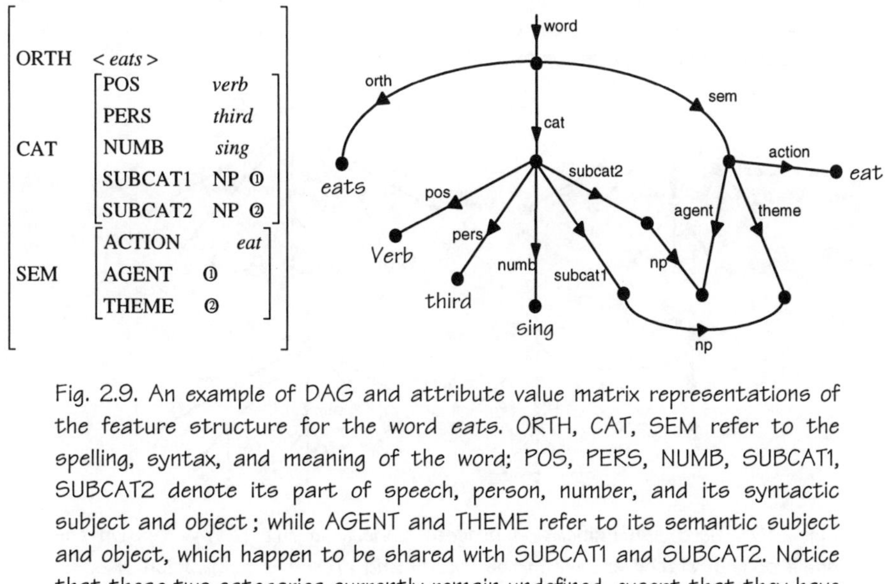

Fig. 2.9. An example of DAG and attribute value matrix representations of the feature structure for the word *eats*. ORTH, CAT, SEM refer to the spelling, syntax, and meaning of the word; POS, PERS, NUMB, SUBCAT1, SUBCAT2 denote its part of speech, person, number, and its syntactic subject and object; while AGENT and THEME refer to its semantic subject and object, which happen to be shared with SUBCAT1 and SUBCAT2. Notice that these two categories currently remain undefined, except that they have to be noun phrases.

[14]Computationally speaking, it can be seen as two variables pointing to the same memory address.

[15]Incidentally, this motivates the use of the PATR notation for a tutorial introduction to lexicons, since it also covers lexical descriptions for nonunification-based grammars.

The PATR notation actually allows to describe PSG-like rules, too, as in

```
Rule X →Y Z :
        <X cat>=S
        <Y cat>=NP
        <Z cat>=VP
        <Y per>=<Z per>
        <Y num>=<Z num>
        <X subj>=Y ,                                (2.17)
```

which says exactly the same as (2.13), except that the order in which features have been declared does not matter, by definition.

A rule like (2.17) can be interpreted as a test on some features (not necessarily all!) of the respective FSs of categories NP, VP, and S to be grammatically correct. It can also, and this is much broader view, be seen as some actions to be performed on the FSs of NP, VP, and S, knowing only some partial information on them (as extracted by previous rules or from lexical entries), for the rule to apply. This graph merging principle is called *graph unification.* A typical case would be the creation of a FS for S given FSs for NP and VP, which would result, as in the case of (2.13), in recognizing NP as the subject of S (see Fig. 2.10). Graph unification is therefore very similar to the *term unification* of DCGs, but since FSs are intrinsically open (given their explicitly allowed unspecification) and since the order of their feature declarations is not relevant, FSs are more convenient to use than Prolog terms.[16]

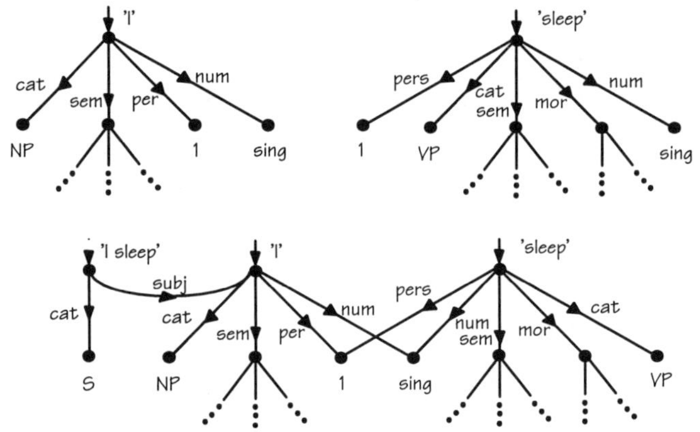

Fig. 2.10. The graph unification process underlying rule (2.17). Top: On the basis of the related entries in the lexicon and after the application of some rules, the categories of words *I* and *sleep* have been hypothesized as NP and VP. Bottom: after the application of rule (2.17), both words have been unified into a sentence level DAG. Notice that even features not specified in (2.17), such as semantic or morphological features, participate in the unification.

[16] It turns out that, for the same reasons, terms allow more efficient computer implementations than FSs. Hence the interest of DCG-like formalisms as well.

2.6. Summary

In this chapter, we have introduced a number of important grammar formalisms, which will be frequently referred to in the following chapters.

Phrase structure grammars and related formalisms have been presented from the simplest to the more complex ones, following Chomsky's hierarchy. Each family has been formally and functionally defined, and inference and parsing algorithms have been mentioned. The formal relationship between regular grammars, hidden Markov models, and multilayer perceptrons has been emphasized. Similarly, context-free phrase structure grammars have been shown to have a stochastic counterpart. Finally, augmented transition networks and definite clause grammars have been mentioned as useful extensions of the context-free formalism.

Unification grammars have also drawn our attention. They progressively incorporate linguistic rules in a lexicon in the form of *feature structures,* so that the only rule that is finally maintained is the very general *unification* principle. We have shown that lexical items can be formally described with the PATR formalism, and depicted as attribute-value matrices or in the form of directed acyclic graphs.

References

BOURLARD, H.A., and N. MORGAN, (1994), *Connectionist Speech Recognition - A Hybrid Approach,* Boston, Kluwer.

BRESNAN, J.W., and R.M. KAPLAN, (1982), "Lexical Functional Grammar", in *The Mental Representation of Grammatical Relations*, MIT press, Cambridge, MA.

CHARNIAK, E., (1993), *Statistical Language Learning*, MIT press, Cambridge, MA.

CHOMSKY, N., (1957), *Syntactic Structures,* Mouton, The Hague.

EARLEY, J., (1970), "An Efficient Context-Free Parsing", *Communications of the ACM,* n°14, pp. 453-460. Reprinted in *Readings in Natural Language Processing*, Barbara J. Grosz, Karen Sparck Jones, and Bonnie Lynn Webber, eds., 1986, pp. 25-33, Morgan Kaufmann, Los Altos, CA.

FORNEY, G.D., (1978), "The Viterbi Algorithm", *Proc. IEEE,* vol. 61, pp. 268-278.

GAZDAR, G, and C. MELLISH, (1989), *Natural Language Processing in Prolog: an Introduction to Computional Linguistics*, Addison-Wesley, Reading, MA.

GAZDAR, G., E. KLEIN, G.I. PULLUM, and I.A. SAG, (1985), *Generalized Phrase Structure Grammar*, Harvard University Press and Blackwell's, Cambridge, MA.

GIBBON, D., (1991), "Linguistic Aspects of Speech Material Complexity", SAM stage report Se.8, UBI-SAM-1/91.

HARRIS, M.D., (1985), *Introduction to Natural Language Processing*, Reston.

HOPCROFT, J., and J. ULLMAN, (1979), *Introduction to Automata Theory, Languages, and Computation*, Addison-Wesley, Reading, MA.

JOHNSON, C.D., (1972), *Formal Aspects of Phonological Description,* Mouton, The Hague.

KAY, M., and R.M. KAPLAN, (1994), "Regular Models of Phonological Rule Systems", *Computational Linguistics,* 20 (3), pp. 331-378.

LARI, K., and S.J. YOUNG, (1990), "The Estimation of Stochastic Context-Free Grammars Using the Inside-Outside Algorithm", *Computer Speech and Language*, n°4, pp. 35-56.

LIPPMAN, R.P., (1987), "An Introduction to Computing with Neural Nets", *IEEE Magazine on Acoustics, Speech, and Signal Processing*, vol. 4, n°2, pp. 4-22.

MICLET, L., (1979), *Inférence de Grammaires Régulières,* Ph.D. dissertation, ENST.

MICLET, L., (1984), *Methodes Structurelles pour la Reconnaissance des Formes*, Eyrolles, Paris.

MOHRI, M., (1994), "Compact Representations by Finite State Transducers", *Proceedings of the 32nd meeting of the Association of Computational Linguistics,* Las Cruces, New Mexico.

MOHRI, M., and R. SPROAT, (1996), "An Efficient Compiler for Weighted Rewrite Rules", *cmp-lg*/9606026[17].

PEREIRA, F., and M. D. RILEY, (1996), "Speech Recognition by Composition of Weighted Finite Automata", *cmp-lg*/9603001

PEREIRA, F., and D. ROE, (1992), "Empirical Properties of Finite-State Approximations for Phrase Structure Grammars", *Proceedings of the International Conference on Spoken Language,* Alberta, pp. 261-264.

PEREIRA, F., and S.M. SHIEBER, (1987), *Prolog and Natural-Language Analysis*, CSLI Lecture Notes N°10.

PEREIRA, F., and D. WARREN, (1980), "Definite Clause Grammars for Language Analysis - a Survey of the Formalism and a Comparison with Augmented Transition Networks", *Artificial Intelligence,* 13(3).

POLLARD, C.J., and I.A SAG, (1987, 1991), *Information-Based Syntax and Semantics*, Vol.1: *Fundamentals*, CSLI Lecture Notes N°13, Chicago University Press; Vol.2: *Agreement, Binding and Control.*

RABINER, L.R., and B.H. JUANG, (1986), "An Introduction to Hidden Markov Models", *IEEE ASSP magazine,* January, pp. 4-16.

RUMELHART, D.E., G.E. HINTON, and R.J. WILLIAMS, (1986), "Learning Internal Representations by Error Propagation", in *Parallel Distributed Processing*: *Exploration of the Microstructures of Cognition*, vol. 1, D.E. Rumelhart and R.J. McClelland, eds, MIT Press, Cambridge, MA, pp. 318-362.

SABAH, G., (1898), *L'Intelligence Artificielle et le Langage, tome 1*: *Représentation des Connaissances, tome2*: *Processus de Compréhension,* Hermes, Paris.

SHIEBER, S.M., (1986), *An Introduction to Unification-Based Approaches to Grammar,* Chicago University Press, Chicago.

WINOGRAD, T., (1983), *Language as a Cognitive Process*, Vol.1, *Syntax*, Addison-Wesley, Reading, MA.

WINSTON, P.H., (1979), *Artificial Intelligence,* Addison-Wesley, Reading, MA.

WOODS, W.A., (1979), "Transition Networks Grammars for Natural Language Analysis", *Communication of the Association for Computer Machinery,* vol. 13, n°10, October.

[17] CMP-LG is the Computation and Language E-Print Archive set up by the consortium for lexical research. It can be accessed through the WorldWideWeb at URL address http://xxx.lanl.gov/abs/cmp-lg. Papers are stored under the name *yymmnnn*, where *yy*, *mm*, and *nnn* refer to its year and month of submission and to its number.

Chapter Three

NLP ARCHITECTURES FOR TTS SYNTHESIS

La grammaire est l'art de lever les ambiguités de la langue;
mais il ne faut pas que le levier soit plus lourd que le fardeau.
(Grammar is the art of removing the ambiguities of a language;
but the lever should not be heavier than the burden).

A. Rivarol, (1784), *De l'Universalité de la Langue*

Text analysis and, better still, text understanding are some of the biggest challenges taken up by artificial intelligence. Over the last thirty-five years, computational linguistics specialists have proposed an impressive number of linguistic formalisms and inference methods to tackle these problems. Is it not surprising, under these conditions, that few of the speech synthesis systems commercialized up to now embody a somewhat complete parser to properly examine the sentences to pronounce? And how do we explain the desperate eagerness of specialists to obtain a reasonable synthetic speech quality with just some simple syntactic heuristics, while research teams in natural language processing (NLP) now focus on the much more complex task of analyzing semantics and pragmatics in the context of automatic understanding, translation, or production of natural language?

Time, memory, and coverage constraints are the main reasons for this paradox. The point is that TTS systems *a priori* have to cope with any text, from simple isolated sentences to complex paragraphs, with possibly embedded phrases, some of them with discursive, if not nongrammatical, structures: whatever the data supplied at the input of a TTS system, speech *must* be generated! Now anyone who has ever been involved in NLP knows that the time required to extract some linguistically relevant information from a text, rapidly goes to infinity as the accuracy of the expected information increases: the more you know, the harder it is to know more (Fig. 3.1). Worse still, obtaining even elementary information becomes questionable as the text to be examined gets more elaborate. This typically originates in the fact that natural languages constantly infringe rules, at any level of the analysis process, so that slightly refining a given analysis or increasing the coverage of a given rule frequently implies augmenting the complexity of the analysis system (by adding rules or lexical entries) in a way that is hard to justify by the refinement it provides. What is more, from the standpoint of the developer, it is also often on a par with substantial changes in the language model and parsing strategy adopted. Hence our quotation from the Comte de Rivarol.[1]

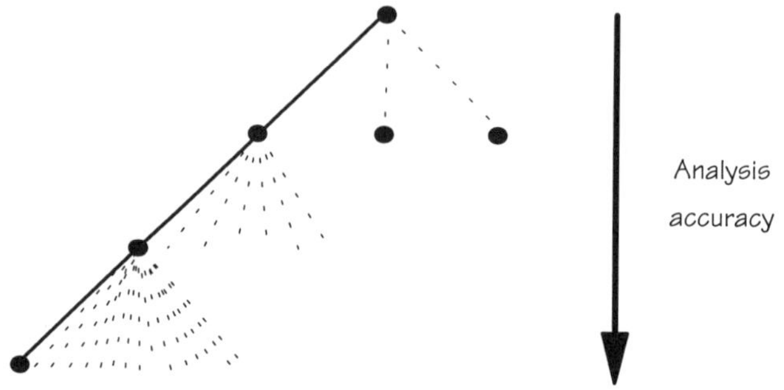

Fig. 3.1. The exponential complexity of the parsing problem: the more you know, the harder it is to know more because the additional amount of information to be processed increases dramatically.

In that respect, some analysis heuristics have succeeded in providing coarse information with a reasonable coverage, at little computational cost and with a reduced lexicon size. Nevertheless, in many cases such optimization-oriented approaches to linguistic problems have also led to describe rules in *procedural* terms (by specifying *actions* to be performed on the data for a rule to apply) rather than in *declarative* ones (by specifying *conditions* on the data, the verification of which is left for a general *inference engine*). This has resulted in hard-coded parsers, the re-usability and extendibility of which is intrinsically limited. In a long-term perspective, a clear separation between the rules of a grammar and the way they are dealt with seems inescapable[2] to us. Likewise, data formalisms should be kept independent from rule specification ones, and should not themselves imply a particular parsing algorithm. These topics will be discussed here.

3.1. Data formalisms

Many early TTS systems made use of linear exchange structures between sequentially organized processing modules (Carlson and Granström, 1976, Hertz, 1982, McAllister, 1987, Allen *et al.*, 1987). The information processed by each unit was merged with its own input character string, and supplied as the input string of the next unit (Fig. 3.2). As a result, intermediate representations become increasingly complex, hence less easy to

---

[1]The same Antoine Rivarol also used to say that "ce qui n'est pas clair n'est pas français!" (What is not clear, is not French!). If only it were true.

[2]More specifically, it is the numerous constraints imposed by rule formalisms, as opposed to programming languages, that is desirable in a text analyzer. After all, as Winston (1979) shows, proceduralists and declarativists secretly agree, as rules and programs are essentially identical things with different levels of constraints.

handle, as processing evolves.[3] A single word like *developments*, for instance, would successively be rewritten as *de-ve-lop-ments*, where - indicates a syllable boundary, as *de-ve-lop/ment/s*, where / indicates a morpheme boundary, and as *de-vE-lop/ment/s*, in which uppercase refers to stress.

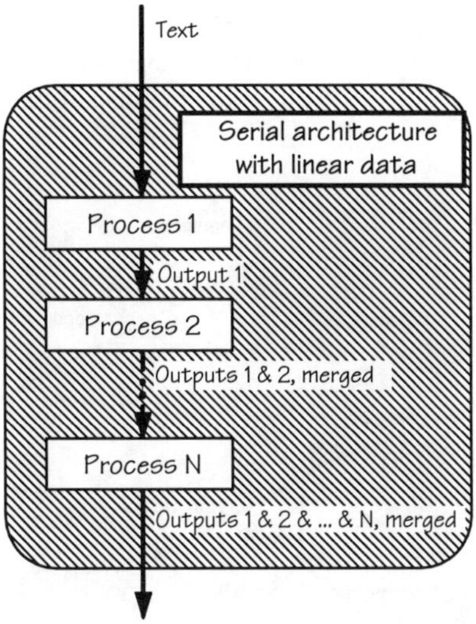

Fig 3.2. An old natural language processing strategy for TTS synthesis: linear exchange structures between sequentially organized processing modules.

In contrast, given the various connections between the many levels of description of a language (phonetics, phonology, morphology, syntax, semantics, and pragmatics), and since the way they can be related to each other is seldom known in advance, it is common practice to organize the data handled by a natural language processing system into an analysis framework that allows relationships to described in many ways—that is, with as many types of grammars as possible—and to be later refined by specifying additional dependencies and possibly additional levels. When trying to catalogue recent TTS systems regarding their internal data organization, two trends emerge. Some benefit from the powerful feature description structures increasingly used by computational linguists—namely, the *feature structures*[4] (FSs) (Fig. 3.3) presented in Chapter Two, in which each level is associated to a given PATR category, as in the work of Cericola *et*

---

[3]The only sensible explanation we see for this is that machine speech in the early days of TTS synthesis was erroneously perceived as mostly a signal processing problem, language processing issues being assumed to reduce to simple transductions. The deep commitment of engineers into TTS system development might well have initiated this bias.

[4]Or, almost equivalently, by terms in a DCG-like approach or any other king of logic programming. In the sequel, we shall systematically refer to FSs because of their increased visual appeal, although most of these so-called FS-based approaches turn out to be implemented with DCGs in current TTS systems—that is, with terms.

al., 1989, in PROPHON (Bäcksröm *et al*, 1989, Ceder and Lyberg, 1992), COMPOST[5] (Bailly and Tran, 1989), LIPSS (Dutoit, 1993) and SVOX (Traber, 1993). Others are based on a so-called *multilevel* (or *multitier*) *data structure* (MLDS) (Fig. 3.4), in which each level appears as an independent description of the sentence, synchronized with the other ones, as in SCYLA (Lazzaretto and Nebbia, 1987), DEPES (Van Coile, 1989), in LIFT (Frenkenberger *et al.*, 1991), SPEECH MAKER (Van Leeuwen, 1989, Van Leeuwen and te Lindert, 1993), PHRITTS (Meyer *et al.* , 1993), and recently in MULTIVOX (Lindström *et al.*, 1993) under the impulse of the DELTA system (Hertz *et al.*, 1985)).[6]

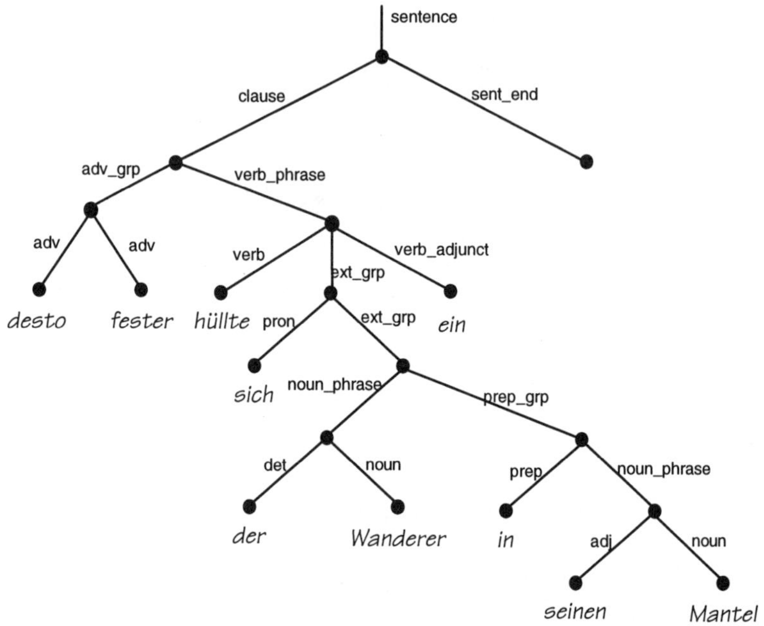

Fig 3.3. An example of a feature structure (depicted as a tree in this case—that is, a DAG with no symbolic sharing of values) when the syntax analysis module has operated on the German sentence fragment *desto fester hüllte sich der Wanderer in seinen Mantel ein* (adapted from Böhm, 1992): "*the more firmly the walker muffles himself up in his coat*").

---

[5]COMPOST actually uses (object-oriented) trees—degenerated DAGs in the sense that trees do not allow features to be *shared* among categories. The rules that can be applied on COMPOST trees, however, include a powerful mark-and-copy mechanism that makes it possible to copy or merge sub-trees through the use of variables. This approach therefore lies somewhere between MLDS and FS-based worlds.

[6]It should be mentioned, however, that some powerful grammar formalisms still make use of flat, linear structures, as it is the case for the interesting constraint grammars proposed in (Karlsson 90), which will be quoted in Chapter Four.

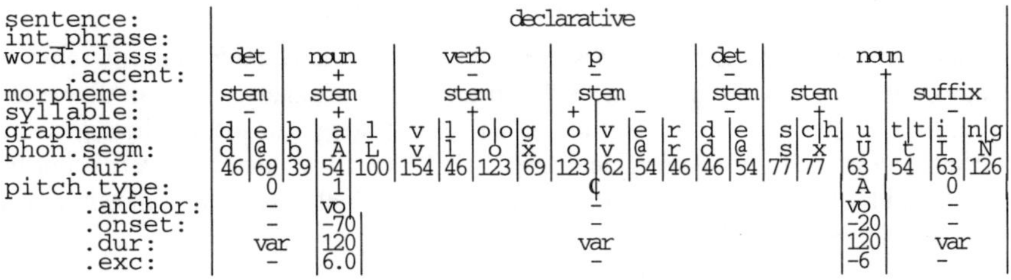

Fig 3.4. An example of a multi-level data structure (MLDS) when all the analysis modules have operated on the Dutch sentence *de bal vloog over de schutting* (*the ball flew over the fence*, after Van Leeuwen and te Lindert, 1993).

Advantages of MLDSs and FSs over a linear organization of information are numerous:

- They naturally increase readability, regarding both data and rules. Indeed, as a result of the intrinsic handling of underspecification with MLDSs and FSs, rules can be underspecified—refer only to the partial information they require and the level(s) or categorie(s) on which they operate.
- Extendibility is greatly enhanced. Since MLDSs and FSs also intrinsically admit unspecification, one can always specify additional layers to MLDSs or categories to FSs to account for new analysis modules, in a way that remains transparent to previously developed ones. This was certainly not the case with linear data structures.
- Debugging is made easier, since information provided by distinct modules is stored at different levels or under different categories. Tracing and understanding the history of data accesses and modifications is therefore facilitated.
- Linear data structures make it hard to exploit additional linguistic knowledge that might be available at the input, as it is the case when speech synthesis is performed from machine-generated concepts (like in dialogue systems; see Youd and Fallside, 1989) rather than from plain text (as in full TTS systems). NLP systems based on MLDSs or FSs offer natural interfaces in all cases: synthesizing speech from concepts simply implies that the MLDS is initially filled with information on other levels than the graphemic one (such as on the *part-of-speech* or the *accent* levels for the MLDS of Fig. 3.4), or that the FSs provided by the lexicon for each word of the sentence are initially linked in some way, which is easy to describe in the PATR formalism.
- Since data is made independent of rule formalisms (see Section 3.2), inter-language portability is much better ensured. No wonder, then, that the introduction of FSs and MLDSs in TTS systems have been initiated by research laboratories with multilingual concerns.
- It is possible to replace a given implementation of a processing block by another implementation of the same functionality but using another rule formalism, another

algorithm, another programming language, and so on. The generality and flexibility of FSs and MLDSs removes burden (from the data representation side, at least).

- Finally, FSs and MLDSs have less implications on process control, in the sense that they do not specifically imply the order, if any, in which analysis modules have to be run. As a matter of fact, the overall text analysis strategy can still be chosen as either serial, hierarchical, or heterarchical (Fig. 3.5).[7] This advantage, however, is not fully exploited by even state-of-the-art TTS systems, in which information mostly remains processed in a sequential way.

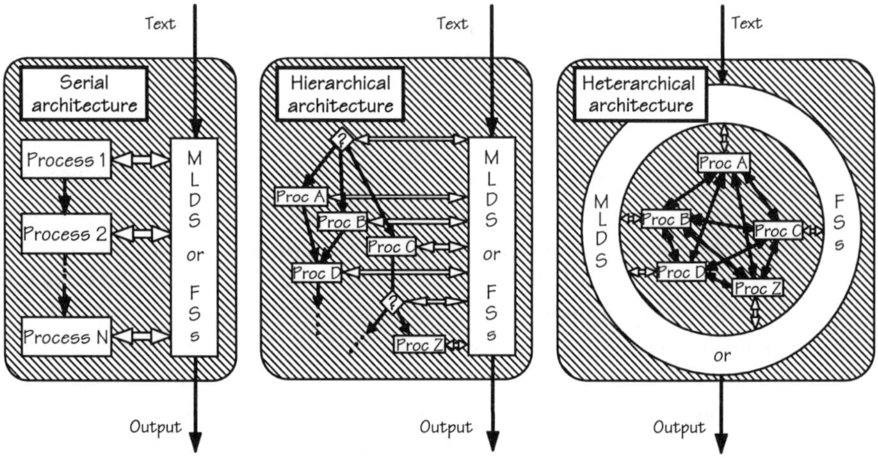

Fig. 3.5. MLDSs and FSs theoretically allow serial, hierarchical, or heterarchical scheduling. In the case of a hierarchical organization, control is still centralized (it is imposed *a priori*, hence the single-headed arrows in the corresponding diagram), so that block firing remains ordered although the exact control path may be data dependent. This is no longer the case for heterarchical architectures, in which control is really distributed and block firing becomes fully data dependent.

3.2. Rule formalisms

Figure 3.6 shows the typical natural language processing module of contemporary TTS systems. It comprises a text analyzer, aimed at providing useful syntactic (if not semantic or pragmatic) information to the subsequent letter-to-sound and prosody generation modules. These tasks will be detailed in Chapters Four, Five, and Six, respectively. For

---

[7] This recalls the blackboard concept adopted in the HEARSAY II system, in the context of the DARPA project for speech understanding (Erman *et al.*, 1980, Engelmore and Morgan, 1988).

the time being, let us restrict ourselves to general considerations about the way information can be handled.

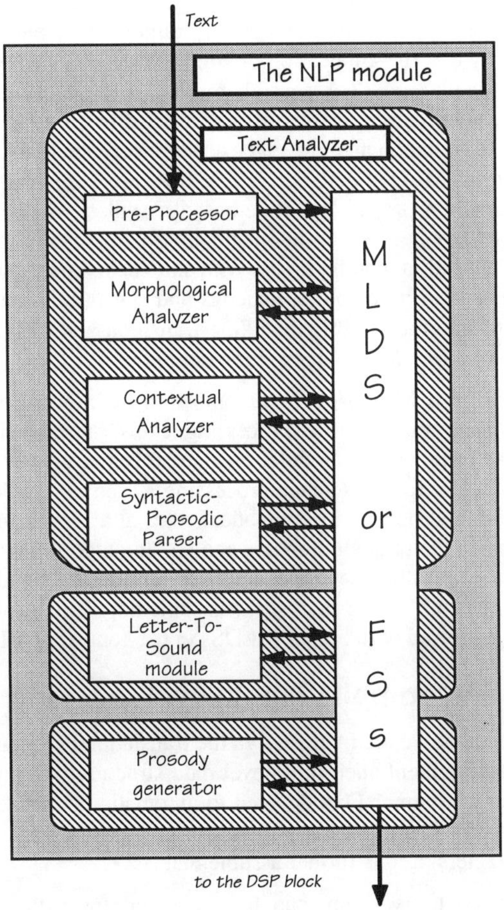

Fig 3.6. The NLP module of a typical text-to-speech conversion system.

The question of rule formalisms is a central one in the context of natural language processing for TTS synthesis. Too tight, they hopelessly confine the set of linguistic events they can describe, or at least they constrain rule writers to use craft to achieve their end; too lax, they bring confusion in the writer's mind. A formalism should be fit to describe a set of facts—not more, not less.[8]

---

[8]To a larger extent, this *principle of parsimony* lies at the very heart of the scientific method itself. As stated by Scharf (1991) in the context of mathematical modeling: "economical models are better than luxury models,

Now it frequently happens that a given natural language processing task can be accounted for by a limited number of quite simple rules, which apply *most* of the time but not *all* the time: a considerable amount of highly intricate rules is then needed as well, the applicability of which is unfortunately much more restricted (as already pointed out in Fig. 3.1). The question is: How should this rather heterogeneous set of rules be handled in such a way that the application of simpler rules does not get penalized by the existence of more complex ones? A possible answer is to adapt rule formalisms to the complexity of each subproblem. After all, a complex problem is always fixed more quickly when specific components are given customized solutions. In that respect, an additional advantage of MLDSs and FSs is that they are sufficiently general to accept many types of information processing. Rule-based modules, whether relying on human knowledge (that is, *expert systems*) or automatically trained with large databases with the adequate inference algorithm, can be mixed with statistical parsers and neural networks, which do nothing else than hard-coding lexicons and rules (to be understood in a broad sense in this section) into their coefficients (the transition and emission probabilities of HMMs and the synaptic weights of MLPs, for example).

However, in addition to a MLDS or FS organization of information, which implicitly imposes the same *data* formalism at all processing levels, the existence of a unified *rule* formalism inside each processing module, and possibly throughout all of them, would also be a desirable feature for a language processing system. It would make it easier for linguists to write, test, and optimize grammatical rules at all levels, and for engineers to write, test, and optimize the related inference and parsing algorithms. The question, then, is also: How should this rather heterogeneous set of rules be handled in a somewhat uniform way? This question, as we shall see, has some implications on the particular rule formalisms that are preferred with FSs or MLDSs in the context of TTS synthesis.

3.2.1. MLDSs and compiled multilevel rewriting rules

Imagine, for instance, that we are interested in the translation of the information stored at a given level of the aforementioned multilevel data structure into another level, as will typically be the case in Chapter Five devoted to automatic phonetization (or grapheme-to-phoneme conversion). The input and output levels, then, are respectively composed of ASCII and IPA characters. Let us further assume that

1. Most of the related transduction can be accounted for with simple FST (hence, regular) rules as (2.4):

   $$a \rightarrow b \,/\, l \,\_\, r \qquad (3.1)$$

   in which l and r refer to the left and right context of a in the *input* level.

2. A number of rules, however, require examination of the left and right contexts of a, as well as the value associated with a itself, on *other* levels (a *syllabification* level, for example: to test the position of a within its syllable; or a *part-of-speech* level: to restrict the application of some rules to particular sequences of syntactic categories). Rules would then take the form:

---

because luxury models often produce spurious effects. These effects are artifacts of the procedure used to identify the model rather than actual characteristics of the system or signal under study".

$$a \rightarrow b\,/\,l\,\_\,r\,/\,\text{level i} : l_i\,\_\,r_i\,/\,\text{level j} : l_j\,\_\,r_j\,/ \ldots \text{level k} : l_k\,\_\,r_k \qquad (3.2)$$

which basically means that if a is found on the input level, surrounded by l and r on the same level, and by l_i and r_i, l_j and r_j, . . . , l_k and r_k on other levels $i, j, \ldots, k$, then b should be produced on the output level.

Such *multilevel rewriting rules* (MLRRs) are obviously best suited to MLDSs (Van Leeuwen, 1993). Within each module of the analyzer, they appear as regular, but their successive use by each module (with different input, output, and context levels each time) typically leads to a nonregular transduction task. Indeed, it recalls in some way the principle of ATNs. Transitions in a regular transition network are submitted to conditions that depend on previous actions. In the particular case of a MLDS, each module produces some output information on a level that can be used later as input or as context for another module. Their power is therefore similar to that of an ATN. What is more, from a formal point of view, uniformity is maintained: simple (regular) and complex rules are written and processed in very similar ways. MLRRs thus appear as the ultimate formalism, in that they provide an efficient answer to our two initial questions.

Things happen to be a little bit more complex in practice. When it comes to performing higher-level syntactic (not to say semantic or pragmatic) analysis, MLRRs face some problems,[9] witness the many extensions that have been proposed to the MLRR formalism of (3.2):

1. Powerful MLRR formalisms include keywords to skip characters or sequences thereof in the left and right context strings. Logical operators (such as *or, and,* and *not.*) and named *sets* (such as vowels, prepositions, liquids) are also useful to increase coverage and compact rules. A rule including

   skip([VOWEL, !u]) (3.3)

   in its left or right context string, for example, could require the MLRR parser to skip all vowels except *u*. Some formalisms even allow calls to procedural languages (like the C language). Others admit the use of variables to transfer some properties of the context to the characters to be rewritten. Regular expressions, like

   word : [-focus] {1,3} *or* segment : [-voice]* (3.4)

   are also in use (which respectively stand for: *at least one, not more than three non-focal words* and *any number of unvoiced segments* (Lindström and Ljungqvist,1994). Such infractions to (3.2) then generally require that MLRRs be compiled into machine-readable code before being used, as in COMPOST or DEPES. If rules remain interpreted, the extra computational load typically prevents parsers from operating in real time, which results in development tools (as opposed to application ones). This was the case, for example, for the early implementation of SPEECHMAKER, which was said to run 20 to 30 times real time on a VAX/VMS 3100 workstation.

2. The weakness of MLRRs does not always originate in coverage limits. More exactly, one should say that *writing MLRRs for extracting higher-level features is not an easy*

---

[9] Besides, ATNs have proved inefficient in such cases.

task for a linguist. Rephrased in such a way, the problem can be understood as emerging from human constraints. This, as we shall see in Chapter Four, is typically the case for part-of-speech disambiguation or *tagging* and for syntactic-prosodic phrasing. It has motivated the introduction, in the contextual analysis and syntactic-prosodic parsing blocks of Fig. 3.4, of stochastic rewriting rules in the form of *N-grams*, of nonlinear multilevel rewriting systems in the form of *multilayer perceptrons* fed with contextual windows, and of corpus-based heuristic analyzers in the form of *classification and regression trees* (CARTs). These approaches indeed provide automatic and efficient training (that is, inference) algorithms. CARTs, in particular, provide an efficient means of inferring MLRRs, as they can handle heterogeneous contextual information.

3. It remains that such HMM, MLP and CART-based analyzers reduce syntax to its lowest levels. Syntax trees, if needed, have to be dealt with by enhanced context-free grammars—that is, mainly with FSs (or equivalently with terms). Besides, although state-of-the-art TTS systems are not involved in semantic or pragmatic parsing yet, it seems clear that one cannot spare unification-based grammars to investigate such intricate areas of natural language processing.

3.2.2. FSs and bottom-up chart-parsed DCGs

The main difference between FSs and MLDSs precisely originates in

- The synchronization of information among levels in the multilevel case, a peculiarity that is neither easy nor even really interesting to maintain with FSs. As we shall see, indeed, FSs have something different to offer.
- The strong relationship between FSs and CF-PSGs, due to the fact that DCG rules are basically CF-PSGs with variables. In contrast, MLDSs are more related to regular grammars, given the regular-like form of MLRRs. This has motivated many TTS system developers to restrain the use of MLDSs to lower-level linguistic tasks known to be accountable for with regular transducers: typically, morphological analysis and letter-to-sound transduction. We show in the sequel that even these simple tasks can somehow benefit from a DCG formulation, which leads to more declarative and more compact rules, while not affecting too much the computational load of the related parsers.

As a matter of fact, MLRRs are generally used as transducers—that is, they actually have to perform two distinct tasks:

1. They have to *parse* the input level—that is, to uncover its inner organization by specifying appropriate left and right contexts.
2. They have to *translate* the input level into the output one, on the basis of the information gained from the contexts.

When both operations are performed simultaneously, rules often get confusing and less readable, and their application becomes increasingly dependent on extralinguistic details such as rule ordering and scanning direction (left-to-right, right-to-left) and scanning hierarchy (from small units to bigger ones, or the converse). If on the opposite they are performed separately, an additional level is needed to store the results of the parse, and the input sequence has to be scanned twice.

Taking again the example of grapheme-to-phoneme transduction, for instance, the left and right contexts of MLRRs are often used to determine which character(s) should be considered as *a* in rule (3.2)—that is, as a means to parse the input string from a *graphotactic* point of view. This is the case for instance with the French graphemic sequence *an*, which is pronounced /ã/ in *bandit* and *banc* but /an/ in *panoplie* and *panne.* Such transcriptions could be accounted for by specifying rules for simultaneous parsing and translation, like

an → ã / _ [CONSONANT, !*n*] (3.5)

a → a (3.6)

nn → n (3.7)

n → n (3.8)

Their interpretation, however, is made difficult by the fact that these rules depend on each other: they are ordered and not mutually exclusive. Their linguistic significance is penalized by the tacit assumption of a left-to-right scanning starting from bigger to smaller units (rules change if another scanning direction or hierarchy is chosen). One could equivalently change (3.5) into a translation-only rule like

an → ã / _ !*n* / syllab : _ I (3.9)

(meaning that *an* is pronounced ã when not followed by *n* and immediately before a syllable break), but this would nevertheless require a previous syllabification (another level of parsing).

In contrast, given the higher level of abstraction they provide, DCGs make it possible to parse and translate simultaneously and in a more declarative way—that is, with rules that do not rely on procedural details (such as the aforementioned scanning direction and hierarchy). Better still, the linguistic significance of DCG rules tends to be left unchanged when rules are considered separately. Traber (1993), for example, describes a declarative natural language processing module for German, which is completely based on DCGs, from syntactic parsing to phonetization, through morphological decomposition. Sentences are analyzed in terms of phrases, words, stems, graphemic sequences, and eventually phonemes with a uniform set of rules.

As already mentioned in Chapter Two, DCGs are built-in facilities with most Prolog interpreters. The point is that they are implemented as top-down, left-to-right, depth-first, nondeterministic parsers (as Prolog itself [10]), so that much processing can be wasted before a dead end is encountered (by confronting hypotheses on the sentence structure with the actual text to parse), and much more before an acceptable parse is found.

This, however, does by no means imply that the DCGs *notation* of rules cannot be used in conjunction with more efficient strategies, leaving the built-in engine as a development tool:

---

[10]Incidentally, Prolog was invented by A. Colmerauer precisely for language processing.

1. Indeed, although we have always understood a rule like

   ```
   nounphrase(np(Det,Adj,Noun)) -->
           determiner(Det,Gender,Number),
           adjective(Adj, Gender,Number),
           noun(Noun,Gender,Number).
   ```

   as

   "to check the existence of a noun phrase, suppose it is composed of a determiner, a noun, and an adjective, check for their existence, and verify that they agree in number and gender,"

   nothing prevents us from reading it as

   "if the string to parse begins with a determiner, check that it is followed by an adjective and a noun that agree in number and gender with it. If it is so, consider the sequence det+ adj+noun as an entity called *np*, and go on parsing—that is, search for rules with *np* as their leftmost right side argument."

   Such an interpretation provides our DCG with a bottom-up, depth-first, left-corner parser, which is obviously a better strategy when there exist many rules with the same left hand side category (*nounphrase* in our example), but few rules with the same first right side category, as it happens to be the case for many natural languages.

2. Another (and complementary) way of improving DCG parsers is to enable them to store already parsed sequences (whether parsing succeeded or not), so as to reduce the computational cost of dead ends by avoiding analyzing the same structure again and again. This is typically achieved with the active chart already mentioned in Chapter Two.

   Better still, chart parsing does not only improve parsing efficiency; it also allows a parser to be robust and deterministic. As a matter of fact, TTS synthesis puts special robustness and determinism demands on a parser, in the sense that whatever sentence is presented, whether legal or not, whether structurally simple or intricate, whether ambiguous or not, one and only one parse *must* be produced. This implies to be able to build a partial parse tree when no full-sentence analysis can be found and to make a choice among competitive parses in the case of an ambiguous sentence. Chart parsing performs well in that respect, as it provides parsed constituents of increasing size as parsing goes by, so that a sentence-level parser automatically changes into a phrase-level parser when nothing else can be done. Notice, however, that combining parsed constituents is not trivial, as some of them may overlap. In Traber (1993), a penalty is associated with each rule and constituents are affected with the sum of the penalties for the rules involved in their parsing. Finding the best partial parse then reduces to uncovering the optimal path through the chart, that is the path with the lowest total penalty. The same principle can be applied to derive the most likely parse in case of ambiguity.[11]

---

[11]Notice that incorporating penalties into rules turns the DCG grammar into something that is close to a stochastic context-free grammar, now being intensively studied in computational linguistics.

3. Finally, many TTS systems have adopted a mixed depth-first breadth-first scheme, by providing the lexicon with all the admissible part-of-speech categories of each word, or by forcing the morphological analyzer to derive all the available decompositions (hence all the admissible part-of-speech as well). In this way, the depth-first bottom-up approach sketched above actually starts with a breadth-first lexical query and may be followed by an early contextual analysis that quickly rejects false candidates and therefore considerably constrains the remaining search space.[12]

Despite these remarks on the computational efficiency of adequately parsed DCGs, it remains that MLRRs are faster than DCGs for problems with more translation than parsing. Consequently, most TTS systems presented at the beginning of this chapter as members of the FSs family (all but SVOX, actually) keep on using a MLDS accessed with MLRRs for low-level, transduction-like linguistic tasks, leaving FSs and DCGs for syntactic (and possibly semantic and pragmatic) analysis, for which MLRRs have proved inefficient.

3.3. Summary

It turns out from the previous discussion that efficient natural language processing modules should be organized into *specific sets of rules* operating on *multilevel data structures*, *feature structures*, or a combination of both, sharing as far as possible a *common formalism*, and preferably a *declarative* one. These goals can be achieved, to different extents, with compiled multilevel rewriting rules and with chart-parsed definite-clause grammar rules. Hence their intensive use in modern TTS systems. The former implement regular-like grammars (although complex grammars can be accounted for by using several output levels); the latter implement context-free like grammars (although complex grammars can be accounted for by using variables). Feature structures inherently lead to a more declarative formulation of linguistic facts. On the other hand, multilevel rewriting rules can be inferred automatically, with hidden Markov models, multilayer perceptrons, and CARTs.

References

ALLEN, J., S. HUNNICUT, and D. KLATT, (1987), *From Text to Speech, The MITALK System*, Cambridge University Press, Cambridge.

BÄCKSTRÖM, M., K. CEDER, and B. LYBERG, (1989), "PROPHON - An Interactive Environment for Text-To-Speech Conversion", *Proceedings of Eurospeech 89*, Paris, vol. 1, pp. 144-148.

BAILLY, G., and A. TRAN, (1989), "Compost, a Rule-Compiler for Speech Synthesis", *Proceedings of Eurospeech 89*, Paris, pp. 136-139.

BÖHM, A., (1992), *Maschinelle Sprachausgabe Deutschen und Englishe Textes*, Ph.D. dissertation, Ruhr-Universität Bochum.

CARLSON, R., and B. GRANSTRÖM, (1976), "A Text-to-Speech System Based Entirely on Rules", *Proceedings of ICASSP 76*, Philadelphia, pp. 686-688.

CEDER, K., and B. LIBERG, (1992), "Yet Another Rule Compiler for Text-to-Speech Conversion?", *Proceedings of the International Conference on Spoken Language Processing*, Alberta, pp. 1151-1154.

---

[12]This will be developed in Chapter Four.

CERICOLA, D., M. DANIELI, M.J. MOLLO, and D. VOLTOLINI, (1989), "Morpho-Syntactic Tools for Speech Processing", *Proceedings of Eurospeech 89*, Paris, vol. 1, pp. 386-389.

DUTOIT, T., (1993), *High Quality Text-To-Speech Synthesis of the French Language*, Ph. D. dissertation, Faculté Polytechnique de Mons.

ENGELMORE R., and T. MORGAN, (1988), *Blackboard Systems*, Addison-Wesley, Padstow, Great-Britain.

ERMAN, L.D., F. HAYES-ROTH, V. LESSER, and D.R. REDDY, (1980), "The HEARSAY II Speech Understanding System: Integrating Knowledge to Resolve Uncertainty", *Computing Surveys*, 12, 2, pp. 213-253.

FRENKENBERGER, S., M. KOMMENDA, and S. MOSSMÜLLER, (1991), "A Multi-Linear Representation and Rule Formalism for Phonetics in Text-to-Speech Synthesis", *Proeedings of the XII'th International Congress of Phonetic Sciences,* Aix-en-Provence, vol. 2, pp. 518-521.

HERTZ, S.S., (1982), "From Text to Speech with SRS", *Journal of the Acoustical Society of America,* n° 72, pp. 1155-1170.

HERTZ, S.R., J. KADIN, and K. KARPLUS, (1985), "The DELTA Rule Ddevelopment System for Speech Synthesis from Text", *IEEE Proceedings. on Acoustics, Speech, and Signal Processing*, n°73, pp. 1589-1601.

KARLSSON, F., (1990), "Constraint Grammars as a Framework for Parsing Running Text", *Proceedings of the Conference on Computational Linguistics*, Helsinki, vol. 3, pp. 168-173.

LAZZARETTO, S., and S. NEBBIA, (1987), "SCYLA: Speech Compiler for Your Language", *Proceedings of the European Conference on Speech Technology 87*, Edinburgh, vol. 1, pp. 381-384.

LINDSTRÖM, A., M. LJUNGQVIST, and K. GUSTAFSONN, (1993), "A Modular Architecture Suppoorting Multiple Hypotheses for Conversion of Text to Phonetic and Linguistic Entities", *Proceedings of Eurospeech 93*, Berlin, pp. 1463-1466.

LINDSTRÖM, A., and M. LJUNGQVIST, (1994), "Text Processing within a Speech Synthesis System", *Proceedings of the International Conference on Speech and Language Processing 94,* Yokohama.

MAC ALLISTER, M., (1987), "The Structural Design of the CSTR Text-to-Speech System", *Proceedings of the European Conference on Speech Technology 87,* Edinburgh, vol. 1, pp. 59-62.

MEYER, P., H.W. RÜHL, R. KRÜGER, M. KUGLER, L.L.M. VOGTEN, A. DIRKSEN, AND K. BELHOULA, (1993), "PHRITTS - A Text-To-Speech Synthesizer for the German Language", *Proceedings of Eurospeech 93*, Berlin, pp. 877-890.

SCHARF, L.L., (1991), "Reduced Rank Signal Processing", *Signal Processing*, 25, n° 2, pp. 113-133.

TRABER, C., (1993), "Syntactic Processing and Prosody Control in the SVOX TTS System for German", *Proceedings of Eurospeech 93*, Berlin, vol. 3, pp. 2099-2102.

VAN COILE, B.M.J., (1989), "The DEPES Development System for Text-to-Speech Synthesis, *Proceedings of the International Conference on Acoustics, Speech, and Signal Processing 89*, Glasgow, pp. 250-253.

VAN LEEUWEN, H.C., (1989), "A Development Tool for Linguistic Rules", *Computer, Speech and Language*, n°3, pp. 83-104.

VAN LEEUWEN, H.C., and E. te LINDERT, (1993), "Speech Maker: a Flexible and General Framework for Text-to-Speech Synthesis, and its Application to Dutch", *Computer, Speech and Language*, n°2, pp. 149-167.

VAN LEEUWEN, H.C., (1993), "Speech Maker Formalism: a Rule Formalism Operating on a Multilevel Synchronized Data Structure", *Computer Speech and Language*, n°. 4.

WINSTON, P.H., (1979), *Artificial Intelligence*, Addison-Wesley, Reading, MA.

YOUD, N.J., and F. FALLSIDE, (1989), "Driving a Speech Synthesizer from Conceptual Input in the Context of a Voice Dialogue System", *Proceedings of Eurospeech 89,* Paris, pp. 514-517.

Chapter Four
MORPHO-SYNTACTIC ANALYSIS

'Twas brillig, and the slithy toves
Did gyre and gimble in the wabe:
All mimst were the borogroves,
And the mome raths outgrabe.

Lewis Carroll, *Jabberwocky*

Chapters Five and Six will highlight the need for some syntactic processing in a high-quality text-to-speech system. Being able to reduce a given sentence into something like the sequence of its parts of speech, and to further describe it in the form of a syntax tree, which unveils its internal structure, is required for at least two reasons:

- Accurate phonetic transcription can be achieved only if the part of speech category of some words is available, and may depend on knowing the dependency relationship between successive words.
- Natural prosody heavily relies on syntax. It also obviously has a lot to do with semantics and pragmatics, but since very little data is currently available on the generative aspects of this dependence, TTS systems merely concentrate on syntax. Yet, as we shall see, few of them are actually provided with full disambiguation and structuring capabilities.

The reader may argue that *syntax* is a rather vague concept. In the context of TTS synthesis, it should be understood as the organization of sentences in the form of dependency relationships between words, examined through their part of speech only. This, in turn, depends on which part-of-speech categories are considered, as well as on what is meant by *word*. Indeed, in the extreme case of one part of speech per word, our definition of syntax remains meaningless. Notice, though, that such a partition is far from being unrealistic. Gross (1986), for example, has classified 12,000 French verbs on the basis of their distributional properties (the peculiar complements they can induce and the way these are formally introduced in sentences), and distinguished no less than 500 syntactic properties, so that the average number of verbs per class is one and a half! The notions of *word* and *part of speech* will be examined in Sections 4.1 and 4.2, respectively devoted to preprocessing and to lexical and morphological analysis. We have also mentioned in Chapter Three that the natural language processing module of most TTS systems is based on a mixed breadth-first depth-first strategy, in which the possible part-of-speech categories of each word are obtained first, and rapidly filtered by a contextual analyzer before being processed by the sentence level parser itself. Contextual analysis will be examined in Section 4.3. Since the sentence-level parser is strongly involved in prosodic considerations[1] (and is therefore termed a *syntactic-prosodic parser*), it will only be introduced in Section 4.4 and further detailed in Chapter Six.

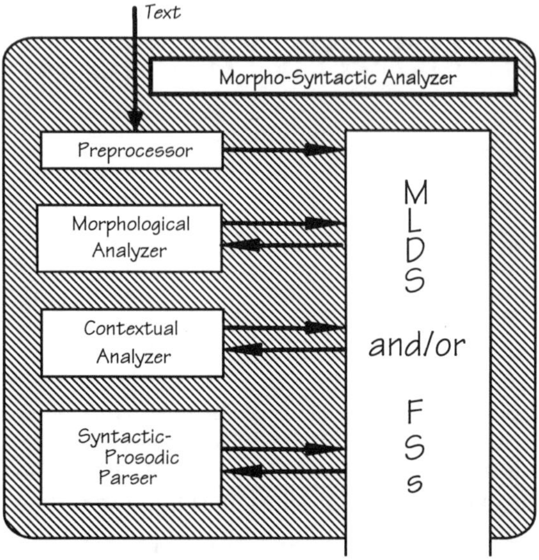

Fig. 4.1. A typical morpho-syntactic analyzer.

The morpho-syntactic analyzer (MSA) of several recent TTS systems is shown on Fig. 4.1. It comprises

- A pre-processing module, the main purpose of which is to serve as an interface between the linear text input and the internal data structure of the analyzer, typically a multilevel data structure (MLDS) or feature structures (FSs). All the words in the current sentence, including numbers, abbreviations, acronyms, and idiomatics, are identified and possibly transformed into full text.
- A morphological analysis module, the task of which is to propose all possible part of speech categories for each word taken individually.
- A contextual analysis module, which considers a word in its context and therefore makes it possible to reduce the list of its possible part of speech categories to a very

[1]Although its results can be put to use for automatic phonetization as well.

restricted number of highly probable hypotheses, given the corresponding possible parts of speech of neighboring words.

- Finally, a syntactic-prosodic parser, which examines the remaining search space and finds the text structure (its organization into clause and phrase-like constituents) that more closely relates to its expected prosodic realization.

4.1. Preprocessing

Texts are presented to the input of the MSA as strings of ASCII characters. They thus have to be transformed into lists of words. Clearly, words are the smallest units on which subsequent morpho-syntactic parsing will be based: they are processed as atomic units by the contextual analyzer and the sentence-level parser and serve as basic material for the morphological analyzer. Their correct detection therefore influences the accuracy of the whole analysis process. Texts, however, do not consist of orthographic words only. In order to complete its interfacing task, the preprocessor is also in charge of detecting the end of the sentence, processing punctuation marks, recognizing acronyms (*K.A.P.*, *TTS*), dealing with abbreviations (*Mr.*, *i.e.*), and transcribing numbers into literals.

4.1.1. Text segmentation

Preprocessing usually begins with a broad segmentation of the input text. McAllister (1989) has proposed a segmentation into sequences of so-called *orthographic islands*—that is, strings of ASCII characters delimited by *white space characters* (either a space, a tab, or a newline) or sequences thereof. Punctuation is then temporarily included into the islands, as in

1,000 words, Dr. Jones. ➔ *(1,000) (words,) (Dr.) (Jones.)*

As a matter of fact, an early incorporation of punctuation marks into syntactically relevant units (leading to their consideration as isolated sentence elements) may be erroneous, given the intrinsic ambiguity of many of them (as for *1,000* and *Dr.* in this example).

In most TTS systems, however, preprocessors isolate punctuation marks by default and later re-associate them with the orthographic island(s) they belong to, if needed. Strings are then typically segmented into *broad segmentation units* (BSUs) taken in the following set:

- A sequence of characters (upper- or lowercase),
- A sequence of digits,
- A single punctuation mark or another special character,
- A sequence of white space characters.

as in

(I)( )(know)( )(1)(.)(000)( )(words)(,)( )(Dr)(.)( )(Jones)(.)

Notice that a little problem arises as early as this segmentation stage, which explains why the preprocessor need not only *detect* but also *memorize* white space characters (hence their consideration as full segmentation units, rather than as simple delimiters): that of the ambiguous hyphens, covering compounds (*high-level*), quotation and cross-line hyphenation (which is sometimes used in printed matter to indicate that a word has been artificially broken due to typographic requirements). Confusing the role of these various hyphens leads to important segmentation errors, in the sense that it provides erroneous words. Correctly recognizing them implies having a look at possibly surrounding white characters (the type of white character is important for cross-line hyphenation, which is always followed by a new-line character).

Preprocessing, as we shall see, reduces in most cases into simple regular rules that rewrite the aforementioned BSUs into a list of word-like units and of syntax bearing punctuation marks (which we shall term as *final segmentation units*, FSU). One should not conclude, however, that segmentation has to be completed first (that is, for the whole text, since the segmentation stage is not able to detect the end of sentences) before starting rewriting its output. On the contrary, it is important to minimize look-ahead by processing segmentation units as soon as possible. In other words, the regular grammars needed to detect FSUs in the sequence of BSUs have to be examined for each incoming BSU, in a left-to-right scan.

4.1.2. Sentence end detection

The far most obvious (and awkward) cases of punctuation ambiguity[2] arise with sentence termination marks (*.*, *?*, *!*, *:*)[3], of which only the question mark is almost unambiguous. Semicolons also appear in hours (*12:30*) and ratios (*1:8*), exclamation marks sometimes (although seldom) refer to the mathematical symbol for *factorial*, and periods are also used in numbers to indicate a decimal point (in English: *1,567.6*) or an exponent of 1,000 (in French: *1.567,6*), and in dates (*12.3.87*). What is worse, is that periods often appear as parts of abbreviations and acronyms, so that disambiguating them does not reduce to a rule of the thumb like

> *Check for the existence of a digit before and after . in the same orthographic island*

nor to detecting capitalized words following periods, since capitalization is used in most languages for names, in some for titles, and even for nouns in German.

On the contrary, accurately recognizing end marks as such first requires parsing the BSUs for abbreviations, acronyms and numbers. However not only periods that have not been accounted for by the related grammars can be considered as sentence ends! Many abbreviations can occur at the end of a sentence (just think of *etc.*). In the tagged Brown corpus of a million words, about 90 percent of periods occur at the end of sentences, 10 percent inside abbreviations, and 0.5 percent in both (Riley, 1990).

Add to this, as pointed out by McAllister (1989), the valid and commonplace practice of quoting direct speech and of inserting personal feelings in texts, as in:

> *The man (and he certainly was one!) just said, "Maybe. I'll see. I can't promise."*

---

[2]A description of punctuation ambiguities is given, for English, in Liberman and Church (1992).

[3]Strictly speaking, semicolons do not end a sentence. In the context of TTS synthesis, however, this is usually allowed, since their left and right parts can be pronounced as if they were isolated, with no important effect on naturalness.

Such sentence embedding shows the complexity of segmenting texts into sentences. It has generally received little attention because of its relationship with discourse (hence with pragmatics), which remains an issue in state-of-the-art TTS systems.

Liberman and Church (1992) have also proposed a probabilistic solution to integrate all aspects of the problem, based on tuning a pattern matching algorithm to data. It has been successfully put to use for Swedish by Lindström *et al.* (1993).

4.1.3. Dealing with abbreviations

Abbreviations are generally read in their full form (*Mr.* reads [mɪstə]); what is more, they often include full stops: two good reasons to take special care of them. Their "grammar" actually reduces to a lexicon that puts abbreviations and related words in correspondence. Notice that the lexicon often includes sequences of BSUs rather than single units:

(n)(°) ➔ *(number)*
(e)(.)(g)(.) ➔ *(for)( )(instance)*

The names of physical units, currencies, or special characters (*kg, $,* +) can also be included in this lexicon, although some of them are ambiguous (mainly the short ones, such as *V*, which stands for *Volt* or for a Roman number). Ambiguity also arises when an abbreviation stands for inflected words, as for the German *tgl.* which corresponds to either *täglich, tägliche, täglichem, täglichen, täglicher, or tägliches* (Böhm, 1992). Finally, some specific but usual abbreviations are intrinsically ambiguous, as in *Dr. Jones lives at the corner of Jones Dr. and St. James St.* (O'Malley *et al.,* 1986). A practical solution to such ambiguity cases is to retain only the most often used acceptation (and possibly to adapt the lexicon to the type of text to read, which may be useful with physical unites) or to design more complex heuristic rules to take limited account of the context. Clearly, the adequate answer would be to leave the transcription for further processing, including syntactic, semantic, and possibly pragmatic analysis.[4]

4.1.4. Recognizing acronyms

As opposed to abbreviations, most acronyms need not be stored in a database, since their pronunciation does not require any extratextual knowledge. Most of the times, it reduces to the sequential reading of all their characters (as for *L.T.S.*) produced individually, or it defaults to usual pronunciation rules when the acronym is pronounceable (*GATT, NATO*).[5] Besides, their rate of generation is too swift to hope to keep up with all of them. They should nevertheless be detected and marked by the pre-processor in order to avoid detecting erroneous sentence ends and to inform the subsequent modules of their peculiar status.

---

[4]Such a strategy would typically imply a hierarchical or heterarchical architecture for the natural language processing system.

[5]The pronunciation of acronyms is itself a complex problem, in which pronunceability criteria play an important part. Spelling them is merely a makeshift.

Acronym surface forms can be accounted for by a regular grammar. Typical FSAs are given in Fig. 4.2. The top automaton describes forms like *to FPMs and he* and *to FPMs. Yesterday, he,* in which the period is correctly interpreted as syntactically bearing. The bottom one is concerned with *to F.P.Ms. and he* but also accepts *to F.P.Ms. Yesterday, he* as an acronym, thereby failing to detect the sentence end. In order to avoid this, the final . of acronyms described by the bottom automaton is not considered as part of the word when followed by an uppercase letter. It should be noticed, however, that this causes first-name initials (*T.M.G. Dutoit*) to be considered as a sentence ends. Such an ambiguity should typically be examined at a higher level, on the basis of syntactic plausibility of the resulting sentences. No current TTS system goes into such details. Notice also that a minimum length is imposed by the bottom automaton, so as to prevent proper names from being considered as acronyms at the end of sentences (*I saw John. He did it.*).

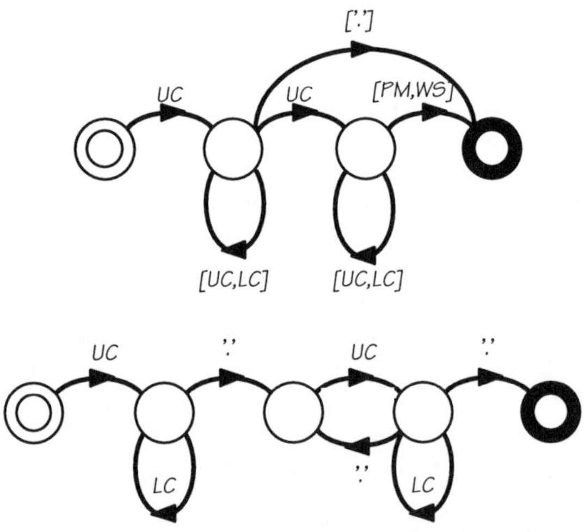

Fig 4.2. Describing acronyms with finite-state automata. UC and LC respectively stand for any upper- and lowercase letter. PM replaces any punctuation mark, and WS stands for a white space character. Notice that the final character in the top automaton (PM, WS, or .) is not considered as part of the acronym. Similarly, the final period in the bottom one is only accepted in the acronym if it is not followed by an uppercase character (this could not be easily shown on the corresponding transition graphs).

4.1.5. Processing numbers

Numbers combine the drawbacks of abbreviations and acronyms: they have to be recognized (by rule, since their number is obviously infinite) *and* transcribed. Again, a simple finite-state automaton like the one of Fig. 2.1 suffices to describe numbers in a formal way. [6] Once detected, numbers are first interpreted as rational numbers (*3.14*), time of the day (*2.16 pm, 13:26*), dates (*08.11.94*), and ordinal numbers (*the 16th*) depending on their context, and translated into the related litterals, following simple but tedious language-specific regular rewriting rules.

4.1.6. Dealing with idioms

Among the 50,000 entries of the *Oxford English Dictionary*, not less than 11,000 are idioms. Natural languages in general are particularly keen on idiomatic expressions, which by definition cannot be analyzed in terms of their constituents but should rather be taken as a whole lexical unit. What is the good of trying to parse *in spite of* or *as a matter of fact*, for example? In some TTS systems, a special lexicon is used to transform the related BSUs into a single FSU as early as in the preprocessing module. In others, idioms are left for the morphological analysis module, since some of them happen to be subject to morphological variations (in French, for instance, *in spite of* translates into *en dépit du* or *en dépit des*, or *en dépit de* according to the number and gender of the following name).

4.2. Morphological analysis

Words are generally classified into two broad categories: *function words* (determiners, pronouns, prepositions, conjunctions) and *content words*.[7] In contrast with function words, which can be stored in a small lexicon, content items theoretically form an infinite set, in the sense that new ones appear every day.[8] What is more, they appear in texts in a multitude of surface forms (masculine / feminine, singular / plural, mode + tense + person for verbs, case). Their existence is therefore often accounted for by rules, in a morphological analyzer.

4.2.1. Function words

Function words make up the skeleton of the sentence. Their identification is of prime importance (some rudimentary TTS parsers have even been proposed, which only focused on function words (Sorin *et al.*, 1987; O'Shaughnessy, 1987). It thus makes sense to keep an exhaustive list of them in a lexicon, which is not cumbersome as their number is particularly small (less than 1,000). What is more, as we shall see in Chapter Five, many function words have an unusual pronunciation (in the sense that it can often not be predicted by rule from their spelling) so that they have to be lexicalized in some way. A typical entry in the function word lexicon therefore includes orthographic (the spelling of the word), syntactic (its possible part of speech categories), and phonological information (its pronunciation). The words *he* and *the* for example, could be represented, in PATR notation (see Section 2.5), by

[6] Yet this one does not account for the scientific notation of numbers (*3.2 e-4*).

[7] Function and content words are sometimes referred to as, respectively, *meaningless* and *meaningful*. This is an oversimplified view : function words obviously contribute to the meaning of the sentence. Suppressing them, for example, introduces ambiguities.

[8] About 12,000 new words have appeared in French in the past twenty years.

word he:
<spel> = he
<syn cat> = pronoun
<syn num> = sing
<syn gen> = masc
<phon> = /hɪ/

word would: [9]
<spel> = would
<syn cat> = modal verb
<phon> = /wʊd/

in which *spel, syn, cat, num, gen,* and *phon* respectively stand for *spelling, syntax, category, number, gender*, and *phonology*.

Besides, the function word lexicon can be used to detect all the word groups that cannot be analyzed on a word basis (what Lindström and Ljungqvist, 1994, term as *collocations*). These include prepositional, pronominal, and conjunctival groups (*so as to, out of*) and idiomatic phrases (*at least, in fact, once upon a time, to and fro*). This, of course, implies that not only single words, but also sequences thereof, are searched in the lexicon (which has to be organized accordingly).

Notice that assigning part-of-speech categories to function words is not a trivial task in the context of TTS synthesis, for which it has proved useful to distinguish more categories than the standard dictionary categories. Indeed, one should always keep in mind that computational and traditional linguistics *a priori* have nothing in common but the object of their study: language. In this connection, the very limited access of state-of-the-art TTS systems to semantics has reinforced the syntactic importance of function words, mainly through their distributional properties. It often happens that many function words with the same traditional part-of-speech category happen to be only partially interchangeable, if at all. Pronouns are the most obvious case: *I see you*, for instance, cannot be changed into *You see I*. This has motivated computational linguists to use many more categories, defined on a distributional basis rather than on a semantic one. In Dutoit (1993), for example, nine classes of pronouns have been distinguished for the French language.

4.2.2. Content words

Morphology[10] is the part of linguistics that describes word forms as a function of a reduced set of abstract semantically bearing units, called *morphemes,* and subsequently separated into *lexical morphemes* (or *lexemes*), *derivational* and *inflectional* morphemes (referred to as *syntactic morphemes*). In *played*, for instance, *play* and *ed* are respectively the written expression of the lexeme *play* and of the syntactic morpheme that accounts, among other things, for the past tense. Morphemes are defined as abstract units in the sense that they may appear in several forms in the words they affect, like in *thought*, which is composed of *think* and of the same syntactic morpheme as above. Morphemes may even not appear at all, as in the past tense of *read.* In practice, when there is a direct

[9]Notice that *person* is not specified, since *would* can be first, second or third person, singular or plural.

[10]Given the lack of space in this textbook and the many areas it covers, we cannot give a precise definition of all the terms used in the context of morphology. Readers interested in more detailed descriptions should refer to Crystal (1985).

mapping between segments of a textually represented word and the underlying abstract morphemes, these segments are termed as *morphs* and are subsequently categorized as *roots* (or *stems*) or as *affixes*. Both the part of speech category and the meaning of surface words are explained in terms of these morphemic constituents. Hence their importance in TTS synthesis.

One generally distinguishes *inflectional*, *derivational*, and *compounding* morphology. Inflectional morphology takes a word and outputs a form of the same word appropriate to a particular syntactic context, due to number, gender, mode, tense or person agreement, or to a particular semantic goal, like in comparisons (*strong, stronger, strongest*) or for case marking in German. Inflections are based on applying affixes, separated into *prefixes* and *suffixes*, on lexeme *stems*. They do not change the word category. Notice that the inflectional morphology of English is quite simple in comparison with that of other languages. English verbs, for example, appear in an absolute maximum of eight forms (as for *be, am, are, is, was, were, been, being*), and most of them (the *regular* ones) in only four forms (as in *love, loves, loved, loving*). In contrast, French regular and irregular verbs exhibit not less than thirty-seven and forty-one forms, respectively. Derivational morphology takes a word and outputs a new word, usually with a different category (*family, familiar, familiarity, familiarize*). What is more, derivation is incomplete: all forms do not always exist for a given word (for instance, no adverb can be derived from *red*). Finally, compounding consists of combining two or more lexemes to form a new lexeme (*so-called, language-specific, morphophonology*).[11] Some languages, like Dutch, German, or Welsh, are particularly keen on such aggregations.[12] It should be noted that not any sequence of morphemes is generally allowed to be a legal word. Just like the ordering of phrases must agree with the *syntax* of a language, the order in which morphemes may appear in a word is described by *morphotactics*.

In the context of multilingual TTS synthesis, a preliminary morphological analysis of the sentences is a must, for many reasons:

- It is the only way to constrain the size of lexicons by storing word roots instead of full forms. Morphology is then used to determine how a word can be analyzed as some form(s) of some root(s) in the lexicon. This in turn requires analyzing words in terms of the stems and affixes of which they are made up, and to find the roots corresponding to the resulting stems.[13] The word *changing*, for instance, should be split into stem *chang* and affix *ing*, and there should be some way to find that *chang* corresponds to the root *change* in the lexicon.
- The way a word can be obtained from its root gives a lot of information on its syntactic category, number, gender, case, conjugation, tense, and so on. Moreover, many words are morphologically ambiguous (in the sense that one and the same inflection, for instance, is used in different cases). This may results in several

[11]The distinction between compounds and small phrases is not always clear.

[12]In Dutch, *exhibition of Hottentot tents* writes *hottentottententententoonstelling'*.

[13]This operation is sometimes termed as *lemmatization*, in contrast to *conjugation* or *declination*.

hypotheses for each incoming word. Morphology therefore has an indirect but strong influence on the prosody of sentences produced by TTS systems.

- Sometimes it happens that the phonetic transcription of a word can best be explained in terms of its morphological structure, as for the German word *Nebenstrasse*, in which *str* is pronounced /ʃtʀ/, although it is read /stʀ/ in *demonstration* because it is not root-initial. Similarly, *th* in the English word *hothouse* is not pronounced as /θ/ because it spans a morph boundary.
- Finally, stress assignment has a lot to do with morphology in so-called *stress-dependent* (or *free-stress*) languages, like English, German, Russian, or Arabic, in which stress can be lexically distinctive, as opposed to languages with *bound stress* (like French) in which it is assigned a fixed position in words. In German, for example, primary stress is often put on the first syllable of the root (e.g. *'Band, Ver'Band*).

Although morphological rules remain highly language-dependent, it is nowadays widely agreed that inflections, derivations, and compounds can be accounted for in most languages by regular grammars, except perhaps for discontinuous morphemes (like in *genomen* in Dutch, where *ge* and *en* simultaneously define a past participle), which could be more conveniently described by context-free grammars (Chomsky type 2).

4.2.2.1. Inflection

In Winograd (1972), rules have been explicitly expressed in a procedural way, in the form of a complete set of ordered tests and actions to be performed on input words as a function of their spelling. Tests examine word endings, and actions progressively suppress them and/or add new suffixes, until the result is found in a root dictionary (not to be confused with a lexeme stem dictionary), as for *plied = pli+ed = pl+i+ed = ply-y+i+ed*. Since morphological transformations are peculiar to word classes, the related derivation or inflection is only accepted if the part-of-speech category of the resulting root is known to accept it. In this way, inflections like *best = bes+t = be+s+t* are rejected, as only adjectives can be given a comparative form and '*be*' is not an adjective. Finally, when the operation performed corresponds to an inflection, the part of speech of the surface word is set to the one of the lexeme found. Otherwise, it is given a category which depends on the particular derivation performed.

Such a procedural view is only conceivable with languages with a small number of inflectional morphemes. A more declarative approach has therefore been followed by Pitrat (1983) to describe inflectional morphology in French, known to be a highly inflected language. He has organized morphological data into *lexeme, suffix group,* and *inflection group* lexicons, as follows:

lexeme: *inflection_group, stem_1, stem_2, ..., stem_N*

inflection_group: Inflection_mode_1, *suffix_group_1, i, j, ..., k*
Inflection_mode_2, *suffix_group_2, l, m, ..., n*
...
Inflection_mode_P, *suffix_group_P, u, v, ..., w*

suffix_group_1: *suffix_11, suffix_12, ..., suffix_1M*
suffix_group_2: *suffix_21, suffix_22, ..., suffix_2M*
...
suffix_group_P: *suffix_P1, suffix_P2, ..., suffix_PM*

in which each line describes a typical entry in each lexicon. Lexemes are associated with an inflection group, which is supposed to describe their inflectional properties, and a series of lexeme stems in order to account for the fact that surface forms are not simply obtained by concatenating a root to a series of suffixes. Inflection groups propose a number of inflection modes (one for nouns and adjectives, as many as the number of tenses for verbs). Each inflection mode refers to a particular suffix group, and gives the indices (denoted as *i, j,...k*) of the stems to be concatenated with the suffixes in the their suffix group. Surface forms for a given lexeme are then assumed to be of the form:

stem_i+suffix_11, stem_j+suffix_12,..., stem_k+suffix_1M,
stem_l+suffix_21, stem_m+suffix_22,..., stem_n+suffix_2M,
...
stem_u+suffix_P1, stem_v+suffix_P2,..., stem_w+suffix_PM,

A verb like *tenir*, for example, whose conjugation is similar to the one of *venir,* would be entered as (Sabah, 1989)

tenir: *venir, tien, ten, tienn, tin, tîn*

venir: *indicatif_présent, suf_ind_prés, 1, 1, 1, 2, 2, 3*
indicatif_imparfait, suf_ind_imp, 2, 2, 2, 2, 2, 2
subjonctif_présent, suf_subj_prés, 3, 3, 3, 2, 2, 3
...

suf_ind_prés: *s, s, t, ons, ez, ent*
suf_ind_imp: *ais, ais, ait, ions, iez, aient*
suf_subj_prés: *e, es, e, ions, iez, ent*
...

which accounts for forms like

je tiens, tu tiens, il tient, nous tenons, vous tenez, ils tiennent
je tenais, tu tenais, il tenait, nous tenions, vous teniez, ils tenaient,
que je tienne, que tu tiennes, qu'il tienne, que nous tenions, que vous teniez, qu'ils tiennent,
...

Clearly, inflection and suffix groups are a way to decrease memory requirements: they are given once for all the lexemes they account for. Notice also that the choice of stems and suffixes can generally be done in different ways (like in *chant/ais, chant/erais* or *chant/ais, chanter/ais),* in which case the choice that minimizes the size of the lexeme dictionary (by minimizing the number of stems) is retained (the first one, in our example).

Using lexeme, suffix group, and inflection group lexicons to parse content words involves the following steps:

1. Get as many segmentations of the word as there are admissible endings at its end. By *admissible*, we understand terminations that can be found in at least one suffix group (possibly many) of the suffix lexicon.

   Then, for each admissible segmentation,

2. Check that the corresponding lexeme stem appears in at least one entry of the lexeme lexicon. If not, reject the current segmentation.

   Otherwise, for all the entries of the lexeme lexicon that contain the current lexeme stem,

3. Check in the corresponding inflection group (in the inflection lexicon) for all the appearings of the current suffix in the suffix groups admitted by the current inflection group. If no match is found, reject the current entry in the lexeme lexicon.

   Otherwise, for every suffix group found,

4. In the related inflection mode, get the index of the lexeme stem corresponding to the current suffix, and check in the current lexeme entry of the lexeme lexicon that the corresponding stem is the current lexeme stem (as obtained from the segmentation). If not, reject the current suffix group. Otherwise, accept the parse, and deduce the morphological features of the word from the current inflection mode and from the current index.

As a result, *tiennent* is recognized as a possible verb, which can be either the present indicative, third person plural, or the present subjunctive, third person plural of *tenir*. It should then also be checked as a possible noun or adjective, with other inflection group, lexeme, and suffix group lexicons accounting for number, gender (for French, for example) and case agreement (typically for German).

4.2.2.2. Compounding

In languages like German or Swedish, inflection is a minor concern in comparison with derivation and, better still, compounding, as shown on Fig. 4.3.

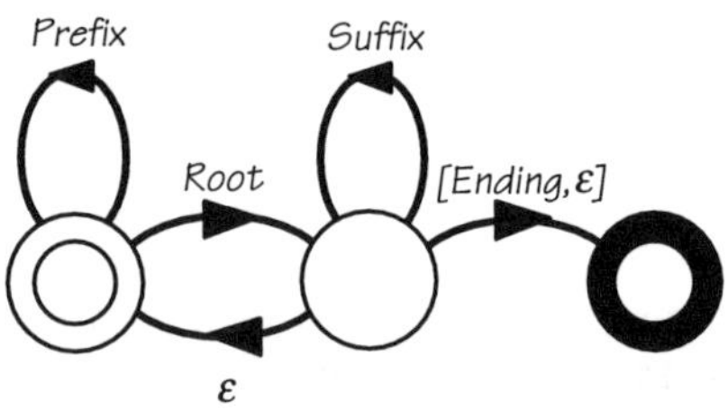

Fig 4.3. The morphological structure of German words (after Böhm, 1992). Notice that ε denotes an empty transition (a transition that produces no output in the surface form). *Endings* account for inflection, *Prefixes* and *Suffixes* allow derivation, and the backward ε generates compounds.

The peculiarity of Fig. 4.3 originates in the existence of several loops (even possibly embedded ones in the case of compounds) that, combined with the fact that affixes are generally short elements, give rise to a higher degree of nondeterminism than when dealing with inflection only (ex: *au-tomaten / auto-maten*). As a result, describing the functioning of a morphological parser as an n step process (in procedural terms!) becomes hazardous. Parsing involves a lot of backtracking. No wonder, then, that German morphological analyzers tend to describe word structure in the form of a true PSG (as induced on Fig. 4.3) and to provide it with adequate parsing strategies in order to minimize the resulting computational load. SVOX (Traber, 1993), for instance, is based on chart-parsed DCGs, Zinglé (1990) uses a customized PROLOG-based expert-system, and Pachuncke *et al.* (1994) exploit FSAs. In Karlsson (1990), such ambiguities are first dealt with by a local disambiguation module based on simple word-level constraint rules, and further processed by a constraint grammar formalism implemented in the syntactic module itself (see Sections 4.3.3 and 6.1.2.2).

Most commercially developed TTS systems that perform some kind of linguistic analysis have adopted morphological analysis principles similar to the two previous ones. See the CNET TTS conversion program for French (Larreur *et al.,* 1989), or the MITALK system (Allen *et al.*, 1987), in which derivational morphology is accounted for by a finite state automaton, which describes legal morpheme sequences. They are generally formulated as MLRRs, FSs, or with DCGs.[14]

Notice, finally, that a unified (and language-independent) formalism has appeared, termed as *two-level morphology*, and implemented in the KIMMO system (Koskenniemi, 1983). Two-level rules have the form

$$a{:}B \Rightarrow ReExp_1 \_ ReExp_2 \tag{4.1}$$

$$a{:}B \Leftarrow ReExp_1 \_ ReExp_2 \tag{4.2}$$

in which a and B are symbols of the vocabulary of the input (or *lexical)* and output (or *surface*) levels, respectively ($a{:}B$ is termed as a *character pair)*, and $ReExp_1$ and $ReExp_2$ are regular expressions over the set of character pairs, related to the left and right contexts of the pair ($a{:}B$). When applied to morphology, the input and output levels respectively relate to the morphemes that make up a word and to the corresponding surface form, both represented as in standard orthography. Rules (4.1) are termed as *context restriction* rules. They state that if a rewrites into B, then it must be that $ReExp_1$ and $ReExp_2$ succeed—that the left and right contexts match the pairs involved. In contrast, rules like (4.2) are termed as *context coercition* rules. They mean that a in the input level produces B on the output level when the above condition applies (but not only then). A combination of both is additionally indicated as $\Leftrightarrow$, and prohibition rules are also allowed (denoted by $/\Leftarrow$), which state that a pair may not appear in a given context. In practice, the input and output levels do not really share the same alphabets: additional abstract symbols, termed as *archigraphemes*, are allowed in the lexical level so as to account for higher abstractions. An archigrapheme N, for example, could map to n or m to account for the fact that *iN* + *sane* = *insane*, while *iN* + *polite* = *impolite*. In order to

[14] The outline of a true FS-based (as opposed to terms with DCGs) implementation can be found in Gazdar and Mellish (1989).

increase the power and the conciseness of two-level rules, more than one context is allowed, and variables may be used and related to each other.

In addition to standardizing morphological analysis formalisms, two-level rules have the advantage that they can be compiled into finite state transducers (Kartunnen *et al.*, 1987), which ensure a high processing speed and a total reversibility of the process (the same system can be used to transduce its output into its input). What is more, they are intrinsically independent of ordering (they could actually be applied in parallel), which somewhat increases their readability, since rules can be examined more independently of one another. Analyzers are now available for a large number of languages. To our knowledge, however, no TTS system currently implements them.

4.2.3. Computational aspects

A number of steps can be taken so as to optimize time and memory constraints. They do not provide more accurate morphological decompositions (some of them even result in additional ambiguities). They are, however, of a prime importance for the computational linguist, and particularly in the context of TTS synthesis, given the *a priori* wide variety of texts encountered.

4.2.3.1. Organizing lexicons into efficient data structures

Lexicons are intensively consulted in the course of the analysis of almost every incoming word, more than once per word on the average since content words, as we have seen, give rise to at least as many lexicon lookups as the number of admissible segmentations. It is therefore important to optimize access times, which mainly condition the analysis speed. A practical way to do this is to organize lexicons into specialized data structures developed by computer scientists: *hash-tables, tries* and *B-trees* (Knuth, 1973, Froideveaux *et al.*, 1990).

Most often, entries in a lexicon contain information indexed by a *key*. The function word lexicon, for instance, informs us about the morphosyntactic features of its keys (the function words themselves). There are basically three ways of storing keys: in a table with fixed length elements, in a list, and in a tree. Choosing a table implies setting the size of its elements as the maximum size of the items to store,[15] which is not efficient in terms of memory, while elements can be accessed very quickly given their fixed size. Alternatively, choosing a list implies scanning the whole list each time a key is searched for. Trees actually combine the advantages of tables and lists. Like lists, they save memory; like tables, their structure allows a fast access to each item. Trees should be arranged so as to minimize the access time to any of their items (in contrast, the time required to insert or delete elements is not really important here). How this can be achieved, however, depends on the physical storage means adopted. If items are to be stored in high-speed memory, then minimizing access time is on a par with minimizing the time needed to decide of the next node in the tree given the current one (when trying to find a way to the leaves). In contrast, storing items on a hard disk implies minimizing

[15]Keys generally have different lengths.

the number of accesses to the disk—that is, the depth of the tree. These two strategies have given birth to tries and B-trees, respectively.

B-trees (which does not stand for *binary trees,* but for *Bayer trees)* are *multiway* trees: they generalize binary trees in that more than one key is stored at each node. All the nodes of a B-tree of order *m* include at most *m-1* keys, and comparing the searched key with each of these node keys indicates which child nodes should be examined next. B-trees are also balanced, which means that the path to each leaf has the same length and that every node, except possibly the root, is at least two-thirds full. As a result, a search for a given key among more than a million keys can be guaranteed, even in the worst case, to require accessing the lexicon only a few times, depending on the number of keys that are stored at each node. One key per node reduces the tree into a list; concentrating all keys on one node has the same result. In most cases optimal speed is achieved with four to eight keys per node. Notice that B-trees are a built in facility in some PROLOG compilers.

As opposed to B-trees in which searching is based on the comparison of entire keys at each node, tries are lexicographic search trees—trees in which each level is concerned with one character of the keys to search for (level one with the first character, level two with the second one, and so on).[16] This leads to at most 26 branches for the first letter, followed by 26 x 26 branches for the second one, and so on. In practice, the exponential growth is only apparent, since many branches do not lead to any key and can therefore be pruned out.

Notice that B-trees and tries find the address of their leaves through a series of steps (node decisions). Another very efficient way to organize data is to directly *compute* the address of leaves by trying to find a mathematical function *F(X)* that associates each element *X* in a table with its address in memory. It is often possible to find good approximations of such a function, known as a *hashing function.* It takes one argument from the data domain (the dimension of which is usually high) and returns an index in the storage table (known as a *hash table*). It is important to handle *collisions* in the table (different elements being associated by the hashing function to the same index). How this is done in practice depends on the particular hash table method chosen.

4.2.3.2. Indexing lexeme and suffix groups

We have presented inflectional morphological parsing in the worst case as a four-step process. Steps 2 and 3, however, are particularly time consuming, since they respectively imply to search for all the lexemes which accept a given stem, and for all the suffix groups which admit a given suffix. Clearly, as suggested by Gazdar and Mellish (1989), lexemes should be indexed on their possible stem values (so as to obtain quickly *tenir* from *tienn*, for instance), and suffix groups could advantageously be indexed by their suffixes (in such a way, for example, that *ent* would immediately hypothesize *indicatif_présent* and *subjonctif_présent*). This in turn can be efficiently achieved with tries or B-trees with stems and suffixes as keys, as in Dutoit (1993).

---

[16]Or conversely, starting from the last character.

4.2.3.3. Unconstraining analysis

One should keep in mind that a TTS system is supposed to read any text, including some ill-formed sentences or misspelled words, or even nonexisting ones (*preconcatenation, artificializing*). After all, we human beings are perfectly able to read sentences such as[17]

Corandic is an emurient grof with many fribs; it granks from corite, an olg which cargs like lange, and grinkles several other tarances, which garkers excarp by glarking the corite and starping it in tranker-clarped storbs. (4.3)

As far as possible, one should thus avoid excessive lexical constraints on the words of the input string or, in other words, take the *input string*, no matter what that input is, rather than accept only the stings that can be successfully described in terms of a set of *linguistic rules*.

Notice, first, that the simple act of decomposing words into morphs is a first step in this direction. As a matter of fact, morphological parsers naturally tend to accept many more forms than strictly needed (unless explicitly specified otherwise, which requires some extra effort), a lot of which are precisely the ones that could occasionally be invented by writers in need of some compact expression of their mental views. Words like *preconcatenation* or *artificializing* are typical examples and should be easily accepted by many TTS systems. It may also be helpful to relax rules explicitly, by allowing exceptions to be handled as regular forms, or conversely, as so as to properly recognize a sentence like *the thiefs have finaly proven sorrowfull.*

The same principle has led many TTS systems to adopt an *open lexicon strategy*, by accepting some morphological decompositions that contain lexeme stems not mentioned in any lexicon (see O'Shaughnessy, 1987, for instance). The most straightforward way to do this is simply to forget about all the stems related to some specific part-of-speech categories, typically nouns, adjectives, adverbs, or all of them, and to allow *a priori* any stem to belong to these categories. In contrast, verb stems are generally maintained: as far as syntax, prosody, and phonetization are concerned, they are by far the most important content words. Their number conditions the number of phrases in the sentence, their position influences prosodic contours, and we shall see in Chapter Five that the verb/nonverb distinction is central to many heterophonic homographs. How such partial information is handled actually depends on morphological analyzers. In the case of Pitrat (1983), for example, it only implies to replace steps 2, 3, and 4 by

2. For all the suffix groups found to accept the current suffix, accept the related morphosyntactic features (obtained from the position of the suffix in suffix groups.

Things become more complex when dealing with compounds. Possible root sequences can then only be segmented with the help of graphotactic heuristics (segmentation rules based on sequences of characters known to be seldom encountered inside a root) as proposed by Rühl (1984) for German.

Independently of the increase of coverage obtained, the open lexicon strategy provides an efficient but rough way of reducing access times and memory requirements, since it

---

[17]This example is taken from Kupiec (1992).

simply eliminates a lot of lexicon lookup. In LIPSS (Dutoit, 1993) for example, only 4500 verb stems have been retained, which cover about 75,000 regular and irregular French verb forms.

Despite these advantages, it has been argued that reduced size lexicons are a major obstacle to high-quality TTS synthesis (Gulikers and Willemse, 1992) since they introduce some arbitrarily high degree of syntactic uncertainty. Taking again the example of LIPSS, every content word can *a priori* be a noun and an adjective, since suffixes are only optional for these categories in French.[18] In the light of sentences like (4.3), we believe (and it has been proved to some extent with LIPSS) that, provided the system is supplied with efficient syntactic disambiguation rules as the ones that will be reviewed in the next section, an important degree of lexical ambiguity can be allowed, although tuning it, is a complex language-dependent task, which deserves particular attention.

4.3. Contextual analysis

When morphological analysis has been performed, the NLP module of a TTS system has really worked out all its lexical knowledge. A number of words, however, and generally many more than initially expected,[19] still have more than one plausible part of speech category. They are:

- All the words that are *intrinsically* ambiguous (that is, even a human reader could not make a choice without examining their syntactic context). These can in turn be divided into

  1. Heterophonic homographs (words that have the same spelling but different pronunciations). In English, for example, there are more than fifty noun/verb ambiguous words that are produced with stress on the first syllable if a noun, and on the second one if a verb, like *record, permit, present.*
  2. All other homophonic content words that do not lead to phonological ambiguities, like *answer* (noun/verb), *kind* (noun/adjective).
  3. A lot of function words, given
     - Function word/function word ambiguities, like for *that* (determiner/pronoun/conjunction/adverb), *in* (verbal particle/preposition)
     - Some function word/content word ambiguities, like for *while* (conjunction/noun), *like* (preposition/verb)

- *Extrinsically* ambiguous words—that is, words whose ambiguity originates in the poor lexical knowledge of prior analysis modules, as when the open lexicon strategy of Section 4.2.3.3 has been followed. This typically covers content words only, of

[18]Notice that in some cases such confusion is in agreement with linguistic facts. French infinitives, for example, can always be used as nouns. Similarly, many nouns can be taken as adjectives, and conversely.

[19]About 40 percent of them in French (taking word frequency into account), according to Larreur *et al.* (1989). Indeed, INALF (1984) reports 4,000 syntactically ambiguous words in French, some of which have a very high frequency. Similarly for English, of the 1 million plus words in the Brown Corpus, about 40 percent are syntactically ambiguous (DeRose, 1988).

which a number can be seen as false heterophonic homographs (when the related artificial ambiguity tends to modify their pronunciation), as it is the case for many English verbs when they are (wrongly) considered as nouns.[20]

At first sight, the number of *a priori* possible analyses of a sequence of isolated words (among which only a few are really syntactically correct) equals the product of all possible part-of-speech categories for each word. In practice, it is even greater, since the syntax of natural languages is nondeterministic, so that a given sequence of parts of speech can give rise to several equally accepted parses (as in *I saw the man with a telescope*). Whatever the language, it thus grows more than exponentially with the sentence length. As a result, for a French sentence as simple[21] as:

> *Le car sortit par la porte sud de la gare de l'est*
> (*The coach departed from the south gate of the east station*)

there are 13,824 possible combinations (among which, surprisingly enough, a human reader immediately finds his way, while being hardly able to explain how).[22] This number goes up to 150,528 when morphosyntactic features (such as mode and tense for verbs, or gender and number for nouns and adjectives) are taken into account.

It follows that some sort of *contextual analysis* is more than welcome before entering the last stage of text analysis (the structural analysis module, if any, whose task is to unveil the internal hierarchical organization of sentences), for two main reasons:

1. Finding the syntactic-prosodic structure of a sentence is typically a declarative problem. Its resolution requires the examination of an important number of possibly nondeterministic rules, the applicability of which must itself be confirmed by other rules, and so on. One often refers to this computationally expensive phenomenon as *combinatorial explosion.* It has often taxed the power even of supercomputers[23].
2. The capability of a syntactic analyzer to disambiguate depends on the restrictions it puts on a language. Remember the particular *unconstrained analysis* principle first introduced in the context of morphology, but also applicable to syntax. Since the aim of the structural analysis module is merely to provide the type of hierarchical description that is minimally required to pronounce a sentence with a natural intonation (rather than to check its adequacy to a restrictive grammar), the syntactic rules that will be introduced in the structural analyzer will inevitably result in a number of accepted parses proportional to the number of proposed ones. Once again, indeed, a TTS system has to take the input string for gospel truth.

---

[20]In French, extrinsic ambiguities do not affect stress (French is a fixed-stress language), but they produce phonetization errors. This is typically the case for verbs ending with *ent*, which are erroneously pronounced with a final [ɑ̃], as if they were nouns.

[21]We admit it was not easy to find a short French sentence with such a high degree of word-level ambiguity.

[22]It is widely accepted, as introduced in Chapter One, that semantics play a leading part in text analysis by humans. Syntax, then, is merely seen as a by-product of semantic analysis.

[23]An old story about computational linguistics reports that when a particularly exhaustive disambiguating program was run at an important IBM center, lights dimmed in the whole building (quoted in Church, 1987).

Contextual analysis, also known as *tagging*, is based on the assumption that, for the specific task of disambiguation, information on extended grammatical structures is seldom necessary: local, low-level information suffices in most cases. It consists, in a sequence of syntactically ambiguous words, of eliminating some hypotheses in the list of possible categories of a word, on the basis of the corresponding lists for the neighboring ones. This strategy is extensively used in speech synthesis and recognition since, as we shall see, it gives very good results in a fairly short time. One can distinguish two types of contextual analysis methods: *probabilistic* ones, which are based on transition probabilities between successive word categories, whether computed explicitly (as with *n*-grams) or implicitly (as with neural networks), and *deterministic*[24] ones, in which a collection of yes/no rules are exploited to accept or reject a combination of syntactic categories.

Clearly, the definition of a set of part-of-speech categories is of prime importance for taggers. If only a small number of syntactic classes are considered, their disambiguation power is intrinsically limited. In contrast, keeping a very detailed set of word classes makes it hard to conceive taggers. Worse still, an arbitrarily large set number of parts-of-speech can introduce unnecessary ambiguity, by erroneously distinguishing words with similar syntactic functions. As a result, an optimal word class set should be composed of the smallest possible number of word classes that allow for the highest possible disambiguation of words. This implies that word classes are chosen so that words with different distributional properties (that is, words that cannot be interchanged without creating syntactic errors in most cases) belong to different classes. In practice, probabilistic taggers make use of a larger set of part-of-speech categories than the standard dictionary categories (typically about a hundred for European languages) (Mastrolonardo and Refice, 1989). Indeed, as opposed to deterministic approaches, which typically make use of declarative rules and are therefore intrinsically capable of handling morphosyntactic features (such as mode, tense, gender, and case) as isolated information, probabilistic approaches artificially include these features into the parts of speech they consider.

4.3.1. *N*-grams

N-grams first appeared in the context of continuous speech recognition (Jelinek, 1976, Jelinek *et al.*, 1992)), as a particular type of Markov model to estimate the probability of a sequence of words $w_1, w_2, \ldots, w_N$ in a given language. An approximate solution to this problem is obtained by implementing a probabilistic finite-state grammar in which, at any point in the analysis, the current state in the corresponding automaton is used as a memory for the n previously encountered words. Several cases arise, depending the complexity of the syntactic relationship one tries to model. *Bigrams* correspond to first-order dependencies (they make use of the probability of a word, given the previous one: n

[24]Notice that *deterministic* in this context should be understood as *nonprobabilistic*. It has nothing to do with the deterministic/nondeterministic feature of grammars. Besides, although a probabilistic grammar is always intrinsically nondeterministic (probabilities are used to find the most likely parse among all the plausible ones), a nonprobabilistic grammar can be deterministic (in the grammatical sense) or not.

= 1), *trigrams* model second-order dependencies, and so on.[25] Trigrams, for example, typically make the assumption that

$$P(w_1, w_2, ..., w_N) \approx P(w_1)\ P(w_2|w_1)\ P(w_3|w_2, w_1)\ P(w_4|w_3, w_2)\ ...\ P(w_N|w_{N-1}, w_{N-2}) \tag{4.4}$$

N-grams (and, more specifically, trigrams) are now widely used in computational linguistics, speech synthesis, and speech recognition because they combine simplicity and efficiency (Garside *et al.*, 1987, Shikano, 1987, Church, 1988, DeRose, 1988, Dumouchel *et al.* , 1988, Mastrolonardo and Refice, 1989, Senders *et al.*, 1989, Cerf-Danon and Elbeze, 1991, Kupiec, 1992, Willemse and Gulikers, 1992). As Jelinek (1993) points out: "That this simple approach is so successful is a source of considerable irritation to me and to some of my colleagues. We have evidence that better language models are obtainable, we think we know many weaknesses of the trigram model, and yet, when we devise more or less subtle methods of improvement, we come up short."

In the context of tagging a given sentence $\boldsymbol{W}=(w_1, w_2, ..., w_N)$, we are looking for the best sequence of tags $\hat{\boldsymbol{T}}$ among all the sequences $\boldsymbol{T}=(t_1, t_2, ..., t_N)$ chosen in the set of admissible tags $\{c_1, c_2, ..., c_M\}$:

$$\hat{\boldsymbol{T}} = \arg\max_{\boldsymbol{T}} P(\boldsymbol{T}|\boldsymbol{W}) \tag{4.5}$$

By Bayes's rule, this is equivalent to finding

$$\hat{\boldsymbol{T}} = \arg\max_{\boldsymbol{T}} \frac{P(\boldsymbol{T},\boldsymbol{W})}{P(\boldsymbol{W})} = \arg\max_{\boldsymbol{T}} \frac{P(\boldsymbol{W}|\boldsymbol{T})P(\boldsymbol{T})}{P(\boldsymbol{W})} \tag{4.6}$$

The denominator of (4.6) in clearly independent of $\boldsymbol{T}$ and can be ignored in the search for $\hat{\boldsymbol{T}}$.

The *n*-gram model for tagging simply makes the following assumptions:

1. The probability of a word given the past depends only on its tag.
2. The probability of a tag given the past depends only on the last *n* tags.

As a result

$$\begin{aligned} P(\boldsymbol{W}|\boldsymbol{T}) &= P(w_1,w_2,...,w_N \mid t_1,t_2, ...,t_N) = \prod_{i=1}^{N} P(w_i|t_i) \\ P(\boldsymbol{T}) &= P(t_1,t_2, ...,t_N) = \prod_{i=1}^{N} P(t_i \mid t_{i-1},t_{i-2},...,t_{i-n}) \end{aligned} \tag{4.7}$$

Hence

$$P(t_1,t_2, ...,t_N|w_1,w_2,...,w_N) = \prod_{i=1}^{N} P(w_i \mid t_i)\ P(t_i \mid t_{i-1},...,t_{i-n}) \tag{4.8}$$

[25]By extension, *unigrams* model the probability of a word given nothing—that is, given the language. A unigram-based contextual analyzer would therefore always retain the most probable part of speech for each word. They only were only used in the early days of TTS synthesis (they give about 80% of correctly tagged words).

Notice that hypothesis one explicitly restricts the scope of n-gram taggers to syntax, while hypothesis two constrains their reach. Both are clearly inexact, but it turns out that they are quite useful.

It is then straightforward to derive the related probabilistic FSA. In a bigram model, for example, a set of M states $\{q_1, q_2, ..., q_M\}$ simply represent the part-of-speech categories considered by the grammar (one state per category: q_1 accounts for c_2, q_2 for c_2, and so on). Transitions are *a priori* allowed between any states q_i and q_j (the model is said to be *fully connected*) and associated with transition probabilities $a(i,j)$. If one assumes that the vocabulary is finite(that is, the set of all possibly encountered words in any sentence to be analyzed comprises a finite number of elements, say L), one can define state-dependent word-emission probabilities $b(j,k)$, which represent the probability that the *kth* word in the lexicon is seen, given that its category is c_j. Provided the elements $a(i,j)$ and $b(j,k)$ of the related transition and emission matrices (of size M x M and M x L, respectively) have been previously computed from a large text corpus, by estimating relative frequencies for instance, the problem of parsing a sentence of N words reduces to trying to find the most probable path of length N in a fully connected FSA with M states, in which *emission* and *transition probabilities* are respectively associated to states and arcs. In practice, since the words to be disambiguated have already been given selected tags through either a lexicon lookup or some morphological analysis, the *a priori* fully connected network automatically, and advantageously, confines itself to a sentence-dependent lattice in which each input word k corresponds to m states, m being the number of admitted categories for it (the number of non-zero elements in $b(j,k)$). An example is given in Fig. 4.4. This problem is solved by dynamic programming with the well-known Viterbi algorithm.

Trigrams are simply an extension of bigrams. A trigram FSA accounting for M categories therefore comprises M^2 states $\{q_{11}, q_{12},..., q_{1M}, q_{21}, q_{22},..., q_{2M},..., q_{M1}, q_{M2},..., q_{MM}\}$, each one corresponding to a couple of part of speech categories (for the current and preceding words: q_{ij} accounts for (c_i,c_j)). Transition probabilities then refer to the probability that a given category is produced given the two previous ones. Notice, however, that the FSA is no longer fully connected (even in the training phase), since a state indexed with tags (c_i,c_j) can only be connected to states (c_j,c_k), for k=1...M. In other words, transition and emission matrices are of size M^2 x M and M^2 x L. In order to avoid the important increase in data storage and computational load resulting from the use of trigrams or wider contextual units, a combination of different length n-grams may be used (see the *polygrams* of Kuhn *et al.*, 1994, and the *n-multigrams* of Deligne and Bimbot, 1995).

In text-to-speech applications, however, the vocabulary is inherently infinite (or at least so large that no training database can claim completeness), which impedes probabilities $b(j,k)$ from being correctly estimated.[26] In order to alleviate this problem, a typical procedure consists of considering content words through a finite number P of *equivalence classes*, while the most frequent words (which naturally include function

---

[26]Notice that this is still a more crucial point when words are not considered via their parts of speech, as it is the case in equation (4.4). Many solutions have been proposed to cope with this relative sparsity of the training data (see Jelinek, 1991). The idea of constraining the size of the FSA by reducing words to tags is only one of them.

words) are maintained as separate lexical entries (Kupiec, 1992). Equivalence classes are theoretically defined as sets of words that are believed to have similar values of *b(j,k)*, for all *j*. These words are not distinguished by the *n*-gram model. In practice, equivalence classes can be chosen as the sets of words with similar ambiguities (for example, all the words that can be either a noun or a verb, for instance, belong to the same class), which implicitly assumes that the distribution of such words into their possible tags is comparable. Emission probabilities are then restricted to equivalence classes and (4.8) becomes

$$P(w_1,w_2,\ldots,w_N \mid t_1,t_2,\ldots,t_N) = \prod_{i=1}^{N} P(EC(w_i) \mid t_i)\, P(t_i \mid t_{i-1},\ldots,t_{i-n}) \qquad (4.9)$$

where $EC(w_i)$ denotes the equivalence class of word *i*.

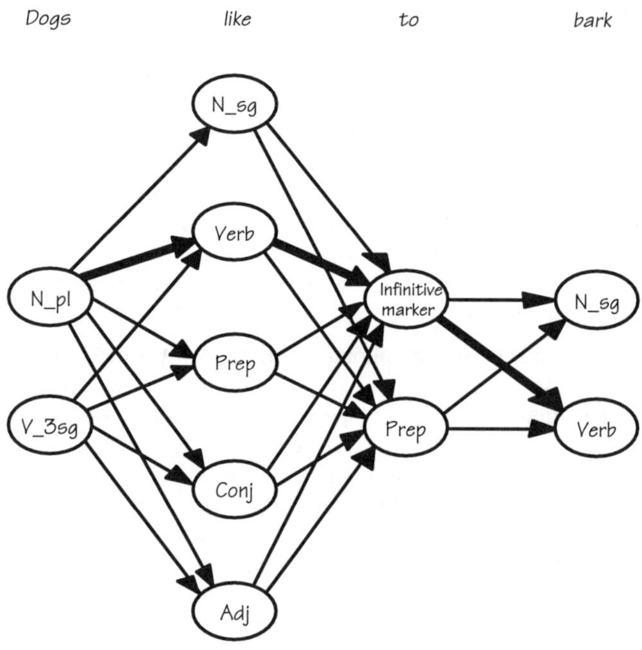

Fig. 4.4. Bigram for the sentence *'Dogs like to bark'* (after Kupiec (1992)).

Recently, a *hidden* Markov model approach has been applied to syntactic category disambiguation with bigrams, by using partially disambiguated sentences, or even a completely untagged corpus, as a training set, thereby leaving the choice for the model to associate ambiguous words to *a priori* any of the states that agree with their possible tags (Kupiec, 1992). It appears that the optimal strategy to build the best possible model is obtained by starting with as much tagged text as possible to compute a first estimation of transition and emission matrices (by computing relative frequencies, for example), and then using as much untagged text as possible to refine the initial model with the forward-backward iterations of the Baum-Welch algorithm (Merialdo, 1991), so as to converge to a local maximum of the likelihood of the unsupervised training data.

The disambiguating power of bi- and trigrams is generally found to be surprisingly high, provided the training database is large enough for probabilities *a(i,j)* and *b(j,k)* to be realistically estimated. Kupiec (1992) reports 94.7 percent of correctly tagged words for English, with a training database of 400,000 words from sentences in the Brown corpus and a hidden Markov model approach combining bi- and trigrams. A very interesting aspect of his HMM-based tagger resides in its ability to tag nonsense sentences like (4.3). This is achieved on the simple basis of an analysis of the possible suffixes of each incoming word whose hypothetical stem cannot be confirmed by a lexeme lexicon (as it typically happens with the aforementioned *open-lexicon strategy*). During training, all the unknown words are associated with a single equivalence class EC_{unk}. When parsing, however, the related equivalence class is further split into Q suffix-based subclasses (Q being the number of suffixes considered:[27] $\{s_1, s_2, ..., s_Q\}$), and its emission probabilities are computed as

$$P(EC_{unk}|t_j) = \max_i P(s_i|\, t_j) = \max_i \left(\prod_{k=1}^{P} P(s_i|\, EC_k)\, P(EC_k \,|\, t_j) \right) \quad (4.10)$$

in which $P(EC_k \,|\, t_j)$ refers to the emission probabilities of the *k*th equivalence class (for all the classes but EC_{unk}) and $P(s_i|\, EC_k)$ can be obtained by measuring the relative frequency of s_i within class EC_k (from even an untagged corpus):

$$P(s_i|\, EC_k) = \frac{P(s_i, EC_k)}{P(EC_k)} \quad (4.11)$$

Notice that *i* in (4.10) refers to all the possible suffixes of the unknown word. Thus, when trying to find the most likely path in resulting word-dependent lattice, the emission probabilities of unknown words are simply computed as their best emission probabilities given their possible suffixes.

4.3.2. Neural networks as taggers

In parallel to the development of finite state probabilistic grammars in the form of *n*-grams, neural networks have been tried as language disambiguating machines. During training, word sequences are presented at the input of a multi-layer perceptron and the disambiguated part of speech category of one of them is given as proposed output (Fig. 4.5). A classical back-propagation algorithm is used the optimize the neuron weights.

A potential advantage of this approach over the previous one is the important data reduction it allows. This theoretically allows multilayer perceptrons to be presented more than three words as input, as in Benello *et al.* (1989), in which a contextual window of six words (four nonambiguous words to the left, the current word, and one possibly ambiguous word to the right) was used as the input of a three-layer perceptron. A total of

---

[27]One of these suffixes being left for the case of a word whose suffix would not be included in the set of the *Q-1* others.

eighty-eight categories were considered for the output layer, but both input and hidden layers contained only 236 units (the four initial words were given a compact representation of word categories, with fifteen units per word), for a total of 80,000 synapses. About 95 percent of correct disambiguation is reported, for a training set of 900 sentences and a test set of 100 sentences, extracted from the Brown corpus. One cannot compare these results with the ones of Kupiec (1992), given the disparity between their respective training and test database sizes, but one can nevertheless conclude that, even though fewer data are necessary with the neural network approach (80,000 weights, in comparison with 88^3 transition probabilities with Markovian trigrams—that is 700,000 numbers), enlarging the contextual window size does not seem to bring significant improvement.

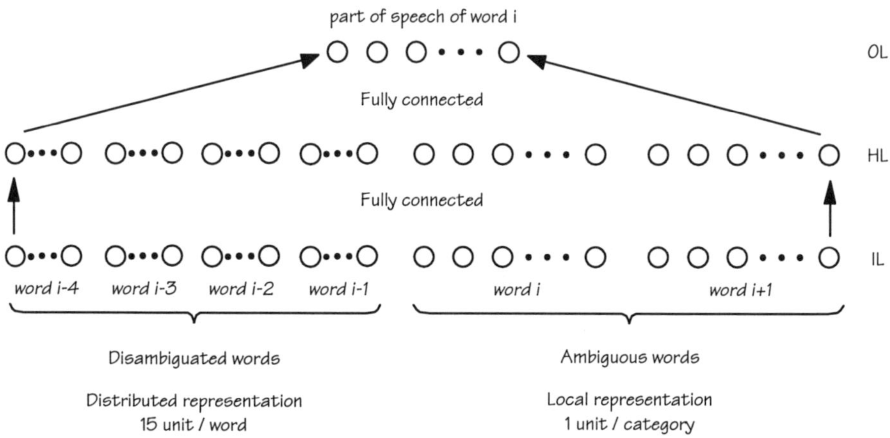

Fig. 4.5. A three-layer perceptron for part-of-speech disambiguation (after Benello *et al.*, 1989). IL, HL, and OL respectively stand for input layer (88 units), hidden layer (236 units), and output layer (88 units). See text for explanation.

Notice that the MLP approach of Benello *et al.* (1989) is intrinsically limited in its disambiguation task, at least in comparison with the *n*-gram models of the previous section. The MLP can, at best, estimate the *part-of-speech* that is most likely to appear in a given context of other parts-of-speech and equivalence classes (of the six input words, the first four were presented disambiguated, while the last two were left ambiguous). In contrast, an *n*-gram uses this probability to find the *sequence of parts-of-speech* which maximizes the likelihood of the whole sentence. For a neural network to have access to this dimension, it should obviously be provided with some means of taking the history of its inputs, or better still of its outputs, into account. This can be achieved by introducing *feedback loops* inside neurons, thereby turning an MLP into a *recurrent network.* However, although these highly non-linear dynamic machines have proved efficient for inferring elementary regular grammars (see Cleeremans *et al.*, 1989) and for parsing the related strings, these tasks can be achieved much more simply with stochastic grammars. To our knowledge, recurrent networks have never been used in the context of part-of-speech disambiguation, due to their intrinsic complexity.

4.3.3. Local nonstochastic grammars

Syntactic disambiguation was studied long before the outgrowth of the powerful automatic inference machines presented above. As early as the 1960s, rule-based expert systems (as opposed to probabilistic ones) were proposed by linguists for many languages, with more or less success, as possible fast solutions to syntactic ambiguity problems. They differ fundamentally from *n*-grams in the sense that, instead of associating probabilities to word sequences in a sentence, they either accept or reject a set of possible categories for each word as a function of a series of tests on its context. The notion of context itself is generally extended, so that useless word sequences can be skipped in the contextual search process.

Consider the word *this*, for example[28]. It can *a priori* be a determiner (*this program works well*), a demonstrative pronoun (*This does not change his opinion*), or an adverb (*He is not this tall*). Two simple rules, however, make it possible to find the right tag in many cases. They are

1. When *this* is a determiner, it cannot be followed by a verb unless the verb is a past or present participle;
2. When *this* is an adverb, it cannot be followed by a noun or by a verb.

In the early days of rule-based disambiguation systems, rules were developed by human experts. In Greene and Rubin (1971), for example, a set of 3,300 rules was proposed for English. Its application to the Brown corpus (more than 1 million words) led to 77 percent of correctly tagged words. Similar scores have been achieved by Blois and Buydens (1968) for French, who have reported 75 percent of correct noun/verb disambiguations (on a 30,000 words test database) with only thirty-six rules.[29] In addition to these poor results, rule-based systems have apparently two important drawbacks: they require experts to design rules (as opposed to the two previous approaches, which are based on a general training procedure) and the rules they implement are obviously language specific, which means that (different) experts are needed for each language considered.

However, even such simple deterministic disambiguation rules should not be ignored, when they are available. They have indeed at least three advantages over *n*-grams and neural networks. First, they remain understandable by human experts, which implies that they can be increasingly refined with time (by combining two sets of rules from different authors, for instance, while combining two n-grams or two neural networks into a single one is an open problem). Second, the complexity of local grammars does not increase exponentially with the size of their right and left contexts, since not all cases have to be

---

[28]This example is taken from Mohri (1994).

[29]The important disparity in the aforementioned numbers of rules simply comes from a difference in rule formalisms. In Blois and Buydens (1968), the use of variables and the possibility of skipping word sequences has led to compact and widely applicable rules.

covered. The CNET TTS conversion system for French, for instance, implements a deterministic contextual analyzer with 1,112 rules that examine the syntactic environment up to five slots to the left and four slots to the right of the current word (Larreur *et al.*, 1989). Similarly, it is relatively easy to specify sequences of words to be skipped for the determination of the right and left contexts, which allows deterministic rules to have a larger scope. And, last but not least, local grammars typically include a large number of simple but error-free rules. As a result, they can truly disambiguate on the word level. In contrast, *n*-grams generally do not produce zero probabilities for those word sequences that have not been encountered in the training set, since they cannot guarantee that the training database is complete. As a result, disambiguation has to be examined on the sentence level, as with the Viterbi algorithm. Simple local grammars can therefore be used as a preprocessing for further probabilistic disambiguation (Tapanainen and Voutilainen, 1994).

There has been recently a revival of nonprobabilistic disambiguation methods, which has led to disambiguation scores comparable to those of *n*-grams and has provided efficient training algorithms. Among the most promising concepts are the *decision lists* of Rivest (1987) and particularly their simplified but highly efficient version provided in Yarowsky (1994), and the *transformation-based error-driven learning* algorithm of Brill (1994). Other interesting approaches have been proposed as a consequence of recent developments in finite state transducing techniques (Koskenniemi, 1990, Mohri, 1994), especially in the framework of *constraint grammars* (Karlsson, 1990).

Decision lists are a typical general-purpose corpus-based method, with minimal presupposed linguistic knowledge, if any. Their power actually originates in the simultaneous availability of large training corpora and of powerful data analysis engines. They are based on expressing the distributional properties of a language in the form of lists of contexts composed of *k* words before and after each ambiguous word, and on further deriving from those contexts the *collocations*[30] that are most useful in solving the ambiguity. Collocations are sorted in the list by degree of evidence (the contexts that have the best disambiguating power are listed first), and disambiguation is obtained by finding, in the decision list of the ambiguous word, the highest-ranking match in the word's context and by choosing the most likely of the word's tags given this context. Decision lists provide a simple and straightforward way of including multiple sources of evidence (words, tags, morpho-syntactic features, and so on). They can therefore *a priori* be used for the resolution of lexical ambiguities (both syntactic and semantic) based on any properties of the surrounding context. They have been successfully applied to the problem of homograph resolution in text-to-speech synthesis, with an average score of 97 percent on thirteen homographs, for which the baseline accuracy was 67 percent (Sproat *et al.*, 1992, Yarowsky, 1994). A more detailed description of such corpus-based analysis techniques will be given in Section 6.1.2.3, devoted to *classification and regression trees* (CARTs) (Brieman *et al.*, 1984).

Similarly, transformation-based error-driven learning has been applied to a number of natural language processing problems, including part-of-speech tagging and syntactic

[30]*Collocation* simply means words appearing adjacent to or near each other (literally, in the same location).

parsing. It is based on first applying a rough disambiguation algorithm (a unigram, for example) to some unannotated text, comparing the results with the corresponding annotated text, and deriving a number of transformation rules that apply on the initial annotation so as to make it better resemble the *truth* (the annotated text). Learning is performed iteratively. Each iteration yields the highest score transformation, which is added to the transformation list and applied on all the training corpus (the unannotated text) before the next iteration. A number of allowable transformation templates are defined, based on the word's context as with decision lists. To learn a transformation, the learner applies every possible transformation, counts the number of tagging errors, and chooses that transformation resulting in the greatest error reduction. Brill (1994) reports a tagging score of 97.2 percent on a 150,000 word test set taken from the Penn Treebank Tagged Wall Street Journal Corpus with only 267 rules, as opposed to the 10,000 contextual probabilities that were learned by the stochastic tagger of Weischedel *et al.* (1993) on the same corpus, and which achieved 96.7 percent. Clearly, with such tools in hand, non-probabilistic rules can simply be used *in place of n*-grams, in which case the very limited number of rules they require constitutes an additional advantage over their probabilistic counterparts.

Finally, constraint grammars have been formalized by Karlsson (1990) as collections of declarative regular constraint rules that take a text file with morphologically analyzed (and possibly ambiguous) word forms as input and basically discard as many alternatives as possible. Each constraint consists of context conditions, an operator, and a target. Context conditions are used to check for the existence (or for the absence) of a given alternative at any position before or after the current word. The target indicates which alternative the constraint is about. Targets and context conditions may always refer to particular readings or to sets thereof. Finally, the operator defines which type of disambiguation to perform, among three possibilities, in order of decreasing strength:

1. The target reading is the correct one if and only if all context conditions are satisfied; all other readings should be discarded. If the context conditions are not satisfied, the target alternative itself should be discarded.
2. The target reading is the correct one if and only if all context conditions are satisfied; all other readings should be discarded.
3. The target reading should be discarded if and only if all context conditions are satisfied; all other readings should be left untouched.

In order to enhance its declarative power, the formalism accepts context conditions with unbounded dependencies by providing a mechanism for searching for given alternatives leftward or rightward of a given word position around the current word, up to the next sentence or clause boundary. There is also a notation for forcing context conditions to apply to nonambiguous contexts only (as opposed to the default operative mode that considers that a condition is fulfilled if one of the readings of the word at a given position satisfies it).

One of the advantages of constraint-based grammars, as we shall see in Section 6.1.2.1, is that their principle can be extended to the determination of intrasentential clause boundaries and to the disambiguation of surface syntactic functions, so as to integrate morphology and syntax within the same formalism. Their efficiency can be illustrated with a simple example, taken from Dutoit (1993). We have followed the deterministic approach of Blois and Buydens (1968) for French, which we have adapted to the particular degree of lexical autonomy of the LIPSS system (given the limited number of entries in its morphological dictionary) and rewritten into thirty-three compact Prolog rules, classified into categories, depending on the type of syntactic constraint they highlight

- Rules based on the existence of a finite set of left neighbors for nouns.

  ex: If *i-1* is an unambiguous determiner or *au, aux, du, des*, and *i+1* is not in the set {adverb, adjective, coordinator, noun}, then *i* is a noun (or is taken substantively), as in: *leurs* politiques *de prestige, mon* pas *de géant*[31]

- Rules based on the existence of a finite set of left neighbors for verbs.

  ex: If *i* is in the set {*le, la, les, leurs, l'*} and is followed by a conjugated verb, a present participle or an infinitive, then it is a pronoun, as for: *l'assemblée* les *choisit, leur adversaire* les *injurie*

- Rules based on verb group limits.

  ex: A word which may be a *noun* or a *conjugated verb* is not a verb if there is an unambiguous conjugated verb on its right within a sequence of words limited by any punctuation mark, coordinator, or subordinating conjunction, as in: *l'institution en* cause *a été transformée*

- Rules based on noun group limits.

  ex: A possible noun is not a noun if there is an unambiguous noun on its right or on its left, within a sequence of words limited by any punctuation mark, coordinator, subordinating conjunction, determiner, preposition, conjugated verb, infinitive, present participle, past participle, or pronoun, as for: *la pièce* joue *dans son logement, le pilote* ferme *la porte*

- Others.

  ex: If an ambiguous word which can be a preposition is encountered before a word included in {*moi, toi, soi, eux, lequel, laquelle, lesquels, lesquelles*}, then it is a preposition, as in: *les avis* suivant *lesquels*

Rules are examined sequentially for each word until a match is found, in which case the corresponding rule is applied and the focus is set to the next word. It should be clear that several successive analyses are profitable, since the disambiguation of a word can help to tag previous ones. The order in which rules are examined has been chosen so as to minimize the number of successive analyses. On the average, no rule can still be applied after two or three cycles. As shown in the example below, this results in a drastic decrease in the number of possible analyses for a sentence:

Le car sortit par la porte sud de la gare de l'est.

---

[31]Notice that, in all these examples, truly ambiguous cases are presented (so that ambiguities can be perceived by human readers too). One should not forget, however, that the number of ambiguous cases for a TTS system can be far greater, given its possibly limited lexical knowledge.

After morphological analysis:

le	DET, masc,sing
	PRONOUN,masc,sing
car	COORDINATOR
	NOUN,masc,sing
sortit	VERB,ind,psimp,1st,sing
	NOUN,unknown gender,sing
	ADJ,unknown gender,sing
par	PREP
la	DET,fem,sing
	PRONOUN,fem,sing
porte	VERB ,ind,pres,1st,sing
	,ind,pres,3rd,sing
	,imper,pres,2nd,sing
	,subj,pres,1st,sing
	,subj,pres,3rd,sing
	ADJ,unknown gender,sing,
	NOUN,unknown gender,sing
sud	NOUN,unknown gender,sing
	ADJ,unknown gender,sing
de	PREP
	DET,unknown gender, plur
la	DET,fem,sing
	PRONOUN,fem,sing
gare	VERB ,ind,pres,person,1st,sing
	,ind,pres,person,3rd,sing
	,impér,pres,person,2nd,sing
	,subj,pres,person,1st,sing
	,subj,pres,person,3rd,sing
	ADJ,unknown gender,sing
	NOUN,unknown gender,sing
de	PREP
	DET,unknown gender,plur
l'	DET,unknown gender,sing
	PRONOUN,unknown gender,sing
est	VERB,ind,pres,3rd,sing
	NOUN,masc,sing

And the application of seven contextual rules (in two parses) reduces it to

le	DET,masc,sing
car	NOUN,masc,sing
sortit	VERB,ind,psimp,3rd,sing
	ADJ,unknown gender,sing
par	PREP
la	DET,fem,sing

porte	ADJ,unknown gender,sing
	NOUN,unknown gender,sing
sud	NOUN,unknown gender,sing
	ADJ,unknown gender,sing
de	PREP
la	DET,fem,sing
gare	ADJ,unknown gender,sing
	NOUN,unknown gender,sing
de	PREP
l'	DET,unknown gender,sing
est	NOUN,masc gender,sing

Hence, the initial 13,824 possible analyses are reduced to 8! To a larger scale, Karlsson (1990) reports that with about 400 context-dependent disambiguation constraints for English, 95 to 97 percent of all morphological ambiguities of running texts can be pruned, depending on the complexity of the text.

4.4. Syntactic-prosodic parsing

At this stage of the analysis process, individual words have been assigned adequate parts-of-speech categories, as the result of the combined contributions of the morphological analysis and contextual disambiguation modules. Yet many of them remain ambiguous, being left with several hypothetical parts of speech. Such situations arise when

- A word is intrinsically ambiguous—when it is referred to as such in dictionaries (like the word *state* which can be either a verb or a noun).
- Morphological analysis has not been able to decompose a word into known stems and affixes, either because it has been misspelled or because it is a new or foreign word or simply because the lexicon is not complete, which typically happens if the open lexicon strategy of Section 4.2.3.3 has been adopted. In the example given at the end of Section 4.3, for instance, the word *sortit* is still considered as a potential adjective (although no such adjective exists in French) because the LIPSS text processing system has no lexicon for names and adjectives, so that every incoming word is *a priori* examined as such.

In either case it is assumed that the neighborhood of the ambiguous word has failed to provide enough restrictive information for its successful contextual disambiguation. Such a disambiguation failure may in turn be due either to the intrinsic weakness of the contextual analysis module, in which case further syntactic parsing may help, or to an ambiguity which can only be solved on semantic or pragmatic grounds. Taking again the above example, both *porte* and *sud* are seen as hypothetical nouns and adjectives, and it is necessary to know that *sud* refers to a cardinal direction to understand that it acts as a qualifier for *porte* (and not the converse).[32]

---

[32]Incidentally, neither *porte* nor *sud* are referred to as adjectives in French dictionaries. The word *sud* happens to be a noun used here as an adjective. Conversely, adjectives are often taken as nouns.

Yet syntax does not reduce to disambiguation. Far beyond the simple determination of part-of-speech categories of individual words, it is the complete organization of sentences into clauses and phrases, and the related functional dependencies ("What does this noun phrase dominate?", "Does this prepositional phrase go with the noun or with the verb?"), which constitute the ultimate object of syntactic analysis. The point is that syntax alone cannot be guaranteed to yield the expected decomposition into interrelated phrases and clauses. Remember sentences like *I saw the man with a telescope* or *time flies like an arrow*, for which several parses are *a priori* acceptable. At least some semantic knowledge is required to make a choice. Add to this that TTS systems are expected to read *any* text, which puts a rather strong and unusual demand on their text analysis modules. Human readers themselves can be confused with the syntax of complex sentences. We cannot always produce an acceptable parse for a sentence, although it does not prevent us from reading it, simply because we understand what we read before being able to analyze it in terms of functional dependencies.

Summing up, it appears that, although syntax has a lot to do with the disambiguation of individual words and with their organization into phrases and clauses, semantics and pragmatics play an important part as well. Therefore neither disambiguation nor dependency relations can be obtained on purely syntactic grounds: syntactic structures, in some way, are a by-product of semantics and pragmatics. As Allen (1992) observes: "At a purely syntactic level, many of the problems of parsing are likely to remain. Compound analysis, prepositional phrase attachment, ambiguous reduced relative clauses, preposition/particle confusions, and the analysis of conjoined constructions present insurmountable difficulties unless the parsing process can be complemented by use of semantic and pragmatic constraints, but these are currently beyond the scope of TTS systems."

How TTS systems currently cope with this paradox, will be detailed in Chapter Six (Section 6.7), devoted to the automatic generation of prosody.

4.5. Summary

In this chapter devoted to morpho-syntactic analysis for TTS synthesis, we have successively examined incoming sentences at the character, word, and phrase levels. We have considered the related analysis modules, respectively referred to as *pre-processing*, *morphological analysis*, and *contextual analysis*, the output of which is eventually processed by a syntactic-prosodic parser so as to describe sentences in terms of hierarchically organized prosodic groups. For all but the last module, we have pointed out which problems should be solved and which NLP inference and parsing techniques can help to do it:

- The pre-processing module organizes the input sentences into manageable lists of words. It identifies numbers, abbreviations, acronyms, and idiomatics and transforms them into full text when needed. An important problem is encountered as soon as the character level: that of punctuation ambiguity (including the critical case of sentence end detection). It can be solved, to some extent, with elementary regular grammars.
- The morphological analysis module generates all possible part of speech categories for each word taken individually, on the basis of their spelling. Inflected, derived, and

compound words are decomposed into their morphs by simple regular grammars exploiting lexicons of stems and affixes.

- The contextual analysis module considers words in their context, which reduces the list of their possible part of speech categories to a very restricted number of highly probable hypotheses, given the corresponding possible parts of speech of neighboring words. This can be achieved either with *n-grams*, which describe local syntactic dependences in the form of probabilistic finite state automata (i.e. as a Markov model), to a lesser extent with *mutlilayer perceptrons* trained to uncover contextual rewrite rules, or with *local, nonstochastic grammars* provided by expert linguists or automatically inferred from a training data set with *classification and regression tree* (CART) techniques.

The organization of syntactic-prosodic parsers will be detailed in Chapter Six.

References

ALLEN, J., S. HUNNICUT, and D. KLATT, (1987), *From Text To Speech, The MITALK System*, Cambridge University Press, Cambridge.

ALLEN, J., (1992), "Overview of Text-to-Speech Systems", in *Advances in Speech Signal Processing*, S. Furui and M. Sondhi, eds., Dekker, New York , pp. 741-790.

BENELLO, J., A.W. MACKIE, and J.A. ANDERSON, (1989), "Syntactic Category Disambiguation with Neural Networks", *Computer Speech and Language*, n°3, pp. 203-217.

BLOIS, J., and J. BUYDENS, (1968), *Problèmes de la Traduction Automatique*, Klinckslieck, Paris.

BÖHM, A., (1992), *Maschinelle Sprachausgabe Deutschen und Englishe Textes*, Ph.D. dissertation, Ruhr-Universität Bochum.

BRIEMAN, L., J.H. FRIEDMAN, R.A. OLSHEN, and C.J. STONE, (1984), *Classification and Regression Trees*, 1984, Wadsworth & Brooks, Monterey, CA.

BRILL, E., (1994), "Some Advances in Transformation-based Part of Speech Tagging", to appear in *Proceedings of the AAAI'94*, also on *CMP-LG*, paper n° 9406010[33].

CERF-DANON, H., and M. ELBEZE, (1991), "Three Different Probabilistic Language Models: Comparison and Combination", *Proceedings of the International Conference on Acoustics, Speech, and Signal Processing 91*, Toronto, pp. 297-300.

CHURCH, K. W., (1987), *Phonological Parsing in Speech Recognition*, Kluwer.

CHURCH, K. W., (1988), "A Stochastic Parts Program and Noun Phrase Parser for Unrestricted Text", Proceedings of the Second Conference on Applied Natural Language Processing, Austin, Texas.

CLEEREMANS, A., D. SERVAN-SCHREIBER, and J.L. McCLELLAND, (1989), "Finite-State Automata and Simple Recurrent Networks", *Neural Computation*, vol. 1, pp. 372-382.

CRYSTAL, D., (1985), *A Dictionary of Linguistics and Phonetics*, Basil Blackwell.

DELIGNE, S., and F. BIMBOT, (1995), "Language Modeling by Variable Length Sequences : Theoretical Formulation and Evaluation of Multigrams", *Proceedings of the International Conference on Acoustics, Speech, and Signal Processing 95*, vol. 1, pp. 169-172.

DeROSE, S., (1988), "Grammatical Category Disambiguation by Statistical Optimization", *Computational Linguistics*, n°14, pp. 31-39.

DUMOUCHEL, P., V. GUPTA, M. LENNIG, and P. MERMELSTEIN, (1988), "Three Probabilistic Models for a Large-Vocabulary Speech Recognizer", *Proceedings of the International Conference on Acoustics, Speech, and Signal Processing 88*, New-York, pp. 513-516.

---

[33]CMP-LG is the Computation and Language E-Print Archive set up by the consortium for lexical research. It can be accessed through the WorldWideWeb at URL address http://xxx.lanl.gov/abs/cmp-lg. Papers are stored under the name *yymmnnn*, where *yy*, *mm*, and *nnn* refer to its year and month of submission and to its number.

DUTOIT, T., (1993), *High Quality Text-To-Speech Synthesis of the French Language*, Ph. D. dissertation, Faculté Polytechnique de Mons.

FROIDEVEAUX, C., M.-C. GAUDEL, and M.SORIA, (1990), *Types de Données et Algorithmes*, McGraw-Hill, Paris, 577 pp.

GARSIDE, R., G. LEECH, and G. SAMPSON, (1987), *The Computational Analysis of English*, Longman, London.

GAZDAR, G., and C. MELLISH, (1989), *Natural Language Processing in Prolog: an Introduction to Computional Linguistics*, Addison-Wesley, Reading, MA.

GREENE, B.B., and G.M. RUBIN, (1977), "Automatic Grammatical Tagging of English", Department of Linguistics, Brown University, Providence, Rhode Island.

GROSS, M., (1986), *Grammaire Transformationnelle du Français I. Syntaxe du Verbe. II. Syntaxe du Nom.*, Cantilène, Paris.

GULIKERS, L., and R. WILLEMSE, (1992), "A Lexicon for a Text-to-Speech System", *Proeedings.of the International Conference on Spoken Language Processing,* Alberta, pp. 101-104.

INALF, (1984), *Dictionnaire des Fréquences: Table de Répartition des Homographes*, CNRS - INALF (institut national de la langue française), Nancy.

JELINEK, F., (1976), "Continuous Speech Recognition by Statistical Methods", *Proceedings of the IEEE,* vol. 64, n°4, pp. 532-556.

JELINEK, F., (1991), "Up from Trigrams!", *Proceedings of Eurospeech 91*, Genova, vol. 3, pp. 1037-1040.

JELINEK, F., R.L. MERCER, and S. ROUKOS, (1992), "Principles of Lexical Language Modeling for Speech Recognition", in *Advances in Speech Signal Processing*, S. Furuy and M. Sondhi, eds., Dekker, New York.

KARLSSON, F., (1990), "Constraint Grammars as a Framework for Parsing Running Text", *Proceedings of the Conference on Computational Linguistics*, Helsinki, vol. 3, pp. 168-173.

KARTUNNEN, L., K. KOSKENNIEMI, and R. KAPLAN, (1987), "A Compiler for Two-Level Phonological Rules". In*:* Daylrimple *et al., Tools for Morphological Analysis,* Report N° CSLI-87-108, Center for Study of Language and Information, Stanford University.

KNUTH, D., (1973), *The Art of Computer Programming*, vol. 2, Addison-Wesley, Reading, MA.

KOSKENNIEMI, K., (1983), *Two Level Morphology: A general Computational Model for Word-Form Recognition and Production*, Ph.D. dissertation, Department of General Linguistics, University of Helsinki.

KOSKENNIEMI, K., (1990), "Finite-State Parsing and Disambiguation", *Proeedings. of the Conference on Computational Linguistics*, Helsinki, vol. 2, pp. 229-232.

KUHN, T., H.NIEMANN, and E.G. SHUKAT-TALAMAZZINI, (1994), "Ergodic Hidden Markov Models and Polygrams for Language Modeling", *Proceedings of the International Conference on Acoustics, Speech, and Signal Processing 94*, vol. 1, pp. 357-360.

KUPIEC, J., (1992), "Robust Part-of-Speech Tagging Using a Hidden Markov Model", *Computer Speech and Language,* n°6, pp. 225-242.

LARREUR, D., F. EMERARD, F. MARTY, (1989), "Linguistic and Prosodic Processing for a Text-to-Speech Synthesis System", *Proceedings of Eurospeech 89*, Paris, pp. 510-513.

LIBERMAN, M.Y., and K.W. CHURCH, (1992), "Text Analysis and Word Pronunciation in Text-to-Speech Systems", in *Advances in Speech Signal Processing,* S. Furui and M. Sondhi, eds., Dekker, New York, pp. 791-831.

LINDSTRÖM, A., M. LJUNGQVIST, and K. GUSTAFSONN, (1993), "A Modular Architecture Suppoorting Multiple Hypotheses for Conversion of Text to Phonetic and Linguistic Entities", *Proceedings of Eurospeech 93*, Berlin, pp. 1463-1466.

LINDSTRÖM, A., and M. LJUNGQVIST, (1994), "Text Processing within a Speech Synthesis System", *Proc. Proceedings.of the International Conference on Spoken Language Processing 94,* Yokohama.

MASTROLONARDO, A., and M. REFICE, (1989), "Measuring the Power of Self-Organized Linguistic Models", *Proceedings of Eurospeech 89*, Paris, vol. 1, pp. 390-393.

McALLISTER, M., (1989), "The Problems of Punctuation Ambiguity in Full Automatic Text-to-Speech Conversion ", *Proceedings of Eurospeech 89,* Paris, vol. 1, pp. 538-541.

MERIALDO, B., (1991), "Tagging Text with a Probabilistic Model", *Proceedings of the International Conference on Acoustics, Speech, and Signal Processing 91*, pp. 809-812.

MOHRI, M., (1994), "Syntactic Analysis by Local Grammars Automata: an Efficient Algorithm", to appear in the *Proceedings of Complex94.* Also on *CMP-LG*, paper n° 9407002.

O'MALLEY, M.H., D.K. LARKIN, and E.W. PETERS, (1986), "Beyond the Reading Machine: What the Next Generation of Intelligent Text-To-Speech Systems Should do for the User", *Proceedings of Speech Technology 86*, pp. 216-219.

O'SHAUGHNESSY, D., (1987), "Specifying Intonation in a Text-to-Speech System Using Only a Small Dictionary", *Proceedings of the International Conference on Acoustics, Speech, and Signal Processing 87,* pp. 1430-1433.

PACHUNKE, T., O. MERTINEIT, K. WOTHKE, and R. SCHMIDT, (1994), "The Linguistic Knowledge in a Morphological Segmentation Procedure of German", *Computer Speech and Language*, vol. 8, pp. 233-245.

PITRAT, J., (1983), *Réalisation d'un Analyseur Lexicographique Général*, rapport de recherche n°79/2, GR22, Institut de programmation, Paris VI.

RILEY, M.D., (1990), "Tree-Based Modeling for Speech Synthesis", *Proceedings of the ESCA Workshop on Speech Synthesis*, Autrans (France), pp. 269-272.

RIVEST, R.L., (1987), "Learning Decision Lists", *Machine Learning*, 2, pp. 229-246.

RÜHL, H.-W., (1984), *Sprachsynthese nach Regeln für Unbeschränkten Deutschen Text*, PhD dissertation, Ruhr-Universtät Bochum.

SABAH, G., (1989), *L'intelligence Artificielle et le Langage, Tome 1: Représentation des Connaissances, Tome2: Processus de Compréhension,* Hermes, Paris.

SENDERS, W., M. KUGLER, and L. BOVES, (1989), "Simultaneous Optimization of Several Variables in a Probabilistic Language Model", *Proceedings of Eurospeech 89,* Paris, vol. 2, pp. 63-67.

SHIKANO, K., (1987), "Improvement of Word Recognition Results by Trigram Model", *Proceedings of the International Conference on Acoustics, Speech, and Signal Processing 87,* pp. 1261-1264.

SORIN, C., D. LARREUR, and R. LLORCA, (1987), "A Rhythm-Based Prosodic Parser for Text-to-Speech Systems in French", *Proceedings of the 11th International Congress on Phonetic Sciences*, Tallin, vol.1, 125-128.

SPROAT, R., J. HIRSHBERG, and D. YAROWSKY, (1992), "A Corpus-Based Synthesizer", *Proc. Proceedings.of the International Conference on Spoken Language Processing 92* Alberta, pp. 563-566.

TAPANAINEN, P., and A. VOUTILAINEN, (1994), "Tagging Accurately -- Don't Guess if you Know", *CMP-LG*, paper n° 9408009.

TRABER, C., (1993), "Syntactic Processing and Prosody Control in the SVOX TTS System for German", *Proceedings of Eurospeech 93*, Berlin, vol. 3, pp. 2099-2102.

WEISCHEDEL, R., M. METEER, R. SCHWARTZ, L. RAMSHAW, and J. PALMUCCI, (1993), "Coping with Ambiguity and Unknown Wordsthrough Probabilistic Models", *Computational Linguistics,* 1994.

WILLEMSE, R., and L. GULIKERS, (1992), "Word Class Assignment in a Text-to-Speech System", *Proceedings.of the International Conference on Spoken Language Processing,* Alberta, pp. 105-108.

WINOGRAD, T., (1972), *Understanding Natural Language,* Academic Press, Edimburgh.

YAROWSKY, D., (1994), "Homograph Disambiguation in Speech Synthesis", *Proceedings of the 2nd ESCA/IEEE Workshop on Speech Synthesis*, New Paltz, NY.

ZINGLE, H., (1990), "Morphological Segmentation and Stress Calculus in German with an Expert System", *Proceedings of the ESCA Workshop on Speech Synthesis*, Autrans (France), pp. 133-136.

Chapter Five

AUTOMATIC PHONETIZATION

Let me take this opportunity of answering a question that has often been asked me, how to pronounce "slithy toves." The "i" in "slithy" is long, as in "writhe"; and "toves" is pronounced so as to rhyme with "groves." Again, the first "o" in "borogoves" is pronounced like the "o" in "borrow." I have heard people try to give it the sound of the "o" in "worry. Such is Human Perversity.

Lewis Carroll, *The Hunting of the Snake*

"Mine is a long and a sad tale!" said the Mouse, turning to Alice, and sighing.
"It IS a long tail, certainly," said Alice, looking down with wonder at the Mouse's tail; "but why do you call it sad?"

Lewis Carroll, *Alice's Adventures in Wonderland*

5.1. From text to phonemes: a long way

Natural languages dispose of a rich vocabulary. Although most natives hardly use more than 5,000 words,[1] the *Oxford Advanced Learner's Dictionary of Current English* (Hornby, 1974) and the French *Petit Robert* (Robert, 1972) both denote more than 50,000 lexical entries. And the latter is merely, with its 2,000 pages, a simplified version of the more exhaustive *Robert* (5,600 pages). Similarly, the *Oxford English Dictionary* contains more than 500,000 entries. Yet only full forms are included. Taking plural, feminine, and conjugated forms into account increases the total number of common words to be transcribed up to at least 1 million, not to mention proper nouns, the pronunciation of which is not always clearly established.

It is well known that the pronunciation of words generally differs from their spelling. Besides, there are few European languages for which the discordance between writing and speech is as important as in English, French being another example. This

[1]The *Centre de Recherches pour un Trésor de la Langue Française* only denotes 4,730 words with a frequency of use greater than 0.01 percent (IMBS, 1971). These figures have been computed on a corpus of 23.5 million words, extracted from a sampling of French texts written in the second half of the twentieth century. For English, Mannel and Clark (1987) report that about 4500 words account for approaching 90 percent of all the words in two big corpora, including the Brown Corpus.

phenomenon partly originates in the natural delay between the spoken language, in perpetual evolution, and the much more rigid written one. As a matter of fact, we now spell French as it was spoken several centuries ago (for example, the French word *'enfant'*, was pronounced /ɛnfant/ in the MiddleAges) (Nyrop 63).[2] No wonder, as Allen (1976) emphasizes, that in these conditions some very usual words are so awkwardly inclined not to conform to well established pronunciation rules. Just think of the fact that *f* is always pronounced /f/ in the 20,000 most frequent words of English, except for the single case of *of*. The word *have* is another striking example, since it does not follow the general rule that a final silent *e* makes the previous vowel long, as in *behave*.

As a consequence, the principle "one character = one phoneme" is frequently not respected. It is sufficiant to notice that

- A single character can correspond to two phonemes, as for *x* /ks/, or none, as for *e* in *surely*[3].
- Several characters can produce a single phoneme, like in the French *ch* /ʃ/, or even none, like for the English *gh* in *though*.
- A character, or a sequence thereof, can be pronounced in different ways as a function of its left and/or right contexts, like for *c*: /s/ in *ancestor*, /ʃ/ in *ancient*, and /k/ in *epic*.
- Inversely, a single phoneme can result from several spellings, like for /ʃ/: *sh* in *dish*, *t* in *action*, *c* in *ancient*.

Besides, a poem of the Dutch high school teacher and linguist G.N. Trenite surveys this problem in an amusing way. It desperately ends with:[4]

> *Finally, which rimes with "enough,"*
> *Though, through, plough, cough, hough, or tough?*
> *Hiccough has the sound of "cup,"*
> *My advice is . . . give it up!*

Fortunately enough, pronunciations of isolated word roots can be found in phonetic dictionaries. Full forms (conjugated verbs, plural and feminine nouns, and so on) are assumed to be pronounced with the help of a list of simple rules described in introductory pages, and forms that infringe these rules are explicitly transcribed.

---

[2]However, this simple delay, common to all languages, cannot be held responsible for the disproportionate problems encountered in French. Actually, the conscious archaism of our ancestors, clerks, attorneys, lawyers, widely contributed to spreading confusion. In the sixteenth century, for instance, scholars began, respectfully in Greek and Latin, to overload their writing with lots of so-called etymological characters, which were often purely ornamental (like *x, y, z*), with the only intention of lengthening lines ! Since then, we have been constrained to write *doigt, sept*, and *pauvre* (from *digitum, septem*, and *pauperem*) instead of *doit, set*, and *povre* as was the case in the eleventh century.

[3]Notice that this remark does not apply to diphtongs and affricates, which are usually considered as a single phoneme.

[4]It is quoted in full in (Witten, 1982).

Let us recall that the letter-to-sound (LTS) module is responsible for the automatic determination of the phonetic transcription[5] of the incoming text. It thus seems, at first sight, that its task is as simple as performing the equivalent of a dictionary look-up[6]! From a deeper examination, however, one quickly realizes that most words appear in genuine speech with several phonetic transcriptions, many of which are not even mentioned in pronunciation dictionaries. What is more, some of these pronunciations seem to be optional (they can be used in place of each other without affecting the intelligibility or the naturalness of speech), while others are obligatory, in which case the decision of using a given pronunciation depends on higher level information than word spelling alone. Here is a rough list of such cases:

- Consonants may be reduced or deleted in clusters, a phenomenon termed as *consonant cluster simplification*, as in *softness* [sɔfnɪs] in which [t] fuses in a single gesture with the following [n].
- *Assimilation* is another source of phonetic variation. It originates in articulatory constraints and leads to a change of some phonological features of a given phoneme, given the corresponding features of some of its neighbors. These changes often result in other recognizable sounds. Voicing assimilation, for instance, changes a voiced phone into an unvoiced one, or *vice-versa*, as in *obstacle* [ɔpstəkl]. French is also apt to *nasal assimilation* (like in *assimilation de nasalité* when pronounced [asimilasiɔ̃nnazalite] instead of the expected [asimilasjɔ̃dnazalite])[7], which principally occurs in fluent speech. They also exist in English, but in such a strong way that they appear in spelling itself (as for *in+possible=impossible*). *Palatization* affects the manner of articulation of [t] and [d] and makes them sound more like the affricates [tʃ] and [dʒ], like in *tree* and *droop*. *Vowel harmony*[8] assimilation also modifies (unstressed) vowels, mainly by a change of their aperture in a given context, even in articulated speech. In French, [e] can be changed into [ɛ] when encountered in a closed syllable due to the muting of an *e*, like in *céderait* [sɛdrɛ] or *événement*

[5]It should be mentioned that the IPA symbols are not very convenient for automatic processing with computers. ASCII codes are preferred, in association with a translation table. Although it is the most compact way to do it, using a single ASCII character per API symbol hinders readability. Phonetization systems for English often refer to the ARPAbet (Shoup, 1980; see also Klatt, 1987) which accepts two characters per API symbol. Lennig and Brassard (1984) have also proposed to standardize a computer phonetic alphabet (CPA) which makes use of one or two characters per symbol, but no clear consensus has emerged. Since then, the standardized SAM phonetic alphabet SAMPA (ESPRIT_SAM, 1992) has arisen, which offers multi-lingual possibilities. In this textbook, however, we shall keep on using the IPA.

[6]Independently of the practical method adopted to do it (whether with a real lexicon or by rule): in this introductive section we are more interested in a functional description of phonetization than in an architectural one.

[7]Incidentally, the French language provides mnemotechnic examples of assimilation, since *assimilation de sonorité* (voicing assimilation) is in turn pronounced [asimilasjɔ̃tsɔnɔrite] instead of [asimilasjɔ̃dsɔnɔrite]—that is, precisely with a voicing assimilation !

[8]Vowel harmony refers to a phenomenon by which the articulation of vowels in a word or phrase influence each other.

[evɛnmɑ̃].[9] Similarly, [ɛ] can be replaced by [e] before [i], [y], or [u] in an open syllable, like in *aimer* [eme].

- *Heterophonic homographs*—words that are pronounced differently even though they have the same spelling, as for *record* (/rekɔːd/ or /rɪkɔːd/)—constitute by far the most tedious class of pronunciation ambiguities. Their correct pronunciation generally depends on their part-of-speech and most frequently contrasts verbs and nonverbs, as for *contrast* (verb/noun) or *intimate* (verb/adjective), although it may also be based on syntactic features, as for *read* (present/past). They are more usual than one might think at first: from the 4,000 French homographs presented in (INALF, 1984), about seventy are heterophonic, some of which having a very high frequency. Just think of *est*, pronounced /ɛ/ (or /ɛt/) when denoting the third person singular of the French verb for *to be* and /ɛst/ when it is the French noun for *east*. They often denote a difference of meaning but can, fortunately, often be detected from syntactic analysis alone. A limited number of ambiguities, however, require a real understanding of the sentence to be read (for example, in French: *les fils* = /lefis/ when understood as *the sons* or /lefil/ when meaning *the threads*; in English: *a bullet wound under his . . .* as opposed to *the cable wound around the . . .*).
- *Phonetic liaisons* affect final consonants of French words immediately followed by a vocalic sound in specific syntactic contexts.[10] This results in the pronunciation of characters that otherwise disappear (*ils aiment* /ilzɛm/) or in a change of their pronunciation (*bon élève* /bɔnelɛv/; this effect is denoted as *unnasalization*). Some are obligatory (*très utile* /trɛzytil/), some are optional (*deux à deux* /dø̆(z)adø/), others are prohibited (*plat exquis* /plaɛkski/). Delattre (1966) gives a comprehensive analysis of their functioning, from which they appear to be mainly conditioned by speaking style and by the degree of syntactic dependency between neighboring words.[11]
- *Schwas* are another source of concern. In French, a simple rule known as the *règle des trois consonnes* states that *e* is pronounced when its removal would mean uttering a sequence of hardly pronounceable consonants (typically three: *table rouge* /tabləruʒ/). In practice however, things are much more complex: [ə] is also maintained before *h* that is not dropped (*le héros* [ləero]), sometimes at the beginning of rhythm groups (*menez-les* [mənele]), or before groups as liquid+semi-vowel: *vous seriez* [vusərje]; rhythmic constraints have an influence when sequences of possibly schwas are encountered (*ce que je ne te redirai pas*), and so on. Lots of variants also exist, the most striking example being the well-known resistance of *e* in southern

[9]This even constitutes a major source of concern for French pupils, desperately trying to write *évènement*, *dussè-je*, or *cèdera*, just as they hear it.

[10]A similar effect arises in English for the pronunciation of *the*, which depends on the following phoneme.

[11]Again, it is interesting to notice that final consonants were always pronounced in ancient French. Phonetic liaisons are thus the survival of old times: they are now applied when the dependency between words has been strong enough to maintain this habit.

regions of France. Similarly, in English, a vowel deletion process termed as *schwa deletion* transforms unstressed vowels into short, central phonetic elements, or simply deletes them. Think of *interesting* [ɪnt(ə)rɛstiŋ] or *thoughtful* [θɔ:tf(ʊ)l].

- Possible *vowel lengthening* can be detected at word level, but it is necessary to know the limits of rhythmic groups inside the sentence to determine whether it is actually maintained or not. The word *three*, for example, is pronounced [θri:] in *I am number three* but rather [θri] in *I saw three of them.* This phenomenon, however, is much more pronounced in French, in which lengthening is strongly related to sentence-level stress (as in *elle est grande, cette grande affiche:* [ɛlɛgrã:dsɛtgrãdafiʃ]).[12]
- *New words* appear everyday, especially in relation with technological advances, although in many cases they can be accounted for by applying compositional and derivational morphology on known morphs. Exceptions include the emergence of foreign words in a language, which is of bigger concern to TTS system developers.
- Although *proper names* can be lexicalized (by including pronunciations into a telephone directory, for instance), one can hardly achieve exhaustivity. Here again, morphology can be of some help, especially for place names (as for *Lucasville* or *Californian*), provided base forms are known. Otherwise, strategies for deriving analogies with known words have to be found.

All these problems are summarized in Table 5.1.

Summing up, associating a single and accurate phonetic transcription with each word of a sentence is not equivalent to a sequence of dictionary look-ups (one per word) for the following reasons:

1. Printed pronunciation dictionaries refer to word roots only. They do not explicitly account for morphological variations, which therefore have to be dealt with by a specific component of morphology, called *morphophonology*.
2. Some words actually correspond to several entries in the dictionary, or more generally to several morphological analyses, generally with different pronunciations. This is typically the case of heterophonic homographs.
3. Pronunciation dictionaries merely provide something that is closer to a *phonemic* transcription than to a *phonetic* one (i.e. they refer to phonemes rather than to phones). As denoted by Withgott and Chen (1993): "while it is relatively straightforward to build computational models for morphophonological phenomena, such as producing the dictionary pronunciation of *electricity* given a baseform *electric*, it is another matter to model how that pronunciation actually sounds". Bridging the gap between phonological descriptions and the realm of speech data cannot be performed without incorporating some knowledge on the many articulatory processes that affect the traits of phonemes underlying the act of pronunciation.

---

[12]In the case of French, whose words include diacritics (é, è, ê, î), phonologists make a distinction between phonemic and phonetic lengthening. The first one is an intrinsic property of a word, whatever the context; it can be deduced from its spelling alone and often conveys lexical information (as in *fête - faites* : [fɛ:t] - [fɛt]). The second one is a natural feature of stressed vowels when placed in given phonetic contexts; it is generally not lexically distinctive (as in the examples given in the main text).

4. Words embedded into sentences are not pronounced as if they were isolated. Surprisingly enough, the difference does not only originate in variations at word boundaries (as with phonetic liaisons), but also on alternations based on the organization of the sentence into nonlexical units—that is, whether into groups of words (as for phonetic lengthening) or into nonlexical parts thereof (many phonological processes, for instance, are sensitive to syllable structure). Hence the need for a sentence-level or *postlexical* phonology.

5. Finally, not all words can be found in a phonetic dictionary: the pronunciation of new words and of many proper names has to be deduced from that of already known words.

Problem	Level	Information
Consonant cluster reduction	word/sentence	reading style, pronunciation of neighbors
Assimilation	word/sentence	reading style, pronunciation of neighbors
Heterophonic homographs	word	part-of-speech, meaning (rare)
Schwa deletion	sentence	syntactic articulation, pronunciation of neighbors, speaking style
Phonetic liaisons	sentence	syntactic articulation, pronunciation of neighbors
Vowel lengthening	sentence	syntactic articulation + prosodic information
New words	word	morphology
Proper names	word	morphology, analogy

Table 5.1. Some of the reasons why the LTS module cannot be reduced to a simple dictionary look-up process. *Level* specifies whether the problem is particular to transcribing words alone, sentences alone, or both. *Information* refers to the information that is needed to choose between variants.

Clearly, points 1 and 2 heavily rely on a preliminary morphosyntactic (and possibly semantic) analysis of the sentences to read. To a lesser extent, it also happens to be the case for point 3 as well, since reduction processes are not only a matter of context-sensitive phonation, but they also rely on morphological structure and on word grouping, that is on morphosyntax. Point 4 puts a strong demand on sentence analysis, whether syntactic or metrical, and point 5 can be partially solved by addressing morphology and/or by finding graphemic analogies between words. However, many aspects of our pronunciation also vary from one person to another, or simply from one moment to another, which made Nyrop (1963) say: "A quoi bon vouloir établir un seul bon usage quand la pratique nous apprend tous les jours qu'il y en plusieurs? (What is the good of trying to establish a single correct French while everyday practice shows us many?)". Consequently, some arbitrary choices have to be made as well, which originate in the *speaking style* of the reader (scholar or casual?, speedy or slow?, and so on).

5.2. Two basic strategies

In the following sections, we will assume that the LTS module of our TTS system is provided with the syntactic information it requires, together with the text to transcribe phonetically. It is then possible to organize its task in many ways (Fig. 5.1), often roughly classified into *dictionary-based* and *rule-based* strategies, although many intermediate solutions exist.

Dictionary-based solutions consist of storing a maximum of phonological knowledge into a lexicon. In order to keep its size reasonably small, entries are generally restricted to morphemes, and the pronunciation of surface forms is accounted for by inflectional, derivational, and compounding morphophonemic rules that describe how the phonetic transcriptions of their morphemic constituents are modified when they are combined into words. Morphemes that cannot be found in the lexicon are transcribed by rule. After a first phonemic transcription of each word has been obtained, some phonetic postprocessing is generally applied, so as to account for coarticulatory smoothing phenomena. This approach has been followed by the MITALK system (Allen *et al.*, 1987) from its very first day. A dictionary of up to 12,000 morphemes covered about 95 percent of the input words. The AT&T Bell Laboratories TTS system follows the same guideline (Levinson *et al.*, 1993), with an augmented morpheme lexicon of 43,000 morphemes (Coker, 1985).

A rather different strategy is adopted in *rule-based* transcription systems, which transfer most of the phonological competence of dictionaries into a set of letter-to-sound (or *grapheme-to-phoneme*) rules. This time, only those words that are pronounced in such a particular way that they constitute a rule on their own are stored in an exception dictionary. Notice that, since many exceptions are found in the most frequent words, a reasonably small exceptions dictionary can account for a large fraction of the words in a running text. In English, for instance, 2,000 words typically suffice to cover 70 percent of the words in text (Hunnicut, 1980). However, the distinction between dictionary- and rule-based approaches goes beyond the simple difference of their respective lexicon size. First, since lexical morphs are no longer extensively lexicalized, morphophonology generally reduces to some aspects of its inflectional and derivational components, termed as *affix stripping*. Second, and perhaps more significantly, although grapheme-to-phoneme rules are often followed by a phonetic postprocessing module as in dictionary-based methods, the distinction between the phonemic and phonetic components of rules tends to be blurred in many rule-based systems. More generally, many of the transformations that are typically accounted for by the morphophonological and phonetic modules of dictionary-based methods are fixed as early as the grapheme-to-phoneme step

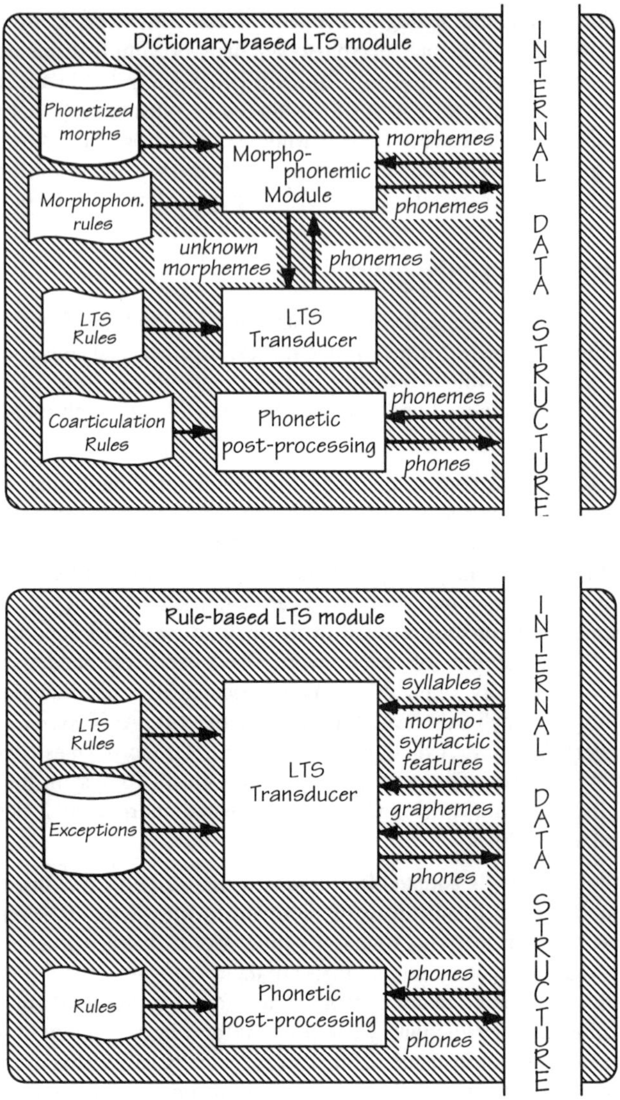

Fig. 5.1. Dictionary-based (top) versus rule-based (bottom) phonetization.

in rule-based ones (which we have indicated on Fig. 5.1 by the fact that the LTS transducer outputs phones instead of phonemes).

It has been argued since the early days of powerful dictionary-based methods that they were inherently capable of achieving higher accuracy than letter-to-sound rules (Coker *et al.*, 1990), given the availability of very large phonetic dictionaries on computers. On the other hand, considerable efforts have recently been made toward designing sets of rules with a very wide coverage (starting from computerized dictionaries and adding rules and exceptions until all words are covered, as in the work of Daelemans and van den Bosch (1993) or that of Belrhali *et al.* (1992). Clearly, some tradeoff is inescapable. Besides, the compromise is language-dependent, given the obvious differences in the reliability of letter-to-sound correspondences for different languages.

In the sequel, we will briefly describe each block of Fig. 5.1—namely, the morphophonemic module, the LTS transducer, and the phonetic postprocessor, independently of their relationship to dictionary or rule-based approaches, and conclude this chapter with a word on the phonetization of proper names.

5.3. The morphophonemic module

Morphophonology is concerned with phonological changes in the pronunciation of morphemes occurring in the process of word formation, in the same way as morphology accounts for the related written changes. Morphemes are thus represented as strings of phoneme segments, in which each phoneme is itself characterized as a set of phonological features. These binary features (having the value + or -) are precisely the ones that are used to define the phonemes of a language, by contrasting meaningful features within minimal pairs. They are related to the manner and place of articulation. The phoneme *p*, for example, could be described as [*-voc, +cons, -high, -back, -low, +anterior, -coronal, -voice, -continuant, -nasal, -strident*]. The aim of the morphophonemic module of a dictionary-based phonetization system is precisely to combine the phoneme strings for each constituent morpheme of a word, so as to produce its genuine pronunciation. Morphemes are assumed to have been obtained from a previous morphological analysis (see Section 6.2), and the related phoneme strings are retrieved from the lexicon or from a LTS transducer for unknown morphemes.

Among the phonological changes to be dealt with by the morphophonemic module, one distinguishes

- Rules for changing phonological features. Typical examples in English are the palatization of alveolar segments before suffixes *ion* and *ure*, as in *completion* or *exposure* or the change in the pronunciation of the suffix *ic* before the front vowels spelled as *e*, *i*, and *y*, as in *electricity*. Similarly, some German and Dutch voiced consonants are devoiced when followed by a morph boundary (for example, in Dutch *bed* and *bedden* give /bɛd/ and /bɛdən/ but [bɛt] and [bɛdən]).
- Rules for deleting or inserting phonemes, as required by the plural of English words ending in a fricative or an affricate, as in *buses*, and by the past of verbs ended in plosive [t] or [d], as in *landed*.
- Rules that account for stress shift in languages such as English or German. Derivation indeed (by either infixation or suffixation) happens to modify the stress pattern of roots, as in *adApt* + *ation* = *adaptAtion,*[13] or in *un* + *schUldig* = *Unschuldig*, although it is not always the case, as for *abOrt* + *ion* = *abOrtion* or *un* + *mÖglich* =

[13]Stressed syllables denoted by an uppercase vowel.

unmÖglich. The decision most often depends on the affix used, which results in categorizing affixes as either *stress-neutral* or *stress-shifting*, but it may also rely on the root with which they appear. More details on stress assignment are given in Chapter 6 (Section 6.8).

Such transformations can be accounted for by regular expressions. These can, however, be formalized in different ways, as examined in the next paragraphs.

Computational morphology was first studied in the 1960s, the golden years of research on word-by-word translating machines. Chomsky and Halle (1968), in their *Sound Pattern of English* (SPE), accounted for morphologically related phonological alterations. Their formalism, known as the *generative phonology* (or SPE: Sound Pattern of English) notation, is based on rewrite rules of the form

$$a \rightarrow b \,/\, l \_ r : c \qquad (5.1)$$

which means that segment *a* is rewritten as segment *b* when surrounded by strings *l* and *r* and provided condition *c* applies (where *l, r, c* can be null). Entries *a, b, l, r* are composed of sets of features and may be underspecified, so as to describe sets of phonemes instead of single ones: [+*voc,-cons*], for instance, would refer to the set of vowels. SPE rules, however, were not presented as a computational method and could not really be used as such. Indeed, the consecutive application of rules led to potential conflicts if more than one rule was applicable at a certain point: *feeding* occurred when the application of a rule produced a string to which another rule could then apply, while *bleeding* resulted from the application of a rule transforming a possible input string to another rule into a sequence to which it could no longer apply.

As a consequence, TTS system developers have often opted for rewrite rules for which the input and output levels are distinct, and more recently for multi-level rewrite rules (MLRRs), which indifferently access the grapheme, morpheme, or phoneme levels in their left and right context specification (as in the Bochum TTS synthesis system, SyRUB) (Böhm, 1992).

A somewhat different approach has recently been made available, as a result of the work of (Koskenniemi, 1983) on computational morphology (see Section 4.2.2.2). Two-level rules like (4.11) and (4.12) are indeed easily adapted to morphophonology, by simply assigning their input and output levels to respectively relate to the morphemes that make up a word and to the corresponding surface form, both represented this time as sequences of phonemes. Just as for morphology, additional abstract symbols, termed as *archiphonemes*, are allowed in the lexical level so as to account for higher abstractions. An archiphoneme /N/, for example, could map to /n/ or /m/ to account for the fact that /ɪN/ + /seɪn/ = /ɪnseɪn/, while /ɪN/ + /pəlaɪt/ = /ɪmpəlaɪt/. Consequently, the input and output levels do not really share the same alphabets. Working rule sets are now available for a large number of languages. They should progressively appear as components of TTS systems.

5.4. The LTS transducer

We now turn to the problems involved in the development of a LTS transducer. As it has been mentioned in Section 5.2, both dictionary- and rule-based LTS modules embody LTS rules. It has also been noted that grapheme clusters generally do not span morph boundaries in many languages (which is why *ea* in *reach* is not pronounced as in *changeable*). As a result, it appears that one should always avoid applying LTS rules to whole words, but rather examine morphs in isolation and reorganize their individual transcriptions into a global word-level one with the help of morphophonemic rules. It should be noted, however, that the letter strings actually passed to LTS rules are significantly different from the morphs obtained by exhaustive morphological analysis. Furthermore, this difference is often not found to be the same for dictionary-based methods as for rule-based ones. As a matter of fact:

- Since dictionary-based systems only invoke LTS rules when no complete morph covering has been found, which almost always originates in the appearing of an unknown lexeme stem (affixes can be exhaustively lexicalized, given their finite number), morphological analysis reduces in these cases to a mere and less reliable *affix stripping*. Compounds, for example, can no longer be analyzed in terms of their constituent roots.
- A similar situation most often arises with rule-based systems, which usually do not involve full lexeme dictionaries. In many such TTS systems, however, affix stripping is not performed explicitly but rather accounted for in practice by the left and right contexts of rewrite rules. Hence, whole words are passed to the LTS rules. Besides, this has motivated the development of training-based phonetization systems which are assumed to take account of morph boundaries in an implicit way, as a by-product of the knowledge they extract from phonetized text corpora (see Sections 5.4.3 and 5.4.4).[14]

With these differences in mind, we will now describe LTS transducers as they are examined by many rule-based approaches. The reader should remember that only affixed stripped words are examined by LTS rules in a dictionary-based framework.

The problem of developing a rule-based system that produces automatic phonetization can be considered as a data-reduction process that is assumed to have generalization properties.[15] Starting from a large database of words and/or sentences and their transcriptions, the system is supposed to end up with a very limited set of fairly consistent grapheme-to-phoneme rules, which are believed to be applicable to words and/or sentences that were not in the analysis database. The leading hypothesis in rule-based systems is that data reduction is on a par with generalization properties.

---

[14]In our opinion, such training-based strategies are often assumed to exhibit much more intelligence than they do in practice, as revealed by their poor transcription scores. We therefore believe that some kind of affix stripping should be performed explicitly anyway.

[15]This is typically an engineering point of view (which does not imply that it is the best one), as opposed to the linguistically motivated approach of dictionary-based solutions.

It is generally accepted that it is possible to account for all (*word, transcription*) pairs of a language with a limited set of regular transcription rules, which implies that it can be achieved with machines as simple as finite-state transducers, without the need to express input strings as hierarchical structures before being able to transcribe them. Deriving the expected set of regular rules then reduces to making the following choices:

1. *In each word, which sequences of characters correspond to which sequences of phonemes?*

   One immediately notices that there is not a one-to-one character-to-phoneme mapping for each word (i.e. each character cannot be associated with a single phoneme, and *vice versa*). Not all characters in *missing* [mɪsɪŋ], for instance, contribute to a distinct phoneme. Likewise, not all phonemes in [taksɪ] originate in a distinct character. As a result, aligning text strings on phonetic ones implies either that *null* characters or phonemes (_) are accepted (*m/i/s/s/i/n/g* = [m/ɪ/s/_/ɪ/ŋ/_], *t/a/_/x/i*=[t/æ/k/s/ɪ]), or that groups of characters and phonemes are accepted as ultimate units[16], (*m/i/ss/i/ng*=[m/ɪ/s/ɪ/ŋ], *t/a/x/i* = [t/æ/ks/ɪ]).[17]

   In the sequel, we shall call the corresponding groups of characters *character clusters* when they involve several characters (as in the table proposed by Lawrence and Kaye (1986) for English, which includes phoneme clusters up to five characters wide), and as *characters* otherwise. It should be clear that, if characters and character clusters are simultaneously accepted, the rule interpreter (or whatever machine is used as a grapheme-to-phoneme transducer) should be provided with a way to recognize them in a word. Why does *missing,* for example, correspond to the graphemic decomposition *m/i/ss/i/ng* rather than *m/i/s/s/i/n/g*? In other words, one needs something like a *graphemic syntax.*

2. *Assuming the above correspondences have been fixed, what type of information will be used to resolve conflicts?*

   Whatever answer given for the previous question, and whatever means adopted to fix the corresponding one-to-one mappings, conflicts will inevitably arise, in the form of identical characters (or character clusters) being transcribed differently throughout the database. Resolving them implies to choose a set of sources of disambiguative information (such as the graphemic context, the syllabic decomposition, morphosyntactic features, etc.) and therefore it conditions the linguistic insight one expects to acquire from the resulting rules.

---

[16]That is, as ultimate *symbols* in their respective vocabularies.

[17]Notice that both choices can be done in a variety of ways. One could *a priori* have chosen to write

missing=[mɪs_ɪŋ_] or [mɪ_sɪŋ_] or [mɪs_ɪ_ŋ] or [mɪ_sɪ_ŋ]

as well as

m/i/ss/i/ng = [m/ɪ/s/ɪ/ŋ] or *m/i/ss/ing* = [m/ɪ/s/ɪŋ] or *m/iss/ing* = [m/ɪss/ɪŋ], ...

3. *What kind of inference and parsing strategy will be used?*

   Once the characters of the languages to be translated into each other have been fixed (that is, the graphemes and the phonemes), after having outlined a grammar to organize character strings into sequences of graphemic units and having selected the discriminant sources of information used to solve potential conflicts, the actual transducing mechanism remains to be chosen. Transduction, indeed, can be performed not only by MLRRs but also with FSTs, MLPs, or whatever automaton known to be good at dealing with regular languages.

As we shall see, the transcription systems developed up to now (including the ones that are proposed by traditional linguistics in phonetization treatises) can be positioned with respect to each other by the particular answers they bring to questions 1, 2, and 3.

5.4.1. Pronunciation treatises

A number of linguists have proposed, and more or less succeeded in establishing a set of transcription rules covering whole languages. It is striking, however, to compare the resulting pronunciation treatises and to notice that they account for the same facts in very different ways. What is more, they often assume important prior knowledge from their potential readers.

All of them group characters into character clusters. Even though many graphemes have been widely accepted (think of the English *th,* for example), a lot of them (generally the less frequently encountered ones) remain questionable. In addition, no graphemic syntax is generally provided explicitly: graphemes are listed, but users are supposed to deduce a parsing strategy by themselves.[18] It can be assumed, for example, that if *oi* and *oin* are both accepted French graphemes, then rules accounting for the transcription of *oin* should first be tested when possible. In case no match is found, then rules for *oi* are the next candidates. Better still, in case two hypothetical graphemes are found to overlap in a word (as for *cc* and *ce* in the French word *accepter* [aksɛpte]), it may be granted that, given a left-to-right transcribing direction, the application of a rule for *cc* prevents *ce* from being examined. Clearly, the underlying graphemic syntax is given as a by-product of rules.

The LTS rules proposed by expert linguists are mostly rewrite rules of the type

$$a \rightarrow [\mathrm{b}]\,/\,l\,\_\,r : \mathrm{condition} \tag{5.2}$$

in which the additional conditions are subject to wide variations from one author to another, depending on the linguistic features chosen to solve transcription conflicts. In the case of French, for instance, Bourciez (1958) mainly develops an etymological description, the purpose of which is to establish what phonological transformations of Latin words have led to our current French vocabulary. Léon (1966) presents phonetization in the form of structured tables, driven by a rigorous list of a hundred graphemes or so. Nyrop (1963) takes an intermediate position: he proposes 232

[18] In that respect, pronunciation treatises are similar to traditional grammars: they address human readers, not machines.

contextual rules, augmented with exceptions lists originating in etymological or syntactic considerations when context does not suffice. Clearly, phonetization treatizes are often not suited for machine implementation.

5.4.2. Expert rule-based systems

Expert rule-based systems are based, as their name literally implies, on rules proposed by expert linguists or phonetic treatises, as the ones that were examined in the previous section. They are the most often adopted approach for TTS synthesis of international languages, since these languages have been extensively studied by linguists far before the outgrowth of computers. Up to now, they have also provided the best results for the automatic phonetization of sentences by rule (while, as we shall see in the next sections, highly accurate word-level phonetization can nowadays be achieved with *trained systems*, without the need for extensive prior linguistic knowledge or explicit rules). Most existing rule-based grapheme-to-phoneme transducers implemented in commercially developed TTS systems are based on this kind of method.

A first approach consists of solving conflicts by inspecting only the graphemic contexts in which they occur. It typically leads to ordered sets of context-sensitive rewrite rules of the same form as the rules used in the context of generative phonology:

$$a \rightarrow [\mathrm{b}]\, /\, l \_ r \tag{5.3}$$

which express that character (or character cluster) *a* is transcribed into phoneme(s) [b] when surrounded by character strings *l* and *r*. There is, however, a major difference between rules like (5.3) and the formalism proposed in (5.1), in that the input and output levels of grapheme-to-phoneme rules are different, which intrinsically guards them against feeding. Only some form of bleeding remains, in the sense that the application of a rule can "consume" letters in the grapheme level that could otherwise have triggered other rules. This has to be faced by a careful rule ordering.

Alternatively, one could use the fact that rules more directly highlight some linguistically pertinent properties of words, such as syllable break positions, part of speech categories, stress patterns, semantic features, or even etymological origins. This can be achieved by either associating conditions to rules (to be interpreted at run time), as in (5.1), or by turning them into the MLRRs of (3.2), so as to restrain their applicability by adding contextual information from other levels.

As a result of the heterogeneity of the linguistic literature on phonetic transcription, experts systems for a given language can be different from each other (they typically use different sets of graphemes, different rule formalisms, and different disambiguation information). Their main contribution with respect to phonetization treatises is the derivation of explicit phonetization procedures, which in turn only implicitly define the graphemic syntax used. Incoming words indeed are never explicitly decomposed into graphemes, which are merely a by-product of the transcription process itself.

In many cases, text is scanned from left to right (even though the analysis direction is more used as part of the implicit graphemic syntax parser than really based on psycholinguistic considerations) with moving focus and context windows. If a match is found, the corresponding phoneme(s) is retained, and the focus window is shifted *N* characters to the left, *N* being the number of characters in the character cluster that triggered the rule. If not, the focus window size is decremented (its first character is dropped), and rules are scanned again, and so on until a match is found (which always happens, as all single characters are graphemic entries for default rules). Each time the focus window is shifted, its size is reset to its maximum value, so that bigger character clusters are tested first (cf. above for *oin* and *oi*). In addition, when several rules exist for the same grapheme in different contexts, rules are tested in the order in which they appear in the rule database. Rule ordering is carefully designed, so that pronunciations are scanned from the most particular to the most general (that is, from the biggest context to the smallest).

There are variants to this general scheme, of course. In the case of English, for instance, it has proved interesting to convert consonant clusters first, so that their transcription can be used as additional contextual information for transcribing vowels (Hunnicut, 1980). Indeed, Klatt and Shipman (1982) have found that consonantal letters are quite regular and amenable to LTS transduction, while vowel letters account for most of the errors. Syllabification is sometimes proposed as an explicit method for segmenting words into graphemes. Choosing between *a/n* and *an* in the French words *ch/an/t/i/er* [ʃ/ã/t/j/e] or *c/a/n/a/r/d* [k/a/n/a:/r/_], for example, can be accounted for by syllabic decomposition: *chan|tier*, *ca|nard*. Syllabification is then described itself in terms of contextual rules (it does itself respect a regular syntax), and a set of exceptions is stored into an appropriate lexicon.

Notice that, just as in the context of the morphophonemic module, the recent availability of the new and coherent formalism of two-level rules (Koskenniemi, 1983) has opened new perspectives for the design of expert systems in phonetization, the advantage being their possible compilation into finite state transducers (Kartunnen *et al.*, 1987). The input and output levels are then respectively related to the graphemic and phonemic surface forms of words. To our knowledge, two-level rules have only been recently tried in the context of phonological rewriting (Williams, 1994). Their main differences with traditional one-level rules are that:

- Two-level rules are not ordered. They can therefore be written somewhat independently from one another, an advantage that can hardly be neglected in the context of rule-based phonetization, given the high degree of intricacy of traditional rules, which itself originates in the existence of possibly ambiguous character clusters.
- Two-level rules can refer to the input string as well as to the output one, which intrinsically enables them to test for the position of the current grapheme within the syllable, without the need to perform a prior syllabification.

5.4.3. Trained rule-based systems

Phonetization treatises or expert linguists are not always available when transcribing national languages (as opposed to international ones) or when developing multilingual TTS systems. *Trained rule-based systems* have thus emerged, for which human experts are no longer required for rule development (although some human-like clear-sightedness is still necessary for data preparation anyway). When provided with some human feedback, in the form of posttraining corrections, they constitute an interesting alternative to the previous ones.

Trained rule-based systems typically lead to more uniformity than expert systems. Most of them are based on single-character graphemes, since, as stated above, designing a phonetization system based on character clusters requires simultaneously defining a reliable graphemic syntax. The first step is to perform the text-to-phoneme alignment of a training corpus composed of words or sentences together with their pronunciation. This operation is not trivial (especially for words with a great difference between spelling and pronunciation). It can be achieved either manually or by rule. Notice that, to some extent, the rules needed to obtain this segmentation can be provided by an expert system, when available. In practice, since the segmentation of a text into its graphemes is only a subproblem of its automatic phonetization, it is also possible to obtain simple alignment rules from the corpus itself and to let a human expert correct the resulting alignment errors. The reader could argue that obtaining such alignment rules automatically is quite similar to inferring the simple graphemic syntax that trained rule-based systems try to avoid. In practice, however, the alignment problem is much simpler, since phonemes are known. What is more, the alignment rules used here need not be absolutely correct, since their result can be checked by human experts before the *training* phase. In contrast, the decomposition of words into their character clusters is performed during the *test* phase, so that decomposition errors inevitably lead to phonetization ones.

Van Coile (1991) has shown that the alignment problem can be solved with simple

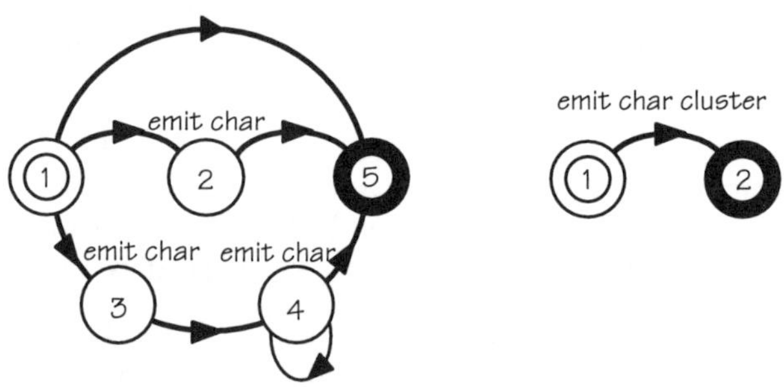

Fig. 5.2. Left: the phoneme HMM used in Van Coile (1991). Only states 2, 3, and 4 are emitting states, and they emit one character at a time. Hence, phonemes are assumed to correspond to null graphemes (upper branch), single characters (middle branch), or a sequence of characters (lower branch). Right: the simpler HMM model of Van Coile (1993).

HMMs for phonemes, whose states are supposed to emit single characters with given emission probabilities (Fig. 5.2). For each word in the training database, a HMM is created by concatenating phoneme models according to the phonetic transcription, and the Baum-Welch or Viterbi algorithms are used incrementally to optimize the emission probabilities so as to maximize the likelihood of the training database. Van Coile (1991) obtained a performance better than 99 percent on a database containing the 10,000 most frequent words of Dutch, of which 7,000 were used for training and 3,000 for testing. Similar results have been presented recently by the same author (1993) for six languages, based on much simpler one-state phoneme HMMs. This time, the probability $P(s|p)$ for each phoneme p to emit one character cluster s chosen in the set of all the possible character clusters up to a given length (including the null grapheme) is modeled by a single transition. As before, these emission probabilities are evaluated with the Viterbi algorithm, which is in turn used for testing.

Once the alignment is performed, character clusters are further decomposed into their letters, and (possibly null) phonemes are associated to each one. Again, this operation is *a priori* not trivial. Intuitively, it should be performed so as to minimize the number of phonemes associated to each letter of the alphabet throughout the training database. In French, for example, character clusters like *an, on, in, un, ...* [ã,ɔ̃,ɛ̃,œ̃,...] should be arranged as *a/n*=[ã/_], *on*=[ɔ̃/_], *in*=[ɛ̃/_], *un*=[œ̃/_], and so on, rather than as *a/n*=[_/ã], *on*=[_/ɔ̃], *in*=[_/ɛ̃], *un*=[_/œ̃], and so on. In many cases, simply mapping the first character of each character cluster to the phoneme, and the other characters to null phonemes gives good results (Daelemans and Van den Bosch, 1993, Van Coile, 1991).

The resulting training database can then be straightforwardly described by a maximal canonical transducer, a transducer version of the maximal canonical grammar that was presented in Section 2.2.1.3. The problem of trained rule-based systems is then confined to the equivalent of applying automatic rule inference procedures to derive a simpler finite state transducer from the trivial one.

In the work of Torkkola (1993), this is done with a simple *dynamic context extension* strategy. The transcription of each character is considered separately. A list of all the possible phonemes found to be associated with it in the training corpus is established and numbered. The *(character, phoneme)* pair that is most frequently encountered is then chosen as reference. All other pairs have to be accounted for by expanding the context one letter to the right or to the left (the expansion strategy is known *a priori*). This context expansion process is applied iteratively for remaining pairs, until no more conflicts are encountered or a maximum context width is reached. The method is clearly suboptimal, in the sense that the data reduction it allows is highly conditioned by an *a priori* context expansion order, which is furthermore fixed for all characters. Its swiftness, however, allows a lot of different strategies to be tested. Similar ideas have been used by Klatt and Shipman (1982) and Lucassen and Mercer (1984), in which decision trees were established on the basis of information theoretic principles, following the ideas examined in Section 4.4.3 about CARTs. Other interesting uses of data-driven algorithms can be found in the *pronunciation by analogy* work of Yvon (1996) and in the *stochastic phonographic transduction* of Luk and Damper (1996).

The advantage of such automatic regular inference approaches is that they are somewhat language independent. What is more, many of them automatically allow a hierarchical description of rules. The transcription of a given character is then searched in a multilevel tree structure in which levels correspond to increasingly wide contexts (that is, a *trie*; see Section 4.2.3.1). In the work of Daelemans and Van den Bosch (1993), a trie is explicitly built from a training corpus. Its first level corresponds to the focus grapheme (again, only single-character graphemes are considered), while the next ones are associated to successive left and right context graphemes (see Fig. 5.3). As with Lucassen and Mercer, the order in which the context graphemes are added to the trie is not randomly determined. It is rather computed from the training database, in such a way that the characters with the strongest disambiguative potential are examined first.[19] By combining this approach with a so-called *similarity-based reasoning* technique inspired by statistical pattern recognition (so as to account for graphemic sequences that might not be present in the trie), they obtained a word transcription rate of about 84 percent (which roughly corresponds to the 94 percent of correct letter transcription of Lucassen and Mercer) with a training database of about 50,000 English words and a test database of 6,000 other words. Besides, they have shown that similar results could be obtained for other languages, with no prior linguistic knowledge. Although very promising, this performance is not yet good enough to compete with expert systems.

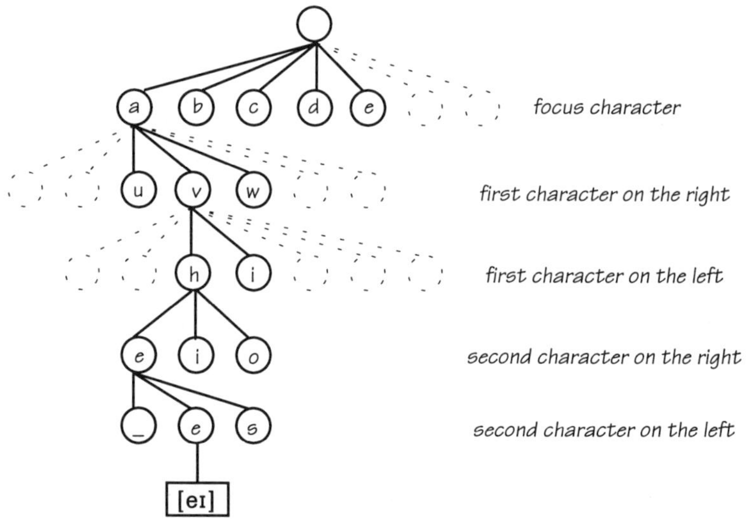

Fig. 5.3. Retrieval of the pronunciation of *a* in *behave* by trie search (after Daelemans and Van den Bosch, 1993).

[19]This is achieved by computing the *information gain* obtained on the training corpus when a context grapheme is known, as a function of the position of this grapheme, and by retaining the position that maximizes this function. The information gain is itself computed as the difference between the original entropy of the information provided by the corpus and the entropy of the information that is left when a context grapheme is known, exactly as in Section 6.7.3.

5.4.4. Neural networks

Neural networks have also been proposed as an alternative to explicit rule-based approaches. Self-organized/connectionist architectures such as NETTALK (Sejnowski and Rosenberg, 1987), NETSPEAK (McCulloch *et al.*, 1987), or the networks of Ainsworth and Pell (1989) and Matsumuto and Yamaguchi (1990) have received considerable attention among the TTS synthesis community. They differ from the previous systems in the sense that rules implicitly appear in the form of a list of numbers (usually the synapses weights in a multilayer perceptron), not easily understandable (nor easy to rectify) by humans.

They typically make use of monocharacter graphemes. Graphemic windows are presented, in a coded way, at the input of a multilayer perceptron, which is trained to produce the phonetic transcription of the central character. In NETTALK, for example, the input window was seven characters wide, each one being associated with input neurons to encode its value among twenty-nine possible values (twenty-six letters and three punctuation marks). A set of eighty hidden neurons and of about twenty-six output neurons were in charge of processing the input information. The output layer actually encoded about forty phonemes in the form of twenty-three articulatory features, and three stress and syllable boundary parameters. The network was fully connected in a feedforward architecture, and the error back-propagation algorithm was used to train the neuron weights (starting from random initial values) on a 20,000 word phonemic dictionary.

As for trained rule-based systems, MLP-based solutions are *a priori* interesting for their multi-lingual capability. Drawbacks, however, are numerous. Since each character corresponds to a single phoneme (including the null phoneme), some of the graphemes traditionally proposed by phonetization treatises are distributed over several characters (think of *ch* when pronounced [k], which has to be examined as [k_]). What is more, automatic handling of syntactic features, as needed to separate heterophonic homographs or to correctly apply phonetic liaisons, is difficult to incorporate. Similarly, the influence of the morphemic organization of words on their pronunciation is only indirectly taken into account, through the graphemic context. Yet a much larger letter context would be needed, for example, to account for graphemic differences as little as *photograph/photography* resulting in such different pronunciations of the first *o* (due to the influence of stress in this case). As a result, the number of errors per phoneme is found to be quite high. Results are so bad that error rates are reported by letter or by phoneme: 92 percent in the best case, which corresponds to a word error rate of approximately 50 percent. This is clearly unacceptable in the context of high-quality TTS synthesis: given the complexity of the task to perform here, MLPs cannot compete with rule-based systems, at least for the time being.

5.5. Phonetic postprocessing

In order to increase the intelligibility and the naturalness of synthetic speech, some kind of phonetic postprocessing is necessary in TTS systems, for the following reasons:

- As they stand, rule-based LTS transducers do not make it easy to express *phonetic* constraints, as opposed to graphemic, metrical, morphological, or syntactic ones, although coarticulatory phenomena are more conveniently described as originating from phonetic influences, as it is typically the case for vowel harmony.
- Neither rule-based nor dictionary-based approaches offer a convenient way of accounting for pronunciation rules that affect a word on the basis of the pronunciation of the following one(s), as in the case of phonetic liaisons or with coarticulatory effects that span word boundaries. Typical examples are the flapping of alveolar stop consonants in selected environments in American English, as in *butter* in which [t] sounds more like a [d], or their replacement by a glottal stop, as for the [t] of *not you*, or the interaction between two plosives, as in *Pat caught cold*, where the explosion of the first plosive is removed. Indeed, although it is possible to design a phonetization system in such a way that these problems are fixed directly in the LTS transduction, by allowing contexts to examine neighboring words, it is more natural to do so in a post-processing step.
- As mentioned in Section 5.1, there may be an important difference between the "received" pronunciation proposed by dictionaries and that observed in genuine speech, due to a number of coarticulatory smoothing phenomena, which moreover depend on the speaking style. Pronunciation is subject to wide variations, from region to region, from person to person, or even for a given speaker in different conditions. It does nevertheless respect some coherency, in the sense that a given coarticulatory effect does not occur (or not) randomly in reading, and should therefore be carefully modeled.

In many cases, these three reasons appear to combine. A description of style-dependent coarticulatory effects can be found in the book of Withgott and Chen (1993) for American English, on the basis of a comparison between the transcription of the TIMIT speech database (Fischer *et al.*, 1987) and the corresponding dictionary transcription. In French, schwa elisions are among the most intricate of such combined difficulties: its pronunciation or elision is based on sentence-level phonetic constraints and submitted to wide speaking-style variations (Dutoit, 1993).

As for the morphophonemic module, the phonetic postprocessor is usually organized into rewrite rules, possibly into MLRRs, so as to allow syntactic restrictions for instance (as needed by phonetic liaisons). It basically examines all word transcriptions as a continuous stream of phonemes and allows modifications to phonemes as a function of their context. As for any other regular transduction problem, it is also possible to design probabilistic transducers or decision trees trained on a phonetized corpus (see Withgott and Chen, 1993; Riley, 1991). This is particularly useful when the rules to infer are too complex for a human expert to obtain them. It has indeed been mentioned in Section 4.4.3 that decision trees provide an efficient means of inferring MLRRs. Withgott and Chen (1993) have designed such context trees in order to account for the difference between phonetic and phonemic transcriptions of the TIMIT corpus. From a predetermined set of possibly useful contextual factors, they have automatically derived by CART techniques a subset of factors that are good at differentiating among the phonetic element distributions. They have found that the most generally useful contextual factors for predicting the variant pronunciations of a phonological symbol are the preceding and following phonological symbol (and not, for instance the neighboring phonetic symbols), and that a small number of additional factors, such as the position of the symbol within the syllable, or the function/content word distinction, can increase the accuracy of the estimates.

5.6. Proper Names

Being able to correctly pronounce proper names is far from being luxury for a TTS system. Indeed, many applications are targeted for public access and involve person names, addresses, and place names. Until recently, however, this fairly specific problem had received, if not little attention, at least little realistic solution.

One of the reasons for this delay originates in the fact that proper names are far more numerous than ordinary words. Liberman and Church (1992) have reported on a database containing 1.5 million names, which is about thirty times the number of words in an abridged dictionary. Yet their relative distribution is found to be particular, too: although a list of the most likely 150 ordinary English words accounts for half the content of running texts, more than 2,300 names are needed to cover half of the surnames in the United States. Better still, it has been shown that, no matter how many names are taken from an American telephone directory, the number of proper names covered in *another* directory reaches an asymptote at about 62 percent (Klatt, 1987).

What is more, the pronunciation of proper names typically cannot be accounted for by the same LTS rules as ordinary words, because it often depends on which language is assumed as the underlying origin of the spelling. Although foreign names are sometimes given nativized pronunciations (unfamiliar phonemes are replaced by their nearest neighbor in the local phonological system and stress is sometimes adapted), it remains the case that similar letter clusters receive different pronunciations as a function of their etymology (Liberman, 1979). Similarly, when it is not simply adapted, stress is distributed for proper names in a way that is reminiscent of their origin (see Section 6.8).

Rule- and dictionary-based strategies have to be modified accordingly:

- Rule-based approaches cannot reasonably spare an etymological classification step prior to the actual phonetization process, so as to enable a language-dependent transduction process. Church (1986) has reported on the determination of etymology from spelling using language-dependent trigram statistics obtained from different corpora. The probability that a proper name is assigned a given origin is simply obtained by computing the related probability for each of its trigrams and multiplying each of these together.[20] Alternatively, Belhoula (1993) has observed language-specific clusters that occur only in names and derived an etymological classification procedure based on mapping these clusters on incoming names and further taking their position into account (final and medial clusters were found to carry more information about the origin of names than initial ones).

---

[20]In other words, he has accounted for graphotactics with second-order Markov chains. The problem, indeed, is not a hidden one since spelling itself is unambiguously known.

- Dictionary-based methods have to be provided with additional means of avoiding general purpose rule-based LTS transduction when names cannot be found in the lexicon. Coker *et al.* (1990) have reported on the use of compounding (*Lucas* + *ville* = *Lucasville*) and derivational (*Jordan* + *ian* = *Jordanian*) morphology, plus a very simple rhyme analogy method that derives the pronunciation of an unknown name by finding a similar one in the lexicon (typically, a name with the same spelling except a few initial letters), obtaining the pronunciation of their common part by analogy, and processing the remainder with LTS rules (as for *Plotsky* = *Trotsky* - *Tr* + *Pl*). They found that the rhyme method covers more names than the more complicated morphological processes.

Although much remains to be done in this field, especially in a multilingual perspective, there has been recently a Europeanwide mobilization in the form of the ONOMASTICA project, established as part of the European Commission Framework Programme—Linguistic Research and Engineering and aiming to create a set of pronunciation lexicons of European names. A total of twelve languages are covered, including Danish, Dutch, English, French, German, Greek, Italian, Portuguese, Spanish, and the Scandinavian languages, with about 1 million names per language and nativized pronunciations for each foreign name in each other language (Schmidt *et al.*, 1993). The results of this project could lay the basis of truly multilingual, dictionary-based name pronunciation systems.

5.7. Summary

After a comprehensive introduction to the difficulty of obtaining the pronunciation of sentences from their spelling, in which we have shown that automatic phonetization does by no means reduce to a sequence of lexicon look-ups (ideally one per word), we have briefly presented the two basic strategies available—namely, the rule-based and dictionary-based ones—and contrasted their philosophical differences. We have then successively examined the organization and the part played by each of their modules:

- The morphophonemic module, which is concerned with phonological changes in the pronunciation of morphemes in the process of word formation (modification of phonological features, insertion or deletion of phonemes, stress shift).
- The LTS transducer, the design of which requires to answer three basic questions: In each word, which sequences of characters correspond to which sequences of phonemes? Assuming these correspondences have been fixed, what type of information will be used to resolve conflicts? What kind of inference and parsing strategy will be used? We have then examined how several transcription systems (as proposed by phonetic treatises, expert rule-based systems, trained systems, and neural networks) can be positioned with respect to each other according to the options they make.
- The phonetic postprocessing module, which provides a convenient way of accounting for coarticulatory phenomena and for variations due to speaking styles.

We have concluded this chapter with a word on the additional problems raised by the automatic phonetization of proper names.

References

AINSWORTH, W.A., and B. PELL, (1989), "Connectionist Architectures for a Text-to-Speech System", *Proceedings of Eurospeech 89*, vol. 1, pp. 125-128.

ALLEN, J., (1976), "Synthesis of Speech from Unrestricted Text", *Proceedings of the IEEE*, vol. 64, n°4, pp. 433-442.

ALLEN, J., S. HUNNICUT, and D. KLATT, (1987), *From Text To Speech, The MITALK System*, Cambridge University Press, Cambridge.

BELRHALI, R., V. AUBERGE, and L.J. BOE, (1992), "From Lexicon to Rules: Towards a Descriptive Method of French Text-to-Phonetics Transcription", *Proceedings of the International Conference on Spoken Language Processing 92,* Alberta, pp. 1183-1186.

BELHOULA, K., (1993), "Rule-based Grapheme-to-Phoneme Conversion of Names", *Proceedings of Eurospeech 93*, Berlin, pp. 881-884.

BÖHM, A., (1992), *Maschinelle Sprachausgabe Deutschen und Englishe Textes*, Ph.D. dissertation, Ruhr-Universität Bochum.

BOURCIEZ, E., (1958), *Précis de Phonétique Française*, Klincksieck, Paris.

CHOMSKY, N., and M. HALLE, (1968), *The Sound Pattern of English*, Harper and Row, New York.

CHURCH, K. W., (1986), "Stress Assignment in Letter-to-Sound Rules for Speech Synthesis", *Proceedings of the International Conference on Acoustics, Speech, and Signal Processing 86,* Tokyo, pp. 2423-2427.

COKER, C.H., (1985), "A Dictionary-Intensive Letter-to-Sound Program", *Journal of the Acoustical Society of America,* suppl. 1, n°78, S7.

COKER, C.H., K.W. CHURCH, and M.Y. LIBERMAN, (1990), "Morphology and Rhyming: Two Powerful Alternatives to Letter-to-Sound Rules for Speech Synthesis", *Proceedings of the ESCA Workshop on Speech Synthesis,* Autrans (France), pp. 83-86.

DAELEMANS, W., and A. VAN DEN BOSCH, (1993), "TabTalk: Reusability in data-oriented grapheme-to-phoneme conversion", *Proceedings of Eurospeech 93,* Berlin, pp. 1459-1462.

DELATTRE, P., (1966), *Studies in French and Comparated Phonetics*, Mouton, The Hague.

DUTOIT, T., (1993), *High Quality Text-To-Speech Synthesis of the French Language*, Ph. D. dissertation, Faculté Polytechnique de Mons.

ESPRIT-SAM (1992), "Standard Computer-Compatible Transcription", ESPRIT Project 2589, *Stage Report Sen.93*, SAM-UCL-037.

FISHER, W., V. ZUE, J. BERNSTEIN, and D. PALLETT, (1987), "An Acoustic-Phonetic Data Base", *Journal of the Acoustical Society of America,* Suppl. 1, 81.

HORNBY, A.S., (ed), (1974), *Oxford Advanced Learner's Dictionary of Current English,* 3rd edition, Oxford University Press, Oxford.

HUNNICUT S., (1980), "Grapheme-to-Phoneme Rules: a Review", Speech Transmission Laboratory, *QPSR 2-3*, Royal Institute of Technology, Stockholm, Sweden, pp. 38-60.

IMBS, P., (1971), *Etudes Statistiques sur le Vocabulaire Français. Dictionnaire des Fréquences.* Centre de Recherches pour un Trésor de la Langue Française (CNRS), Marcel Didier, Nançy, Paris.

INALF, (1984), *Dictionnaire des Fréquences: Table de Répartition des Homographes*, CNRS - INALF (Institut National de la Langue Française), Nancy.

KARTUNNEN, L.,, K. KOSKENNIEMI, and R.M. KAPLAN, (1987), "A Compiler for Two-Level Phonological Rules", *Report n° CSLI-87-108*, Center for the Study of Language and Information, Stanford University.

KLATT, D., and D. SHIPMAN, (1982), "Letter-to-Phoneme Rules: a Semiautomatic Discovery Procedure", *Journal of the Acoustical Society of America*, suppl 1, n° 72, S.48.

KLATT, D. H., (1987), "Review of Text-to-Speech Conversion for English", *Journal of the Acoustical Society of America,* 82, 3, pp. 737-793.

KOSKENNIEMI, K., (1983), *Two Level Morphology: A General Computational Model for Word-Form Recognition and Production*, Ph.D. dissertation, Department of General Linguistics, University of Helsinki.

LAWRENCE, S.G.C., and G. KAYE, (1986), "Alignment of Phonemes with their Corresponding Orthography", *Computer Speech and Language*, 1, pp.153-165.

LENNIG, M., and J.P. BRASSARD, (1984), "Machine Readable Phonetic Alphabet for English and French", *Speech Communication*, 3, pp. 165-166.

LEON, P.R., (1966), Prononciation du Français Standard, Didier, Paris.

LEVINSON, S.E., J.P. OLIVE, and J.S. TSCHIRGI, (1993), "Speech Synthesis in Telecommunications", *IEEE Communications Magazine*, pp. 46-53.

LIBERMAN, M.J., (1979), "Phonemic Transcription, Stress, and Segment Durations for Spelled Proper Names", *Journal of the Acoustical Society of America*, Suppl.1, 64, S. 163.

LIBERMAN, M.J., and K.W. CHURCH, (1992), "Text Analysis and Word Pronunciation in Text-to-Speech Synthesis", in *Advances in Speech Signal Processing*, S. Furuy, M.M. Sondhi, eds., Dekker, New York, pp.791-831.

LUCASSEN, J.M. and R.L. MERCER, (1984), "An Information Theoretic Approach to the Automatic Determination of Phoneme Base Forms", *Proceedings of the International Conference on Acoustics, Speech, and Signal Processing 84*, 52.5, pp. 42-45.

LUK, R.W.P., and DAMPER, R.I., (1996), "Stochastic Phonographic Transduction for English", *Computer Speech and Language*, n°10, 133-153.

MATSUMUTO, T. and Y. YAMAGUCHI, (1990), "A Multi-Language Text-to-Speech System Using Neural Networks", *Proceedings of the ESCA Workshop on Speech Synthesis*, Autrans, pp. 269-272.

MANNEL, R., and J.E. CLARK, (1987), "Text-to-Speech Rule and Dictionary Development", *Speech Communication*, n°6, pp. 317-324.

MC CULLOCH, N., M. BEDWORTH, and J. BRIDLE, (1987), "NETSPEAK - a re-implementation of NETTALK", *Computer Speech and Language*, 2, pp. 289-301.

NYROP, K., (1963), *Manuel Phonétique du Français Parlé*, Gylendal, Paris.

RILEY, M.D., (1991), "A Statistical Model for Generating Pronunciation Networks", *Proceedings of the International Conference on Acoustics, Speech, and Signal Processing 91,* pp. 737-740

ROBERT, P., (1972), *Le Petit Robert*, Société du Nouveau Littré, Paris.

SCHMIDT, M., S. FITT, C. SCOTT, and M. JACK, (193), "Phonetic Transcription Standards for European Names (ONOMASTICA)", *Proceedings of Eurospeech 93*, Berlin, pp. 279-282. (see also URL address : `http://guagua.echo.lu/langeng/en/lre1/onomas.html`)

SEJNOWSKI, T., and C.R. ROSENBERG, (1987), "Parallel Networks that Learn to Pronounce English Text", *Complex Systems*, n°1, pp. 145-168.

SHOUP, J. E., (1980), "Phonological Aspects of Speech Recognition", in *Trends in Speech Recognition,* W.A. Lea, ed., Prentice-Hall, New-York, pp. 125-138.

TORKKOLA, K., (1993), "An Efficient Way to Learn English Grapheme-to-Phoneme Rules Automatically", *Proceedings of the International Conference on Acoustics, Speech, and Signal Processing 93*, pp. II.199-II.202.

VAN COILE, B., (1991), "Inductive Learning of Pronunciation Rules with the DEPES System", *Proceedings of the International Conference on Acoustics, Speech, and Signal Processing 91,* Toronto, vol. 2, pp. 745-748.

VAN COILE, B., (1993), "On the Development of Pronunciation Rules for Text-to-Speech Synthesis", *Proceedings of Eurospeech 93,* Berlin, vol. 2, pp. 1455-1458.

WILLIAMS, B., (1994), "Welsh Letter-to-Sound Rules: Rewrite Rules and Two-Level Rules Compared", *Computer Speech and Language*, n°8, 261-277.

WITHGOTT, M. M., and F.R. CHEN, (1993), *Computational Models of American English,* CSLI Lecture Notes, , n°32, Center for the Study of Language and Information, Stanford University.

WITTEN, I.H., (1982), *Principles of Computer Speech*, Academic Press.

YVON, F., (1996), "Grapheme-to-Phoneme Conversion of Multiple Unbounded Overlapping Chunks", *CMP-LG*, paper n° 9608006[21]

[21]CMP-LG is the Computation and Language E-Print Archive set up by the consortium for lexical research. It can be accessed through the WorldWideWeb at URL address http://xxx.lanl.gov/abs/cmp-lg. Papers are stored under the name *yymmnnn*, where *yy*, *mm*, and *nnn* refer to its year and month of submission and to its number.

Chapter Six

AUTOMATIC PROSODY GENERATION

"But if you HADN'T done them," the Queen said, "that would have been better still; better, and better, and better!" Her voice went higher with each "better," till it got quite to a squeak at last.

Lewis Carroll, *Through the Looking Glass*

The impression of naturalness that issues from a TTS system is a function of the richness of the melodic contours and the quality of the rhythmic patterns it is able to apply to the speech it produces. The prosody generator is responsible for these two aspects. We shall now examine its requirements and examine how its goals can be partially achieved, after some definitions and a few words on the difficulty of deriving a consistent model of prosody.

6.1. What is prosody?

The term *prosody* refers to certain properties of the speech signal such as audible changes in pitch, loudness, and syllable length. For some authors the set of prosodic features also includes (other) aspects related to speech timing such as rhythm (mostly determined by the timing of *stressed*[1] syllables) and speech rate.

Because prosodic events appear to be time-aligned with syllables or groups of syllables, rather than with segments (sounds, phonemes), they are also referred to as *suprasegmental* phenomena (or simply *suprasegmentals*).

The term *intonation* is used by some as a synonym for prosody. It is restricted to the tonal (melodic) aspects of prosody by others (it will be the case here, too); finally it is sometimes reserved for the structural interpretation of prosodic phenomena.

---

[1]Stress is defined in Section 6.3.

6.2. Levels of representation of prosodic phenomena

As for other properties of the speech signal, prosodic events can be studied at various levels of representation (see Table 6.1):

- First, the *acoustic* level: the acoustic manifestation of prosody (fundamental frequency, amplitude, and duration[2]) can be measured directly, using specialized hardware or algorithms (such as pitch determination algorithms).
- Second, the *perceptual* level represents the prosodic events as heard by the (average) listener. As for spectral properties of speech sounds, acoustic characteristics that can be measured are not always perceptible. The perceptual representation is accessible to the individual listener, but this mental representation can hardly be measured. Alternatively it can be computed with a fair amount of precision on the basis of our knowledge about psychoacoustics.
- Finally, the *linguistic* level represents the prosody of an utterance as a sequence of abstract units (signs, symbols), some of which have a communicative function in speech, while others may just fulfill syntactic requirements. The linguistic structure of prosody is not some hidden code that simply can be revealed using some standard procedure. The linguistic model is a structural interpretation of the data, which results from the distributional or functional analysis of prosodic data by a linguist. There can be, and usually there are, alternative linguistic representations of the same data. Linguistic representations can not be "measured"; they can not be verified or falsified beyond the level of mere descriptive adequacy. However, it is possible to build recognition systems that generate a transcription of the prosody of an arbitrary utterance according to some specific linguistic model of prosody.

Acoustic	Perceptual	Linguistic
Fundamental frequency (F0)	Pitch	Tone, intonation, aspect of stress
Amplitude, energy, intensity	Loudness	Aspect of stress
Duration	Length	Aspect of stress
Amplitude dynamics	Strength	Aspect of stress

Table 6.1. Links between levels of representation of prosodic phenomena.

Given the different nature of these representations, it is important to keep them apart. It can be helpful to have the terminology reflect the level of representation. For instance,

[2]Notice that even duration is not strictly an acoustic feature, since it depends on the prior determination of segmental boundaries.

measuring loudness does not equal measuring signal energy. It is obvious that the perception of loudness is not exclusively related to the amplitude at one point of the signal, but also dependent on the duration of a speech fragment (the loudness of which we are measuring), and relative to the loudness of other parts in the signal.

As one moves away from the acoustic level towards the perceptual and/or linguistic levels, the measurement of some given prosodic property will progressively involve segmentation (for example, into syllables), context (such as relative prominence), and structural information (the linguistic interpretation of a syllabic tone, for example, often depends on whether the related syllable is stressed or not, which requires a prior analysis of the segmental layer).

6.3 Major components of prosody

Among prosodic events, the most overt are changes in *pitch* (in the order of five semitones[3] or more), which together constitute the pitch contour of the utterance. Distributional analysis of sentence-level pitch contours shows that the pitch contour of longer utterances can be broken down to a sequence of elementary contours, which can further be divided into syllabic contours or even static pitch targets associated with syllables or parts of syllables.

Stress is a prosodic property that has been described since the very first work on prosody in phonetics. It was said to be related to loudness and phonatory force. Both these characterizations refer to the perceptual form of prosody: the syllable carrying stress is prominent with respect to the surrounding syllables, either due to its loudness, or to its dynamic properties. In the 1950s and 1960s, increased acoustic research showed that is was impossible to find a (simple) acoustic correlate of stress (see Mol and Uhlenbeck, 1956, for instance), and consequently the concept was abandoned in most acoustic research on prosody. Some models of prosody claim there is no such thing as stress, and that prosody reduces to pitch change. Others have tried to capture stress in terms of pitch and syllable length, in which case it is sometimes referred to as *accent*.[4] However, in linguistic analyses stress is inevitable because it is a necessary concept for explaining the relation between prosodic structure (such as elementary contours) and segmental structure (such as word stress). Recent research on stress correlates has investigated other possible cues to stress related to syllabic timing (P-centers, see Section 6.8.1.1) or to spectral properties (Sluijter *et al.*, 1995). Nowadays it has become clear that the rejection of stress as a pertinent prosodic feature was based on a simplistic view of (absolute) acoustic correlates.

Syllabic length plays a role in the indication of boundaries and of stress. Although the absolute duration of an event is easily measured, this is only a trivial aspect of the measurement of length. First of all, it is not obvious which events should be measured: entire syllables, voiced parts, syllabic nuclei, and so on. Second, the duration of syllables

---

[3]The musical scale is logarithmically divided into *octaves* and *semitones*. An octave corresponds to a pitch doubling and is composed of twelve *semitones*, which correspond to $2^{1/12}$ octave.

[4]In this textbook, we use both words interchangeably.

and speech sounds depends on several (dependent or interdependent) factors such as speech rate, rhythm, phonetic nature, and so on.

The acoustic parameters related to prosody (mainly *F0*, *intensity*, and *duration*), exhibit short-term variations that can be measured (provided very accurate instrumentation is available), but that cannot be perceived (as such) and clearly have no function in intonation (although they may contribute to segmental identification). These variations are referred to as *microprosody*. For instance, the *F0* contour usually shows a very small dip during nasal and liquid consonants; following unvoiced consonants, the first two or three vibration periods typically are considerably shorter. Most of these phenomena can be explained by the aerodynamics of speech production. When using acoustic analysis for the study of prosody, one has to be aware of these phenomena. It would be a major error to treat these parts of the *F0* contour as meaningful pitch rises or falls.

6.4. The meanings of prosody

Prosodic features have specific functions in speech communication (see Fig. 6.1). The most apparent effect of prosody is that of focus. For instance, certain pitch events make a syllable stand out within the utterance, and indirectly the word or syntactic group it belongs to will be highlighted as an important or new component in the meaning of that utterance. The presence of a focus marking may have various effects, such as contrast, depending on the place where it occurs, or the semantic context of the utterance.

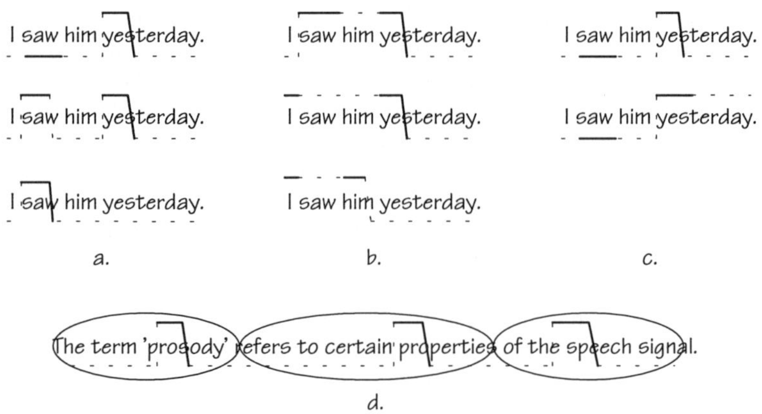

Fig. 6.1. Different kinds of information provided by intonation (lines indicate pitch movements; solid lines indicate stress).

a. Focus or given/new information;

b. Relationships between words (saw-yesterday; I-yesterday; I-him)

c. Finality (top) or continuation (bottom), as it appears on the last syllable;

d. Segmentation of the sentence into groups of syllables.

Although maybe less obvious, there are other, more systematic or general functions.

Prosodic features create a segmentation of the speech chain into groups of syllables, or, put the other way round, they give rise to the grouping of syllables and words into larger chunks.

Moreover, there are prosodic features that indicate relationships between such groups, indicating that two or more groups of syllables are linked in some way. This grouping effect is hierarchical, although not necessarily identical to the syntactic structuring of the utterance.

The well-known declarative or sentence-final intonation, marked in many languages by a very low pitch near the bottom of the speaker's pitch range, is often seen as a specialized prosodic marker indicating the sentence is closed. A high pitch at the end of the utterance indicates question intonation. In terms of the prosodic links we just talked about, these phenomena can be seen as side-effects of the grouping hierarchy: a prosodic structure is completed once a sentence final tone is encountered, and a closed structure is a declaration; when the low tone is not encountered by the time we reach the end of the utterance, the structure is unfinished, or left open by the speaker, indicating a link with what is to follow.

All these aspects of intonation can be grouped under the header of *linguistic* (functional) *aspects* of intonation. They are part of the structure of language (and specific to any given language) in the same way as morphology and syntax are. There is, however, another range of phenomena that are also expressed by prosodic means (such as pitch), but are independent from the functional ones. They could be called *emotional aspects* because they convey information about the emotional or physical state of the speaker. Angry people, for example, usually have faster pitch changes, a larger pitch range, and a larger dynamic amplitude range; whereas depressed people typically show the opposite trend. But while the pitch range may be affected by such emotional factors, the basic functional pitch shapes and configurations remain unaffected. The emotional state does not alter the linguistic code; it merely affects its realization. This is why these aspects are sometimes called *paralinguistic aspects*, together with other phenomena such as voice quality, for instance (Crystal and Quirk, 1964).

6.5. Intonation models[5]

A number of intonation transcription formalisms and methods have been developed over the years. Following the levels of description introduced in Section 6.2, they can be classified as either acoustic, perceptual, or linguistic.

[5]In this section, we focus only on the melodic component of prosody. Duration models are mentioned in Section 6.8, as their understanding is less critical than that of intonation models. As for the energetic component, it appears that the intensity of a given phoneme is approximately constant throughout its appearings in a message that is read with no special emphasis. We shall thus consider, as it is usually the case in TTS synthesis, that it is an intrinsic feature of phonemes. This does not imply, however, that phonemes have a fixed loudness, as loudness is a perceptual variable that is influenced by phoneme length, for instance.

6.5.1. Acoustic models of intonation

Just as it is possible to describe the segmental features of speech in the form of a sequence of spectral vectors related to an acoustic model (see Chapters Eight and Nine), it is reasonably possible to obtain a numerical representation of prosody, in the form of a sequence of *F0* values (one every 10 ms, for instance), with the help of pitch analysis algorithms[6] (see Fig. 6.2).

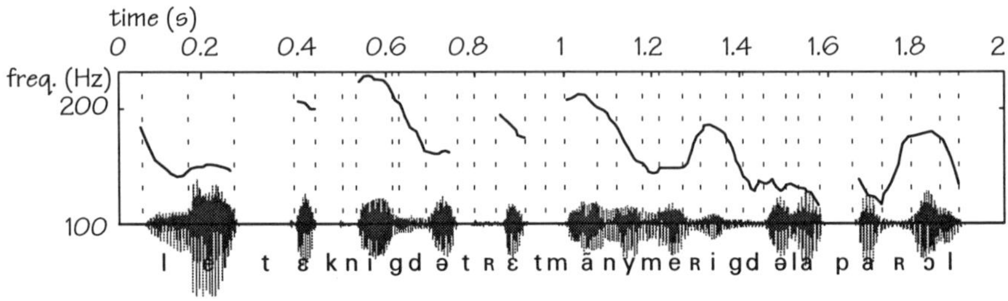

Fig. 6.2. A low-level acoustic description of the prosody of the French utterance: *Les techniques de traitement numérique de la parole...* (*speech processing techniques...*): amplitude (time waveform), phoneme durations (between vertical dotted lines), and *FO* curve. Fundamental frequency is plotted on a logarithmic scale and set to zero in unvoiced segments.

Such a description, however, has limited usefulness. In the context of text-to-speech synthesis, it can at best be used as the prosodic input of a *TTS copy synthesizer*, the aim of which is to repeat utterances produced by a human speaker with a synthetic voice, while keeping the original prosody.[7]

A full acoustic representation is not very constrictive either, in that the range of acoustic events it is able to account for is much too large for a description of human prosody. Not all pitch curves correspond to natural nor even to physically pronounceable data. As a result, it does not easily lend itself to prosodic modifications with a copy synthesizer and exhibits, *a fortiori*, limited interest for prosody generation. As a matter of fact, modifying the prosody of an utterance typically means modifying several *F0* values homogeneously. This can only be achieved, to a degree, with the help of a full acoustic intonation *model* (as opposed to the *description* mentioned above), which examines prosodic data at a higher level and accounts for *F0* curves with a limited number of *parameters*, a change of which makes it possible to browse a wide range of prosodic effects. Such models actually exist, known as *Fujisaki's model* and *acoustic stylization methods.*

---

[6]For a review of pitch determination algorithms, see Hess (1983, 1992).

[7]Such systems are very useful to assess the *segmental* quality of a TTS system—that is, its ability to produce natural sounding speech when provided with natural prosody.

6.5.1.1. Fujisaki's model

Fujisaki's model (Fujisaki , 1992) is a continuation of Ohman's work on the prosody of words (Ohman, 1967). It is based on the fundamental assumption that intonation curves, although continuous in time and frequency, originate in discrete events triggered by the reader that appear as a continuum given *physiological* mechanisms related to fundamental frequency control. Fujisaki distinguishes two types of discrete events, termed as *phrase* and *accent* commands and respectively modeled as pulses and step functions. These commands drive critically-damped second order linear filters whose outputs are summed to yield *F0* values (see Fig. 6.3). Fujisaki (1992) argues on the choice of this transfer function and of two such types of commands, on physiological grounds.

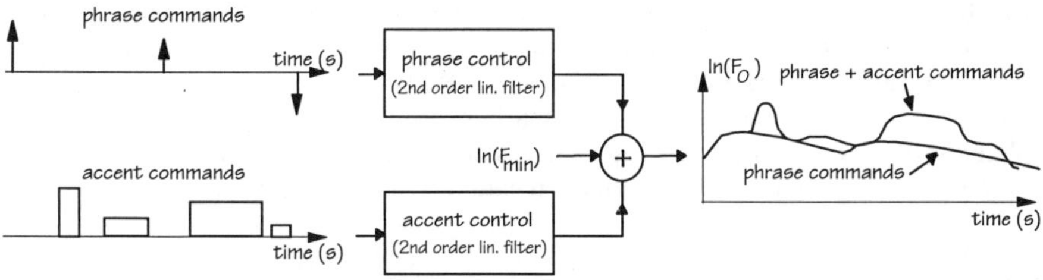

Fig. 6.3. Fujisaki's production model of intonation (modified from Ljungqvist and Fujisaki , 1993).

Assuming the parameters of the linear filters are fixed for a given speaker (since they are related to physiological constraints), the important parameters of this model are relatively few:

- A bias frequency level F_{min} on which all accent and phrase commands are superimposed;
- The number of phrase commands and two parameters for each command: its amplitude and timing;
- The number of accent commands and three parameters for each command: its amplitude and onset and offset timings;

These parameters (including the filter parameters) are obtained by first estimating the number of phrase and accent commands from a visual inspection of the analyzed *F0* contour, and then minimizing the mean squared error between this contour and that of the model through a search in the parametric space. A pitch analysis algorithm based on this model can be found in (Geoffrois, 1993), which additionally derives the number of phrase and accent commands automatically.

The model has been applied to several languages (including Japanese, English, Mandarin Chinese, and German) and good approximations of *F0* contours were obtained. Notice, however, that arbitrarily good approximations can always be achieved provided the number of phrase and accent commands is sufficiently high. Some extra constraint is necessary.

6.5.1.2. Acoustic stylization methods

Another means of presenting the information embedded in F0 curves in a comprehensive way is to highlight the *acoustic invariants* they might contain, by computing *declination lines* and/or approximating curves by a sequence of *target points*.

It has been noted in many languages that *F0* curves tend to evolve around or between average values that decrease with time (Vaissière, 1983). Or, the other way round, if one computes the average fundamental frequency in an utterance, *F0* values tend to be greater than this mean value in the first part of the curve and lower in the second one. This general tendency is termed as *declination.* How it can be accounted for mathematically, however, is another matter. A simple idea would be to compute *declination lines* as the best linear regression fit of all the local extrema (*peaks* and *valleys*) of an *F0* curve (a *topline* and/or a *baseline*: see Fig. 6.4). This task is made more difficult in practice by the fact that speakers tend to *reset* their vocal register after a particularly low *F0* value has been produced (typically after pauses separating major phrasal constituents) ('t Hart *et al.*, 1991). One also often observes very low *F0* values at the beginning and end of utterances, which bias the computation of the baseline. The topline itself does not always appear clearly, since *F0* maxima depend on accent values. Better estimations can be obtained after perceptual stylization (see Section 6.5.2).

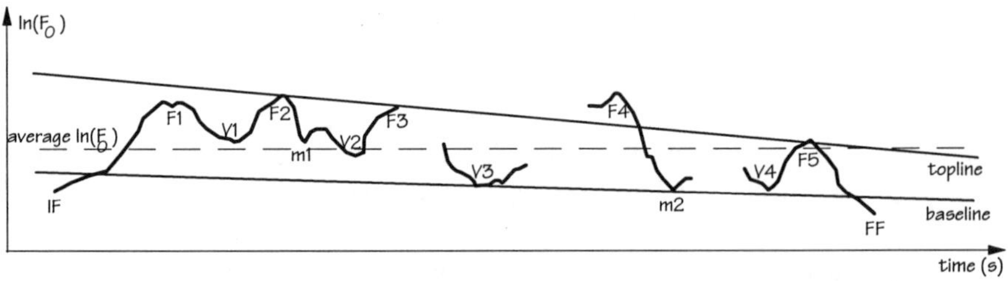

Fig. 6.4. Declination lines obtained from an acoustic analysis (after Beaugendre, 1995). *IF, FF, Vx, Px,* and *mx* respectively, refer to initial frequency, final frequency, valleys, peaks, and micromelodic dips.

Independently of the computation of declination lines, *F0* curves can be expressed as sequences of target points, assuming that between these points, transitions are filled by a (usually monotonic) interpolation function—like straight lines (Bagshaw, 1994), quadratic functions as in the acoustic part of Pierrehumbert's model (1981) or of Taylor's RFC model (Taylor, 1994), or quadratic splines (Hirst *et al.*, 1991)[8] (see Fig. 6.5). This approach is sometimes referred to as *acoustic stylization.* The parameters of the interpolation functions are again either imposed *a priori* or adjusted through linear regression. The position of pattern points is obtained by computing the correlation between successive *F0* values, setting up a pattern point each time it drops under a specified value (see Scheffers, 1988). Stylization may be broad or narrow, depending the

[8]Interpolation functions are much less critical than target points ('t Hart, 1991).

allowed maximum density of target points. Aubergé (1991) and Larreur *et al.* (1989), for instance, choose narrow stylization by modeling the intonation of each vowel with three target points (interpolated with straight lines), *a priori* positioned at its onset, extremum or medium, and offset. In contrast, Bagshaw, Hirst, Taylor, or Pierrehumbert 's stylizations are broad ones (hence, more ambitious, in terms of data compression).

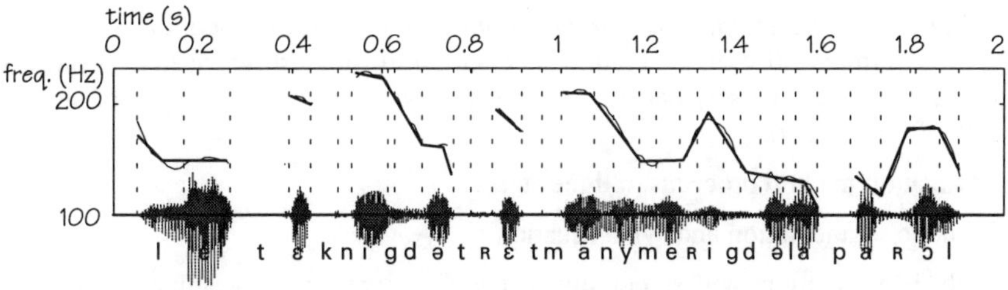

Fig. 6.5. A straight-line acoustic stylization of the example of Fig. 6.2.

6.5.2. Perceptual models of intonation

Acoustic models can hardly be used as such for a linguistic, functional study of prosody, which should at least take perception into account. They provide approximated *F0* curves, but since the parameter estimation algorithms take the acoustic measures as input, they cannot guarantee that the details that are smoothed out are not actually audible, and conversely that the ones that remain can really be heard.[9] Perceptual models on the other hand aim to yield a quantitative but compact description of the prosodic attributes of the signal that are perceived. We shall briefly examine two such models: the *IPO* model and a recent *automatic perceptual stylization* model.

6.5.2.1. Automatic perceptual stylization

Mertens (1987) and more recently d'Alessandro and Mertens (1995) have proposed methods for automatically stylizing pitch contours using a model of tonal perception. It is based on the following hypotheses on tonal perception:

- There is some evidence from studies on perception that *F0* contours should not be interpreted in isolation, but rather in conjunction with other phonetic and prosodic characteristics of the speech signal (Kohler, 1991). In particular, *syllable-sized units* often appear to act as elementary units for intonation perception.
- The perceived pitch contours within a syllable can be decomposed into elementary contours called *tonal segments*, the number of which is expected to be limited by some language-specific properties, in particular by the inventory of the syllabic pitch contours for the language. One should not confuse, however, the perceptual segments studied by d'Alessandro and Mertens and the linguistic tonal segments proposed by many phonological theories of intonation (see Section 6.5.3): the number of perceived tonal segments can be larger than the number identified in the linguistic analysis, in

[9]Besides, it is highly dependent on the accuracy of the pitch determination algorithm.

which case several perceptual segments can be seen as free variants of a single abstract intonation unit (in the same way as phonemes can be realized as several allophones).

- Not all pitch changes can be heard. A number of studies have reported on the existence of an absolute threshold of pitch change, or *glissando*[10] *threshold*, expressed in semitones per second. Similarly, not all glissandi can be perceived as distinct. The *differential glissando threshold* (also expressed in semitones per second) is the minimum difference in slope necessary to distinguish between two successive glissandi, more precisely between two successive tonal segments (see 't Hart *et al.*, 1991, for a review).

The stylization method proceeds in three steps:

1. Phonetic segmentation and syllabification.
2. Pitch determination, and smoothing of the *F0* curve with a weighted time average model that replaces each *F0* value by a time average of the *F0* pattern viewed through a data window taking voicing into account as well. This operation is supposed to account for the integration characteristics of the auditory system.
3. Stylization of syllabic pitch contours. The syllabic pitch contour is first decomposed into a sequence of tonal segments on the basis of the differential glissando threshold and the glissando threshold. This step is followed by the actual stylization, in which pitch targets are assigned to each tonal segment.

The method of d'Alessandro and Mertens has been extensively tested for French using re-synthesis and has been reported to preserve a high level of perceptual equality. Mertens's method (1987) has already been used for comparison with a linguistic model of intonation (see Section 6.5.3) for French and Dutch and applied to the automatic recognition of their intonation (Mertens, 1989).

6.5.2.2. The IPO model of intonation

The intonation model developed since the 1960s, at the Institute for Perception Research (Instituut voor Perceptieonderzoek, IPO) is based on an *analysis-by-synthesis* method, in which utterances are analyzed and resynthesized with a stylized pitch composed of sequences of straight lines (on a logarithmic scale), so as to verify the perceptual equivalence between original and stylized intonation ('t Hart *et al.*, 1991). At first sight, it thus looks like a manual equivalent of the automatic stylization procedures of the previous section, or like a perceptually oriented version of the target points approach of Section 6.5.1.2. This similarity is further reinforced by the fact that IPO stylization also makes use of declination lines (defined this time on perceptual grounds).

The IPO model does, however, go one step further in the parametric approach of pitch contours, by approximating stylized contours (sequences of straight lines) as sequences

---

[10]A *glissando* is an audible pitch change.

of patterns chosen in a limited set of *standardized* patterns.[11] All stylized straight lines (as obtained from perceptual stylization) have their own amplitude, slope, duration, and position with respect to the syllable in which they appear. The idea is that, since not all these stylized movements can be perceived as different, it is possible to establish classes of perceptually equivalent movements, each class being represented by a *standardized movement.* While pitch stylization approaches assume that pitch curves are completely determined by the *position* and *amplitude* of *target* points, pitch movements are mainly distinguished by their *timing, slope,* and *ambitus.* The IPO approach jointly defines a formal grammar to account for the possible grouping of pitch movement into standardized pitch contours (or *patterns*). Both the set of standardized pitch movements and the related grammar are language-dependent (see 't Hart *et al.*, 1991 and Collier, 1991, for a review).

6.5.3. Linguistic models of intonation

Extracting the phonological nature of prosodic data (by confusing data that convey the same information) is much more problematic than deriving acoustic or perceptual characteristics. Prosodic information is intrinsically relative (as opposed to segmental information). It does not correspond to absolute *F0,* duration, or intensity values nor even to absolute pitch, loudness, length, or syllable strength values. Besides, identical acoustic or perceptual data can be understood very differently by the listener, depending on prosodic context.

As a result, laying linguistically oriented foundations of a transcription system for intonation implies a simultaneous definition of some prosodic *vocabulary* to account for the intonative events revealed by acoustic or perceptual analysis and of some *grammar* to solve the ambiguities that arise in the use of these symbols for different purposes. In a word, one needs to define a comprehensive linguistic theory of supra-segmentals, similar to that for segmental data but not necessarily related to it. Supra-segmentals (and more particularly intonation) need not be systematically analyzed in terms of the segmental organization of sentences into phonemes, morphemes, words, phrases, clauses, and so on. Selkirk (1984), for instance, presents a phonological hierarchy containing syllable, foot, prosodic word, phonological phrase, and intonational phrase. Such an autonomous analysis can be achieved in many ways. It typically reveals the existence of functional prosodic units (denoted by the symbols of the aforementioned vocabulary) organized into phrases and clauses (whose syntax is accounted for by the aforementioned grammar). Exactly as for the segmental level, functional prosodic units (one could call them *prosodemes*) have no independent existence: they constitute a structured set in which each element can be contrasted with all the other ones—changing one element into another in a prosodic string may lead to a change in the understanding of that string (such as changing the position of the focus, for instance). The syntax of prosodic strings, which restricts the set of admissible prosodic strings for a given language, is itself described with the rules of a grammar. Such a grammar is by no way unique. Its rules depend on the general formalism adopted for the description and, more importantly in this case, on

[11]Mertens (1987) also defines standardized tonal units, but these are derived from a functional, linguistic analysis of contours obtained by auditory transcription, while the IPO patterns are established on perceptual grounds.

the linguistic features one wants the parser to highlight[12] as well as on the ones that are assumed to be available at its input.[13]

The point is that intonation is very much related to semantics and pragmatics.[14] We have also mentioned in Section 6.4 that it exhibits emotional, para-linguistic aspects. As a result, even though the International Phonetic Alphabet (IPA) appeared more than a century ago, its prosodic counterpart has not yet been settled down. *A fortiori*, no universally adopted prosodic syntax has been established. It is nevertheless a fundamental requirement for the development of linguistic theories and for the outgrowth of speech applications (both recognition and synthesis).

Phonological models of intonation can be classified in many ways, mainly depending on the degree of segmentation they imply (to which parts of an utterance they associate prosodic symbols), the status of stress (whether or not it is explicitly used), and the hierarchy, or *prosodic domains*, they define. We shall now briefly mention two important approaches, which respectively analyze intonation in terms of *pitch contours* and *tone sequences*.

6.5.3.1. Pitch contour theory

According to some authors, the intonation of an utterance originates in a sequence of elementary pitch contours taken from a limited set, programmed by the speaker and approximately realized by his or her vocal apparatus. Each contour is seen as an basic intonation unit, which cannot be itself decomposed into smaller parts.

The application of this principle to British English has led to a theory known as *tonetics* (also known as the *British school*), proposed by Crystal (1969) (see also Halliday, 1967, and O'Connor and Arnold, 1973). Speech is divided into *tone groups*, each composed of, at most, four components: *prehead, head, nucleus* and *tail*. The accented part of the tone group forms the nucleus, which is the only obligatory part and carries primary stress on its first syllable. Tonetics associates nuclear syllables with a limited number of *nuclear tones* and some combinations thereof, termed as *compound nuclear tones*. Crystal, for instance, distinguishes four nuclear tones: a *fall*, a *rise*, a *fall-rise*, or a *rise-fall*, and four compound tones: *rise* or *fall-rise* plus *fall*, and *fall* or *rise-fall* plus *rise* (Halliday proposes five nuclear tones and two nuclear tones; O'Connor and Arnold distinguish seven tones and only one compound). The tail consists of additional syllables on which the nuclear pitch movement is completed. The head is composed of the syllables from the first accented syllable of the tone group up to, but not including, the first syllable of the nucleus. Accented syllables in the head receive secondary stress. This theory, however, is strongly language dependent, and can be criticized for its inability to capture a hierarchical prosodic structure (see Ladd, 1986).

---

[12]Let us recall that a parser is most often used as a transducer that not only accepts or rejects strings but also outputs their description in terms of the rules of a grammar (see Chapter Two).

[13]Some models, for instance, will try to use the segmental syntactic organization as input information; others will only refer to lexical information (see Section 6.7)

[14]This will be developed in Section 6.6.

Delattre (1966) was one of the first to apply similar ideas to French. He analyzed utterances in terms of *ten fundamental intonations* (See Fig. 6.6) defined on a grid of four levels covering approximately one octave (level 2 being the unaccented level). It is assumed that these pitch contours can be formally distinguished by their slope and curvature and by the initial and final levels. The speaker adopts a particular contour on the basis of syntactic and semantic criteria. An important difference between Delattre's approach and tonetics resides in the fact that utterances are segmented into contours on the basis of the melodic curve itself rather than as a result of a prior analysis in terms of stressed/unstressed syllables, although it appears in practice that pitch contours cannot be unambiguously obtained without information on stress (see Mertens, 1987).

A more elaborate model has been proposed by Martin (see 1982, for instance). Similarly to tonetics, he analyses French utterances in terms of rhythmic groups (each rhythmic group contains one and only one stressed syllable) and reduces their intonation to a single basic contour on their stressed syllable. He distinguishes ten basic contours (four

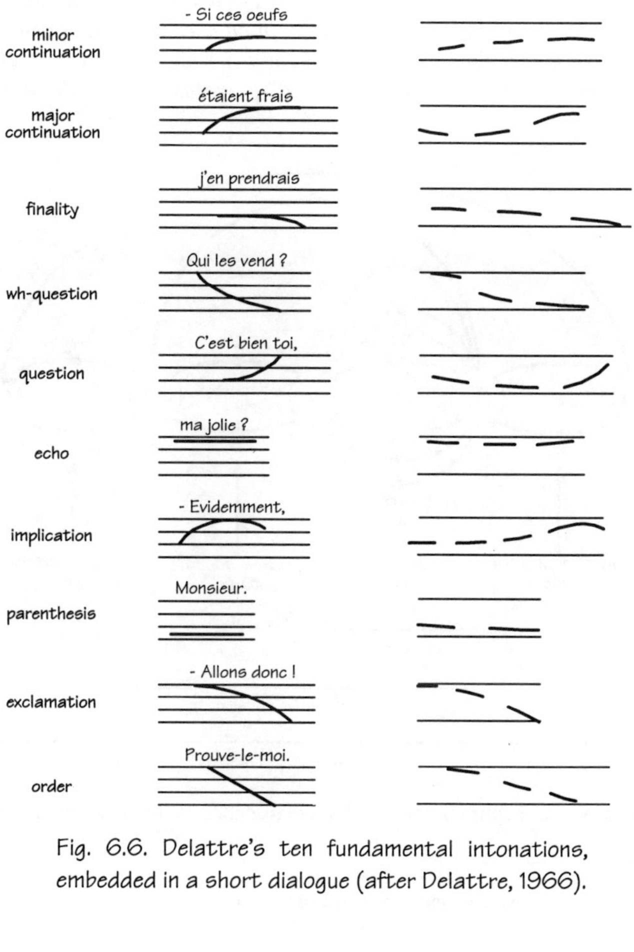

Fig. 6.6. Delattre's ten fundamental intonations, embedded in a short dialogue (after Delattre, 1966).

sentence-final ones and six sentence-internal ones) and additionally defines dependency relationships between them (a grammar), which enables him to derive hierarchical descriptions of intonation from sequences of contours. Martin has also studied the connections between his hierarchical descriptions and syntactic and semantic organization (see also Rossi *et al.*, 1981).

6.5.3.2. Tone sequence theory

Unlike pitch contour specialists, who consider absolute curves, exponents of the tone sequence theory describe melodic curves in terms of relative *tones*. Following the pioneering work of Pierrehumbert (1980) for American English, tones are defined as the phonological abstractions for the target points obtained after broad acoustic stylization (see Section 6.5.1.2).

Pierrehumbert distinguishes only two tones, a high tone (**H**) and a low tone (**L**), which are contrasted against each other: **H** is higher in the speaker's range than **L** would be in the same place.[15] Sequences of **H** and **L** tones are restricted by a finite-state grammar (Fig. 6.7), which in turns distinguishes four categories of tones on the basis of their distributional properties: *initial boundary tones*, *pitch accent tones*, *phrase accent tones*, and *final boundary tones*. This grammar explicitly introduces a three-level hierarchical description of intonation, in terms of *prosodic words*, *intermediate phrases,* and

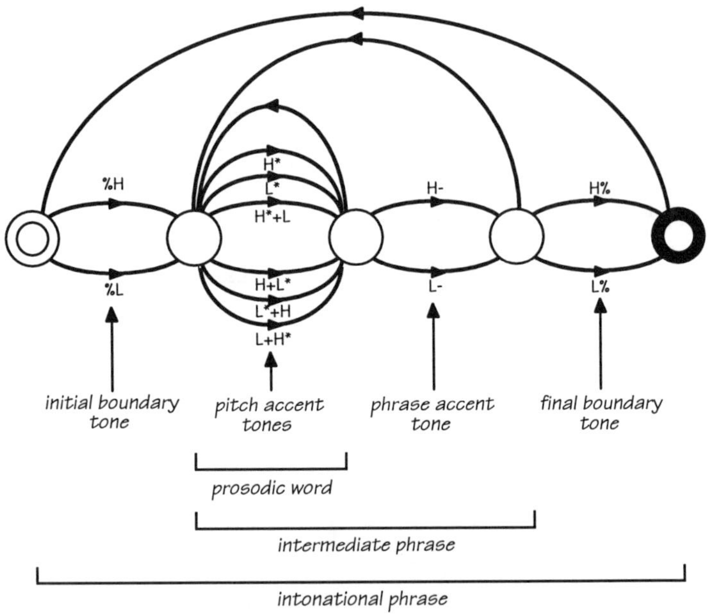

Fig. 6.7. Finite-state grammar for H/L tone sequences (after Pierrehumbert, 1980).

[15]Hence, **H** and **L** do by no means correspond to absolute pitch values.

intonational phrases, which moreover respect the *strict layer hypothesis*: an intonational phrase is composed of a number of intermediate phrases that are in turn made up of prosodic words. Utterances are basically decomposed into prosodic words, which receive one and only one pitch accent. Pitch accent itself appears as either a single tone (**H*** or **L***) or two tones[16] (**H*+L, H+L*, L*+H,** or **L+H***), where the tone associated with the accented syllable is denoted by a star diacritic (*). Prosodic words combine into intermediate phrases which themselves constitute intonational phrases. The offset pitch of intermediate phrases is controlled by a phrase accent tone denoted by a hyphen diacritic (-). The onset and offset pitch of intonational phrases are imposed by initial and final boundary tones denoted by a percent diacritic (%). This theory has been more deeply formalized into the *ToBI (Tones and Break Indices)* transcription system (Silverman *et al.*, 1992).

[[NA]		AI]	[NA]		AF	[NA]	
l h	l h	H L	l h	l h	L- (L-) H+ (H+) H L- H L	l-	l-
					L H H (H)	l- h	l- h
					H (H) L (L)		

Table 6.2. A regular grammar of French syllable sequences on the basis of the tones they accept (after Mertens, 1990).

For French, Mertens (1990) basically distinguishes four basic tones: high (**H** or **h**)[17] and low (**L** or **l**), plus two tones to account for the existence of very large pitch jumps: extrahigh (**H+** or **h+**) and extralow (**L-** or **l-**). Mertens's tones somewhat differ from Pierrehumbert's ones in that the former are associated to syllables and account for their observed pitch levels while the latter are restricted to accented syllables and phrase boundaries and influence the realization of pitch on several syllables (since they are related to pitch targets). Having defined his syllabic tones, Mertens examines the occurrence of each tone on each syllable in a representative corpus and notices that many combinations would result in unnatural intonation. He then classifies syllables as a function of the tones they accept and obtains three types of syllables for French: initial accent (AI), final accent (AF) and unaccented syllables (NA) (Table 6.2). On the basis of this classification, he establishes a segmentation of utterances into *intonational groups* (IG) obeying a simple regular grammar:

---

[16]Double tones are sometimes considered as a single tone, termed as *dynamic* (as opposed to *static* tones).

[17]Uppercase characters denote tones that occur in stressed syllables.

$$IG \rightarrow [\,[NA]\,AI\,]\quad [\,NA\,]\quad AF\quad [\,NA\,] \tag{6.1}$$

Being strongly based on an analytical examination of speech corpora, with limited influence of segmental information, Mertens's model lends itself to automatic analysis. Fig. 6.8 gives an example for Dutch (following a similar model developped by Mertens, 1989) and shows how a tone sequence is obtained from speech data, by first computing acoustic parameters (F_0, intensity, zero-crossing rate) from which syllable nuclei are spotted (and a first estimation of stress is obtained). A perceptual stylization of intonation is then achieved, which is interpreted in terms of tones, and the result is matched with

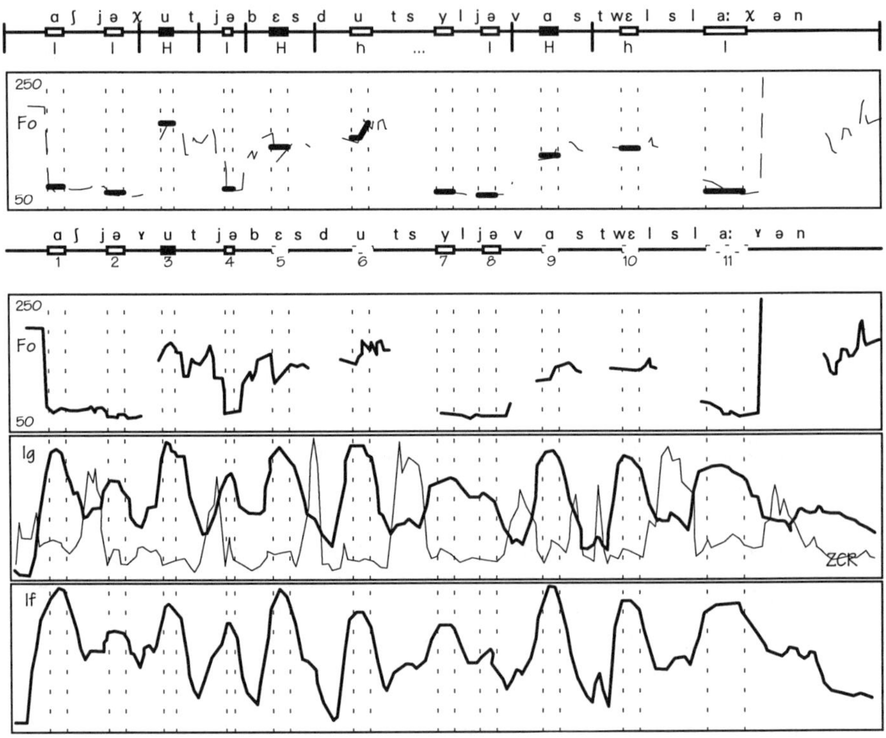

Fig. 6.8. Automatic analysis of the intonation of the Dutch utterance *als je goed je best doet, zul je vast wel slagen* in terms of a tone sequence (after Mertens, 1989). From bottom to top: intensity of the band-pass filtered speech signal (*If*; between 500 and 3000 Hz), intensity of the speech signal (*Ig*), zero crossing rate, fundamental frequency (*Fo*), syllabic nuclei (indicated as black rectangles for syllables detected as stressed, white rectangle for unstressed syllables, dotted rectangles for unknown stress), stylized pitch (taking perception into account), and final tone sequence (together with a complete stress detection; tones on stressed syllables appear as uppercase characters) as obtained by matching the data with a grammar of tones sequences.

(6.1) to obtain a final estimation of stress.

Tone sequence theories have been successfully applied to the analysis of several other languages (see Gussenhoven, 1984 for Dutch; Grice, 1992 for Italian; Pierrehumbert and Beckman; 1988 for Japanese; Bruce and Grandström, 1993 for Swedish; see also the INTSINT system adopted in the European MULTEXT project, Hirst *et al.*, 1994).

6.6. Relationships between prosody and other aspects of speech

After having established acoustic, perceptual, and linguistic models of prosody, it is important to examine its interaction with the lexicon, with syntax, and with semantics and pramatics, since the main task of the prosody generator component of a TTS system is to derive an acoustic representation of prosody from these linguistic levels.

This relationship is more complex than it appears at first sight. Indeed, it is often wrongly assumed that the intonation we apply on the sentences we read is only a function of their punctuation. Punctuation marks, however, are merely the visible sign of the strong structuring of a language, for which intonation acts as the echo, the acoustic trait that should help the listener to perceive all its subtlety. Prosody, as we shall see, makes demands on each and every level of linguistic competence of the reader.

6.6.1. Lexicon and prosody

In a great number of languages, known as *tone languages* (like Chinese and most African languages), the pitch level carried by a word is an essential feature of the meaning of that word. In Beijing Mandarin Chinese the word *ma* when pronounced in a level tone means *mother* and in a falling-rising tone means *horse* (two out of four possible tone contrasts in that language). In order to account for this, the IPA includes tone symbols (the rise, fall, high tone, low tone, rise-fall, and fall-rise).

It is not the case for *intonation languages* like most European ones. Intonation languages, however, can be classified as either *free-stress* (English) or *fixed stress languages* (French). In free stress languages, the position of the stressed syllable within a word is lexically distinctive (it must be stored in the lexicon). In contrast, word stress is predictable from word structure in fixed stress languages (this will be further detailed in Section 6.8).

Accenting and intonation contours, however, depend on the organization of words into higher level units, termed stress groups, intonational phrases, intermediate phrases, and so on. We shall see in Section 6.7.1 that many TTS systems decompose sentences into such groups with the help of simple heuristic rules based mainly on lexical information.

6.6.2. Syntax and prosody

Given the relative simplicity of syntactic analysis in comparison with the many (and yet mostly unsolved) problems raised by automatic handling of semantics and pragmatics, the possibility of describing prosody as a function of syntactic relationships between the words associated into groups and phrases is of high importance for TTS synthesis.

Intuitively, it makes sense to assume that syntax plays an important role for prosody. After all, melodic phrases often coincide with morpho-syntactic groups, and pauses are seldom encountered between strongly associated words.[18] It was suggested in Section 1.3.3, however, that speakers derive the syntax they choose to express a given idea from the idea itself, so that it is rather difficult for us to imagine what our speech would be like if only syntactic information were available. The question is: is it possible to read a text without really understanding it? From the point of view of the prosody peneration module, the absence of meanings would just result in a complete interchangeability between words that are classified in the same syntactic category. A similar result can be obtained by replacing, in a given sentence, every word by a dummy word indicating its part-of-speech category and by apply to the substituted forms the inflection of the original ones. Clearly, a new sentence appears, which has lost most of its meaning, but little of its syntax.[19] Our initial question then turns into: Is it possible to give a natural prosody to sentences like this:

> *We shall adverbially verb its nouns and verb how its nouns can be adverbially verbed, after some nouns and a few nouns on the noun of verbing an adjective noun noun.* (6.2)

Obviously, the answer to this question is yes, in most cases. Finding how this can be achieved is the main task of the prosody generation module.

6.6.3. Semantics, pragmatics and prosody

The aforementioned substitutions, although related to syntax, sometimes require semantic information, as it is typically the case for the example of Section 1.1.6: *Time flies like an arrow*. Clearly, the difference in syntactic alignment will advantageously be contrasted in the pronunciation, especially if the most unexpected interpretation is meant, in order to help the listener in his task. As Steedman (1991) writes:

> *Thus, whatever problems for parsing written text arise from the profusion of equivalent alternative surface structures . . . , these spurious ambiguities seem to be resolved to some extent by prosody in spoken language.*

In order to solve such an ambiguity, it is necessary to know where the main verb is. A high-quality TTS system should be able to detect such ambiguous situations and try to solve them in some way, either by finding a common and somewhat neutral prosody for all cases or by applying semantic rules to make the right decision.

Focus is still more important for prosody. The many possible interpretations of the example of Section 1.1.6 (*John did not come to New York by plane*) can easily be contrasted by choosing the right syllable lengths and intonation. Or, conversely, applying a wrong focus can result in serious misunderstanding. In order to avoid it, the perfect TTS system should be able to understand not only isolated sentences but also complete texts.

---

[18] This is further examined in Section 6.7.

[19] In practice, things happen to be more complex than that, as not all words with the same part-of-speech can be freely interchanged. It was mentioned in Section 4.3 that a much larger number of part-of-speech categories must be used.

Last but not least, sentences repeatedly refer to a knowledge of the world shared by the writer and the listener. The prosodic patterns produced by a given speaker therefore depend on his (or her) mental representation of a state of affairs. They are related to numerous features that have, strictly speaking, nothing to do with syntax or semantics, such as the position of the reader in relation to his (or her) reading, the aesthetic dimension he (she) wants to confer on the text to be read, or even his (her) own personality. Just think of the many registers available to an actor when reciting a poem. Similar options are available in spontaneous speech. Obviously, the choice of a particular prosodic scheme is not operated by a mere chance: it conveys some information by itself. Returning to the previous example, one might also mean:

Please, stop repeating "John came to New-York by plane": It is the third time I tell you he did NOT. (6.3)

The simple idea that John could come to New York by plane is ridiculous: he hates flying! (6.4)

and many others, by carefully selecting intensity, rhythm, and pitch patterns.[20] Pragmatics is the part of linguistics that studies such extratextual phenomena. It accounts, among other things, for the prosodic processes that allow speakers to be involved in their speech.

6.7. Syntactic-prosodic parsing

Summing up the last two sections, the absence of a universal prosody transcription framework, together with the implication of prosody in many levels of linguistic description makes it very hard to build a widely accepted theory of intonation.[21]

So what? Does this mean that TTS systems are doomed to a mere robot-like intonation until a brilliant computational linguist announces a working semantic-pragmatic analyzer for unrestricted text—that is, not for a long time? There are various reasons to think not, provided one accepts an important restriction on the naturalness of the synthetic voice—that its intonation is kept *acceptably neutral* (Monaghan, 1989):

> Acceptable intonation must be plausible, but need not be the most appropriate intonation for a particular utterance: no assumption of understanding or generation by the machine need be made. Neutral intonation does not express unusual emphasis, contrastive stress or stylistic effects: it is the default intonation which might be used for an utterance out of context. (...) This approach removes the necessity for reference to context or world knowledge while retaining ambitious linguistic goals.

---

[20]And, to some extent, facial expressions. This still appears more clearly when deaf-mute persons express themselves in sign language: prosody has no other communication channel at its disposal.

[21]In that respect, it is striking to notice how different the respective approaches of linguists and engineers are. Most linguistic research on prosody tends to focus on very specific points, for which a deep study is performed and a number of narrow but accurate conclusions are drawn. Their reorganization into a general view, however, is far from being simple. In contrast, engineers often adopt a more pragmatic behavior, by trying to build up general but unavoidably simplistic and somewhat erroneous prosody generation modules. Clearly, advances in this intricate area require a constructive partnership between both parties.

The key idea is that the "correct" syntactic structure, the one that precisely requires some semantic and pragmatic insight, is not essential for producing such a prosody (see also O'Shaughnessy, 1990)[22]. As explained in Section 6.4, human readers basically vary prosody for two reasons:

1. To partition the utterance into shorter parts, so as to help the listener to parse the utterance.
2. To cue certain meanings, either by using contrastive stress to set the *focus* on the information that is at the center of the reader's interest or to emphasize words that bring new (as opposed to given) information, by prosodically expressing some particular relationship between words, or simply by using high or low tones to indicate whether an intonation group is assumed to be followed by another one or not.

The latter is by far the most complex one. In contrast, in the simple case of a TTS system that would merely try to achieve goal 1 (like most current systems), thereby concentrating its efforts to the production of the aforementioned acceptably neutral prosody, it is widely accepted that the organization of the short text segments that act as prosodic groups (also termed as *prosodic domains*) is much simpler than the related syntactic dependencies. In other words, the prosodic structure of a sentence is flatter than its syntactic one. Hence, there must be levels of structure in syntactic analyses which are less relevant to prosody (what Monaghan (1990a) refers to as a *many-to-one syntax-to-prosody mapping*) and omitting these levels in the syntactic-prosodic analysis or failing to account for them (by avoiding access to semantics or pragmatics, for instance) will result in minor degradations in the prosodic contour.

With these considerations in mind, it is not surprising that commercially developed TTS systems have emphasized coverage rather than linguistic sophistication, by concentrating their efforts on text analysis strategies aimed at a segmentation of the *surface structure* of incoming sentences, as opposed to their syntactic, semantic, and pragmatic underlying *deep structure.*[23] The resulting syntactic-prosodic descriptions, as we shall see, organize sentences in terms of prosodic groups strongly related to phrases (and therefore also termed as *minor* or *intermediate phrases*), but with a very limited amount of embedding, typically a single level for these minor phrases as parts of higher-order prosodic phrases (also termed as *major* or *intonational phrases,* which can be seen as a prosodic-syntactic equivalent for clauses) and a second one for these major phrases as parts of sentences,[24]

[22]This does not mean, however, that prosodic models accounting for semantic or pragmatic cues (see Rossi, 1993, for instance) are useless in speech synthesis. In a class of applications, termed as *synthesis from concepts* (as opposed to text-to-speech synthesis), this information is indeed readily available to the computer. This is typically the case of a synthesizer in a man-machine dialogue system.

[23]In that respect, commercially developed TTS systems should be contrasted with laboratory systems. As Monaghan (1990a) notes: "Almost every conceivable combination of parsing techniques has been applied to the problem of analyzing unrestricted text. Until recently, however, the criteria for deciding what parsing techniques would be implemented in a given TTS system had more to do with researchers' interests in syntax than with the requirements of text-to-speech conversion."

[24]It is found in practice that clause-level parsing is much more difficult and computationally expensive than phrase-level analysis, so that many TTS systems use phrase-level parsing only.

to the extent that the related major phrase boundaries can be safely obtained from relatively simple text analysis methods. In other words, they focus on obtaining an acceptable segmentation and translate it into the continuation or finality marks of Fig. 6.1.c but ignore the relationships or contrastive meaning of Fig. 6.1.a and 6.1.b.

In the next sections, we have classified these syntactic-prosodic analysis methods as either based on hand-derived heuristics, on grammars or on automatic, corpus-based methods.

6.7.1. Hand-derived heuristics

When reading a sentence, we are usually able to start pronouncing it far before we reach its end. It must be then that we organize the sentence into simple prosodic groups which can be obtained from a local, phrase-level analysis. This simple observation has led many TTS system developers to try to avoid the computational load and complexity of full-fledged grammars by developing low-level heuristics to uncover prosodic groups.

Following the widely spread fallacy that we organize the prosody of a sentence as a result of its punctuation alone, heuristics have first attempted to build up prosodic phrases on the simple basis of the position of commas, semicolons, hyphens, and so on. Colons and semicolons are then seen as sentence delimiters for prosodic purposes. Left marks (quotes, parentheses, brackets, braces) and the corresponding right marks are assumed to introduce phrases. Dashes and commas partition the sentence into phrases and clauses (O'Shaughnessy, 1990). It should be clear that such an elementary analysis is doomed if not to failure at least to limited success. For one thing, punctuation merely is one of the apparent and occasional marks of some deeper syntactic organization. Obviously, not all prosodic phrases nor even all clause boundaries happen to be delimited by punctuation marks. Major syntactic breaks indeed can appear as conjunctions, or simply not appear at all (as in *I found the book I wanted to read*). On the other hand, not all punctuation marks are prosodically relevant, the case of a list of adjectives being a well-known example in English (as in *the old, dark, red house* for the prosody of which it makes little difference whether the commas are present or not).

As a result, the division of a sentence into syntactic-prosodic phrases is often assumed to take place not only at punctuation marks but also at the boundary between lexical words and grammatical words, the latter being understood as phrase delimiters. With some slight refinements, however, this apparently plain strategy has proved very successful in practice for right-headed languages—languages whose function words tend to pile up at the left edge of constituents (which happens to be the case for most European languages). Liberman and Church (1992), for instance, have recently reported on such a very crude algorithm, termed as the *chinks 'n chunks* algorithm, in which prosodic phrases (which they call *f-groups*) are accounted for by the simple regular rule (Fig. 6.9):

a (minor) prosodic phrase = a sequence of chinks followed by a sequence of chunks (6.5)

in which chinks and chunks belong to sets of words that basically correspond to function and content words, respectively, with the difference that object pronouns (like *him* or *them*) are seen as chunks and that tensed verb forms (such as '*produced*') are considered as chinks. They show that this approach produces efficient grouping in most cases, slightly better actually than the simpler decomposition into sequences of function and content words, as shown in the example below:

Function words / Content words	Chinks / Chunks
I asked	*I asked them*
them if they were going home	*if they were going home*
to Idaho	*to Idaho*
and they said yes	*and they said yes*
and anticipated	*and anticipated one more stop*
one more stop	*before getting home* (6.7)
before getting home (6.6)	

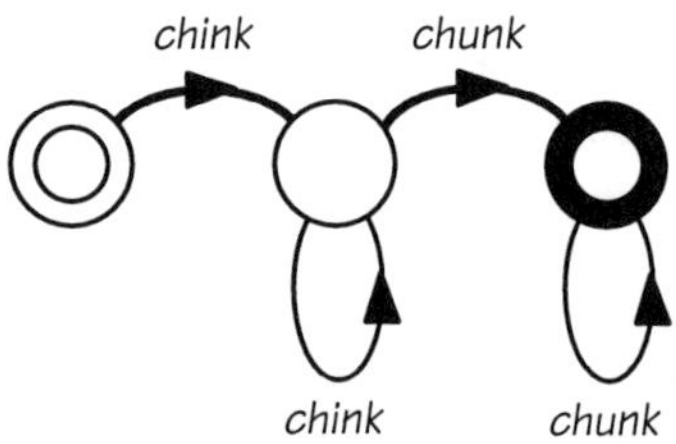

Fig. 6.9. A FSA for the simple but efficient chinks 'n chunks algorithm.

Similarly, Quené and Kager (1989) organize sentences into phonological phrases that "consist of a lexical head (i.e. noun, verb, adverb, adjective), its left-hand specifiers, and all non-lexical words (i.e. function words, as prepositions, complementizers, conjunctions, and copulas) to the left up to the next lexical head". For French, Larreur *et al.* (1989) and more recently Keller *et al.* (1993) define the former prosodic parsing approach of their TTS system in the same way. So do Balestri *et al.* (1993) for Italian.

Even though the segmentation of a sentence into minor prosodic phrases is an important aspect of syntactic-prosodic analysis, it does not yet make it possible to account for the focused acceptably neutral prosody. Obviously, not all prosodic phrases are equivalent. Better still, their peculiarity does not simply result from their position within a sentence. Just try to read example (6.7) with

- The same prosodic pattern on each prosodic phrase, with a pitch rise at the end of the phrase, except for the last one which ends with a pitch fall;
- The same prosodic pattern on every second prosodic phrase, with a pitch rise at the end of even phrases, and a a pitch fall at the end of odd ones;
- The same prosodic pattern on each prosodic phrase, with a pitch rise at the end of the phrase, except for every third phrase, which ends with a pitch fall; and so on.

None of these rules of the thumb works satisfactorily. If one of them lead to an acceptable prosody, it would nevertheless be easy to find lots of sentences for which the apparently winning rule would fail horribly. Clearly, one needs some sort of a hierarchical organization of these minor prosodic phrases into larger clause-like units (major phrases) that moreover cannot be accurately determined without some prosodic phrase hierarchy. Although this should be achieved by examining the syntactic components of these phrases (see Mertens, 1993a; 1993b), it can be performed, although with much less success, with heuristics. In Quené and Kager (1989), for instance, some constituents obligatorily form an intonational phrase, such as displaced syntactic constituents, parentheticals, and nonrestrictive relative clauses,[25] as in

[(This sentence) (is divided) (into intonational phrases)] [(which are divided) (into phonological phrases)] (6.8)

in which *()* and *[]* denote phonological and intonational phrases. Complex noun phrases and subordinate clauses form intonational phrases as much as possible, depending on their length and on the style of speech. Intonational phrases, however, are not obtained, as one might expect at first, from a prior syntactic analysis, but rather directly from the sequence of parts of speech of which the sentence is made. A number of simple rewrite rules identify subordinate clauses by inserting intonational phrase boundaries between adjacent words which cannot belong to the same clause. Other rules delimit intonational phrases on the basis of punctuation. Similar approaches have been reported for French by Larreur *et al.* (1989) and Bailly (1989) and by Frenkenberger *et al.* (1994) for German. Larreur *et al.* use rewrite rules, about 140 of them, and additionally allow incoming words to be only partially disambiguated, by specifying sets of parts of speech as context instead of single ones. Their parser also inserts the related rule number with a syntactic-prosodic boundary, so that further prosodic processing can discriminate among various types of phrases and clauses. Bailly inserts prosodic markers chosen in an ordered set of eight elements on the basis of the syntactic dependency between adjacent words and derives phrases by deleting markers on the basis of marker priorities and rhythmic constraints.

Notice that a rather different approach could also be conceived, as the result of the work of Karlsson (1990), already mentioned in Section 4.3.3, who has proposed organizing syntactic-level heuristics into the coherent framework of *constraint grammars.* These basically differ from the above strategies in that all plausible syntactic information is *a priori* assigned to each incoming word (can it be the head of some phrase? which type of word or phrase can it modify? can it be followed by a clause boundary? which one? etc.), which results in a highly ambiguous initial data to be progressively disambiguated by heuristic constraint rules. The first step is to provide as many syntactic labels as possible in the lexicon. Then some morpho-syntactic mapping is applied to all words that do not yet have any syntactic function label, by examining its morphological features (and possibly the syntactic labels of its neighbors) and deriving its range of syntactic labels. In step three, syntactic constraints discard as many superfluous syntactic labels as possible. In other words, constraint grammars organize phrase and clause-level syntactic parsing

---

[25]Relative clauses are classified as restrictive or non-restrictive, as in *the man who is behind you* and *my friend, who is behind you,* respectively. This distinction is also present in prosody, since only non-restrictive relative clauses tend to be preceded by a pause.

by associating syntactic labels to words in a very similar way to that of morphological and contextual analysis for parts-of-speech.

Heuristic approaches toward syntactic-prosodic analysis have a number of advantages over more complex parsers. Most of them attempt left-to-right, real-time[26] analysis, instead of examining the entire sentence before producing parsing results. What is more, their organization into a first and simple phrase-level segmentation followed by a more complex and risky clause-level grouping enables them to output relatively fail-safe low-level information in all cases, even when examining agrammatical sentences. State-of-the-art heuristics, however, are not error-free. They heavily rely on the accuracy of part-of-speech tagging. Misidentifying verbs, for instance, can be disastrous for clause-level analysis. They also inevitably fail to account for higher-level linguistic phenomena, such as finding the beginning of a clause that does not start with a specific function word or the end of a clause that is not terminated by a comma. How are they to detect tags and parenthetical material, which should be marked as such since they are typically pronounced in a lower tone? How are they to parse complex nominals when they arise? Even phrase-level heuristics are not guaranteed to succeed, as in the example provided in (Liberman and Church, 1992):

(the outspoken conspiracy theorist) (will not pass) (up a chance) (to speak) (6.9)

In order to cope with such deficiencies, many TTS systems provide simple mechanisms for a user to indicate phrase and clause boundaries in the incoming text, by surrounding them with specific markers (parentheses and brackets, for instance; see Emerard, 1977).

6.7.2. Grammar-based systems

As already introduced in Chapter Two, many grammar formalisms have been proposed as extensions to the popular but limited context-free one (Chomsky type 2). They are principally augmented transition networks (ATNs), definite-clause grammars (DCGs), and unification grammars (UGs). A number of TTS systems have attempted to formalize their syntactic-prosodic analysis in these ways, thereby often implicitly assuming that both clausal structure and predicate-argument relations are important in determining prosodic phrasing. The problem, then, is to find an implementation of such grammars which achieves robustness and flexibility. Real sentences are often much more complicated than those found in linguistic texts (as emphasized in Tomita, 1988).

ATNs have mainly been proposed for phrase-level syntactic analysis, given the intrinsic simplicity of this task. Allen *et al.* (1987), for instance, have described noun and verb groups with relatively simple ATNs. Their phrase-level analysis, however, is still based on sentence-level heuristics, in the form of a global parsing strategy aimed to constrain the order in which ATNs are matched with the input data. Starting from the first word of the sentence as current word, they first look for the longest noun group they can find. If no noun group is found, they attempt to find the longest verb group starting at the current word. If neither type of group can be found, they simply move to the next word and the

[26]In the context of TTS synthesis, real-time implies that the parser must be able to analyze any text at a speed that exceeds the average speaking rate.

process begins again. O'Shaughnessy (1987) has reported on the use of a much more complex procedural analyzer. His parsing algorithm decomposes the input sentence into noun phrases, verb groups, and so-called independent phrases (including adverbial phrases and parenthetical phrases) with a detailed regular-like grammar which also accounts for number agreement. Clauses are described in terms of admissible phrase sequences. The parser marks any deviation from the expected order of events as a potential syntactic break, assuming that deviations are occasions where a speaker tends to mark syntax prosodically.

DCGs and UGs offer a very flexible framework for real language processing. They implement context-free grammars augmented with variables, which naturally enables them to account for various kinds of agreement rules (such as gender, number, and case), and to naturally transduce input sentences into their syntactic-prosodic representation by instanciating variables during the parsing process. As such, DCGs and UGs can be seen as recursive transition networks (RTNs) whose arcs are labeled by complex categories (including variables) instead of by atomic symbols. Unlike ATNs, however, they are fully declarative, and they easily lend themselves to chart parsing, a very powerful concept which allows the parser to record already parsed sub-structures into a table or a chart, so as to avoid analyzing them again when backtracking. One of the most advanced DCG-based parsers for TTS synthesis is probably the one of Traber (1993). After the work of Russi (1992) and Tomita (1986), who have compared various unification-based parsing strategies (bottom-up and top-down, with variants), he has derived a DCG for a substantial part of German syntax in which rules are augmented with penalty values, used when building the optimal parse tree: the nodes of the tree inherit the *penalty* (or *score*) values of the corresponding grammar rules, and the parse tree with globally minimal summed penalty is chosen as the final result. When such penalties are used for beam-search pruning (so as to process only a limited number of best-scored edges in the chart), linear parsing time can be achieved with respect to sentence length (Cericola *et al.*, 1989). Another advantage of scoring parses from word to sentence-level arises when considering the robustness of the analyzer. As a matter of fact, when no full-sentence parse can be found, a partial parse can straightforwardly be obtained by retaining the best scored constituents, thereby smoothly turning the sentence-level parser into a phrase-level one (Traber, 1993). Without scoring, robustness can still be achieved to a lesser extent by choosing the largest single structure in the chart and concatenating it to the remaining words (Ceder and Lyberg, 1992).

Once a syntactic structure has been derived, the gap remains to be bridged between prosodic and syntactic phrasing. This is traditionally achieved by first inserting major and minor phrase boundaries between words as a function of their highest common ancestor in the syntactic tree (the higher the ancestor, the stronger the phrase boundary; see Fig. 6.7). It should be recalled, however, that the automatic assignment of prosodic markers (such as the determination of minor and major intonational phrases) from syntactic grounds only is a somewhat risky business. As a matter of fact, misalignments frequently occur between the two levels of phrasing (see Bachenko and Fitzpatrick, 1990 for a review), as with embedded phrases like (see Fig. 6.10)

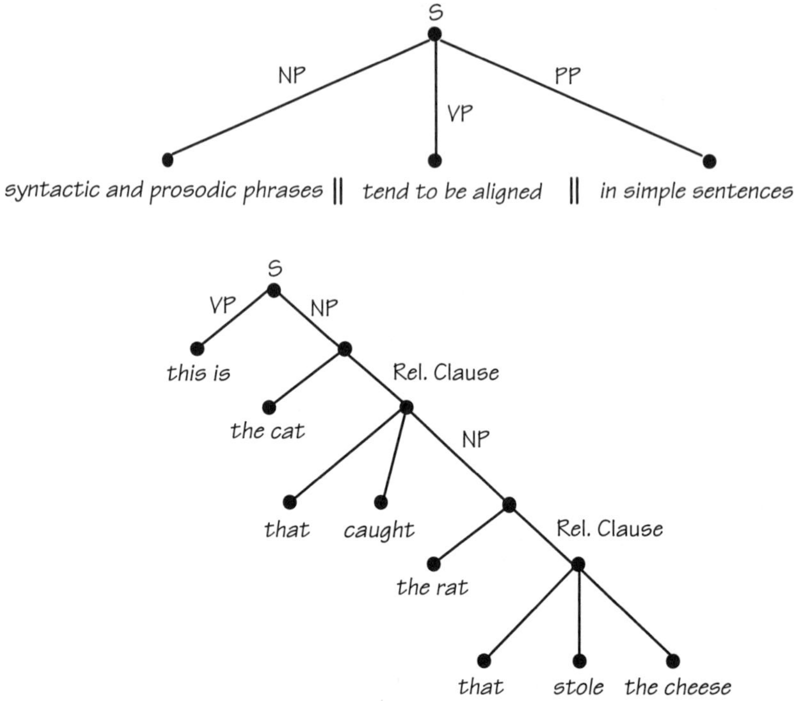

should be : *this is* ||*the cat* | *that caught the rat that stole the cheese*
(if one assumes a strong mapping between syntax and prosody)
observed : *this is the cat* ||*that caught the rat*|| *that stole the cheese*

Fig. 6.10. Deriving prosodic phrases from syntactic ones, with more or less success. Major and minor phrase boundaries are denoted as double and single vertical bars, respectively.

This is the cat that caught the rat that stole the cheese. (6.10)

in which the prosodic structure obviously is much flatter than the syntactic one. In order to cope with such mismatches, *readjustment rules* are often invoked, which produce a flattened syntactic structure. Other mismatches between prosody and syntax can be related to specific nonsyntactic factors, the most important of which being constituent length, which interacts with grammatical structure and sometimes overrides it (see Wenk and Wioland, 1982 for French). Hence the difference in the assignment of prosodic boundaries in:

One of my friends || *has recently resigned.* (6.11)

My friend has told me || *about his resignation.* (6.12)

Syntax-based prosodic phrasing algorithms approximately account for this by first setting prosodic boundaries as determined from a flattened syntactic tree and associating them with *strength* or *salience* values and by subsequently deleting boundaries in the order of their strength, under the condition that the length of constituents does not exceed a given value (Traber, 1993). A more complete algorithm for English can be found in Bachenko and Fitzpatrick (1990).

A practical consequence of such problems is reported by Meyer *et al.* (1993), for the design of a TTS system for German. When confronted with the dilemma of having to choose among two German syntactic-prosodic parsers for the assignment of prosodic boundaries and the determination of accents, a "conservative" one and a more elaborate one (termed as such in connection with the number of phrase boundaries they provide), they opt for an intermediate solution, by keeping the accents provided by both parsers (which achieve similar accentuation) and maintaining only the boundaries marked by punctuation, following the observation that "it is more acceptable to leave out boundaries than to insert e.g. pauses where they are not expected." Similarly, although Lindström *et al.* (1993) once opted for a DCG-based approach, they have now abandoned it in favor of a simpler part-of-speech tagging scheme followed by rules for grouping of prosodic words and prosodic phrases (see Ljungqvist *et al.*, 1994), as described in the previous section and refined in the next one.

We are convinced, however, that a lot remains to be done to better account for the congruency between syntax and prosody. An intonation model like the one that is proposed by Mertens (1993a; 1993b) for French, for instance, is particularly appealing in that its multilayered description of the prosodic structure precisely explains the relations between intonation, syntax, and morphology.

6.7.3. Automatic, corpus-based methods

One should not conclude from the end of Section 6.7.1 that basic heuristics are of no practical use. On the contrary, as denoted by Liberman and Church (1992):

> Comparing the extremely simple chinks 'n chunks approach to the output of more complex parsers, we have found that the division into small f-group-sized units is nearly the same, while complex parser's decisions on grouping of larger units are wrong too often to be relied on. . . . We believe that the current research in text analysis, combining sensible models of linguistic structure with techniques for learning from very large text corpora, is likely to produce techniques that work well enough to be used safely in TTS applications.

Indeed, as mentioned at the beginning of Chapter Three, increasing the coverage of a set of rules (whether heuristic or not) often implies increasing the complexity of the analysis system greatly without expanding the functionality significantly. As a result, improving heuristics can often no longer be achieved by using human expertise alone. Big text corpora and powerful rule finding, editing and assessment tools become indispensable. Statistical inference algorithms are often invoked. The problem can be stated as follows: given a large text corpus in which phrase boundaries (minor, major, and whatever other boundary type is expected to be useful) have been positioned by hand, how can we derive automatically a set of heuristic rules that account for most of the boundaries?

A first and straightforward means of trying to solve it would be to express each sentence as a Markov chain in which states would be associated to words, and arcs to *a priori* possible boundaries. In practice, since even a very big corpus would not make it possible to reliably estimate the probabilities of appearing of each boundary type between each pair of words of the language, words would be grouped into equivalence classes, as a function of their part-of-speech, for instance, which would lead to associating states to parts-of-speech (Fig. 6.11) (Ejerhed, 1988). Although this approach could be refined by taking larger contexts into account, it should be clear that the error rate could not be indefinitely lowered, for several reasons:

- Prosodic phrase boundaries cannot be set independently of each other. Phrase length has often been suggested as a determiner of boundary location, such that prosodic phrases roughly have the same length (see Martin, 1987, and Bachenko and Fitzpatrick, 1990, for instance).
- It has also been proposed that syntactic constituent boundaries may be more or less likely to coincide with intonational boundaries. Unfortunately, there is no one-to-one relationship between a local sequence of parts-of-speech and the syntactic structure they belong to. Again, this could be solved by accepting very large contextual sequences, but it would quickly be impossible to estimate the related probabilities.
- It is likely that the location of prosodic boundaries depends on the length of the sentence, as well as on their position within the sentence (i.e. on the number of words or syllables from its beginning to its end).

In short, many contextual factors should *a priori* be considered when designing a strategy for positioning prosodic boundaries. So many, indeed, that one could hardly derive a probabilistic FSA to account for all of them simultaneously. In other words, the process of finding a way to assign prosodic boundaries can be seen as equivalent to inferring multi-level rewrite rules (MLRRs) (see Chapter Three) with many *a priori* useful input levels. Fortunately, a highly efficient corpus-based analysis method has been derived, which organizes heuristics into so-called *decision trees* obtained by *classification and regression tree* (CART) techniques (Brieman *et al.*, 1984; see also Bahl *et al.*, 1989, and Withgott and Chen, 1993, which have inspired the following presentation), and allows

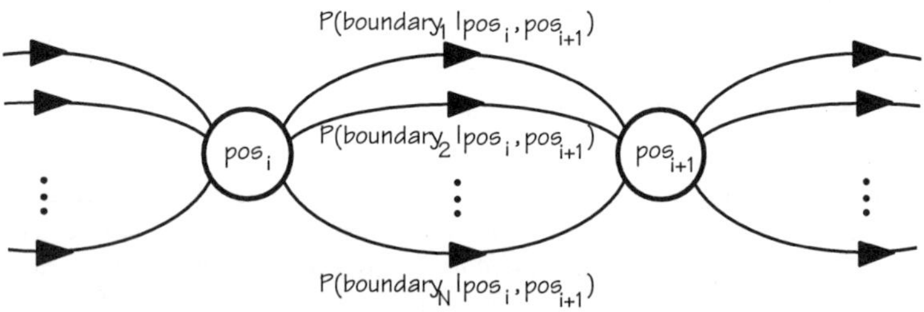

Fig. 6.11. Transitions allowed between two states of a simple Markov chain accounting for the presence/absence of prosodic boundaries (chosen among *N* possible ones, one of these being the *null* boundary) within a sentence. Words are considered as parts-of-speech only.

the most significant contextual factor to be statistically selected using a greedy algorithm.[27] CARTs permit both categorical and continuous features to be considered (the former take their values in a finite set of atomic values, while the latter can take any value in a given range) and allows human interpretation and exploration of the results. They naturally provide a structured, tree-based presentation of a set of automatically inferred MLRRs.

Basically, CARTs express the labeling process as binary decision trees.[28] At each non-terminal node, there is a question requiring a yes or no answer about the value of the contextual factor associated with the node, and for each possible answer there is a branch leading to the next question (see Fig. 6.12 for an example). Questions relate to contextual factors assumed to be useful. Questions about categorical factors partition their values into two mutually exclusive subsets; questions about continuous factors split their range in two by setting a cut-point. Terminal nodes, or *leaves*, are associated with a label (a prosodic boundary in our case, including the *null* boundary).[29]

In order to build such a tree, one needs a *training set* composed of tokens associated with decisions. From this set, relations between contextual factor values and decisions are used to determine predictors for these decisions. At start-up, all the training data is assigned to a first *parent* node. The tree is then built by recursively splitting the data in a parent node into subsets that form descendant or *child* nodes. Each node encodes the distribution of the training data in a given context. Central to CARTs is the node splitting algorithm, based on the minimization of the *entropy* of the training data. Entropy is an information theoretic concept that can be though of a measure of the "randomness" of the data. It is measured in *bits* and computed as the negative of the mean of the log-likelihood of all labels *l* in the set of admissible ones *L*:

$$H(L|\text{node}) = -\sum_{l \in L} P(l|\text{node}) \log_2 P(l|\text{node}) \quad (6.13)$$

This value differs from node to node, as the distribution of labels at a node is influenced by all the choices that have been made on contextual factor from the very first parent node. If, for instance, L contains 2^N labels seen as equally probable at a given node position, formula (6.13) reduces to $H(L|\text{node}) = -\log_2(2^{-N}) = N$—precisely the number of bits required to encode any label *at that node position.* In case labels are not equally probable, which corresponds to a reduction of the "randomness" of L, $H(L)$ decreases, and any label can be coded with less than N bits. Each splitting of a parent node, based on the partition c_i^j of the values of a contextual factor c_i into two subsets C_1^j and C_2^j, produces two child nodes and the resulting average entropy is given by:

$$H(L|\text{child}_1, \text{child}_2) = H(L|\text{child}_1)P(\text{child}_1) + H(L|\text{child}_2)P(\text{child}_2) \quad (6.14)$$

[27]A greedy algorithm makes optimal decisions at each step, without regard to subsequent steps. It aims to build a tree which is locally optimal, but very likely not globally optimal.

[28]Strictly speaking, *n*-ary trees are also covered. We will only consider binary trees, as in the recent implementations for TTS synthesis.

[29]They could also be associated with a continuous variable and given a numerical value.

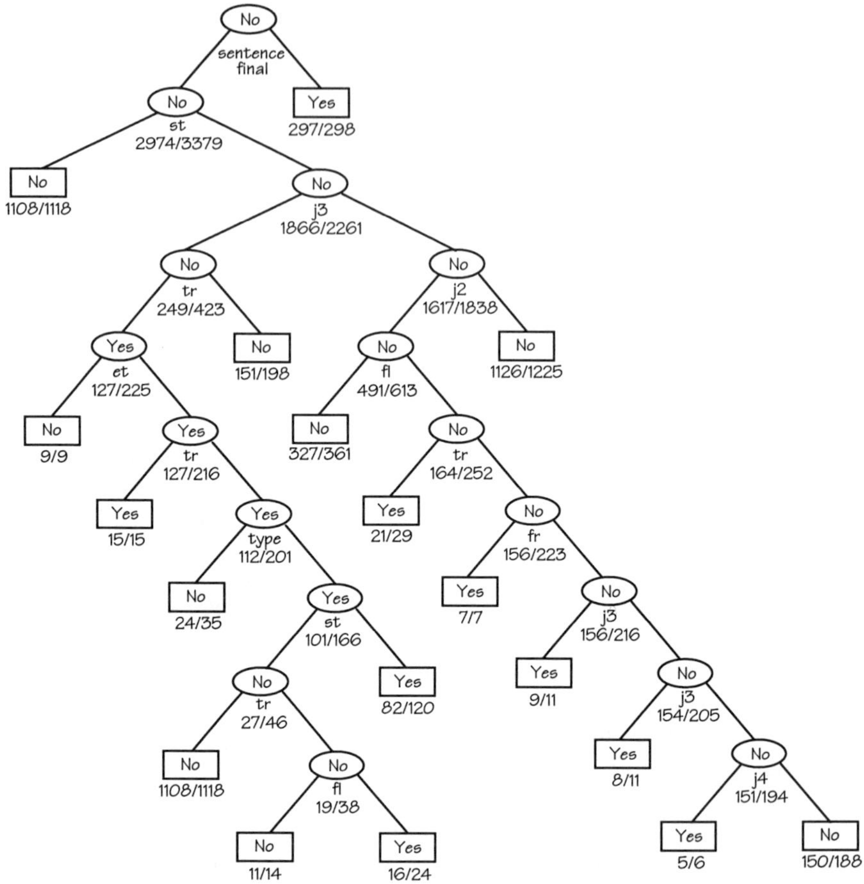

Fig. 6.12. A yes/no decision tree for predicting prosodic boundaries in texts, using text-based information alone. Only contextual factors are indicated at each branch (see text for interpretation of node labels); values are omitted. Each node indicates the number of correctly classified samples (adapted from Wang and Hirschberg, 1991).

where $P(\text{child}_1)$ and $P(\text{child}_2)$ are the probabilities of visiting the child nodes—the probabilities that c_i falls in C_1^j and C_2^j, respectively, given their parent node[30]. Indices i and j respectively refer to contextual factors (in the set of *a priori* useful ones) and to partitions of their values. The key idea is that the best splitting $c_{i,best}^{j,best}$ is the one that maximizes the difference between the entropies before and after splitting. This

[30]As a matter of fact, these probabilities cannot be estimated *a priori*: they have to take account of the choices that have previously been made to lead to the parent node.

difference is defined as the *average mutual information* $I(L,\ c_i^j)$ between the labels to predict and splitting c_i^j:

$$I(L,c_i^j) = H(L|\text{parent}) - H(L|\text{child}_1,\text{child}_2) \qquad (6.15)$$

It can be obtained by first examining each contextual factor c_i and finding the partition $c_i^{j,best}$ that maximizes $I(L,\ c_i^j)$, and then maximizing $I(L, c_i^{j,best})$ over i:

$$c_i^{j,best} = \arg\max_j I(L,c_i^j) \qquad (6.16)$$

$$c_{i,best}^{j,best} = \arg\max_i I(L,c_i^{j,best}) \qquad (6.17)$$

The node splitting algorithm is iteratively applied on child nodes, and branches of the tree are stopped when maximum average mutual information falls below a threshold, in which case a further reduction of entropy is seen as not significant. Notice that in practice entropies have to be estimated from the training data, by computing relative frequencies. This clearly biases the tree growth, as the entropy of a finite set of samples can always be made arbitrarily small, by increasing the number of leaves. Hence, the lower the threshold of our stopping condition, the more biased the decision tree. Typically, such decision trees would tend to achieve extremely high prediction scores for their training data by modeling all their peculiarities rather then capturing the expected generalizations on the related process. As a result, it would fail to account for other data originating in the same process. In order to avoid such "over-training", nodes are only extended when the number of data samples they account for is greater than a specified threshold. In addition, the data is generally initially divided into two parts: one for *training*—for estimating (6.16) and (6.17) on each node and for deciding which splitting to choose—and the other one for *cross-validation*—for checking the accuracy of the decrease of entropy estimated with the training data. When the estimate obtained from the cross-validation data falls below a threshold, tree growth is stopped, even if the estimate obtained from the training data suggests that further splitting would significantly improve the prediction.

CART techniques have been successfully applied to the problem of syntactic-prosodic phrasing in English by Wang and Hirschberg (1991) and Hirschberg (1991). Their training corpus consisted of 298 sentences from the DARPA Air Travel Information System (ATIS), labeled by hand, for location and type of intonational boundary (a major, minor, or null boundary was assigned to each of the 3,677 pairs of adjacent words $<w_i,w_j>$, although major and minor phrases were finally collapsed into a single category). The contextual factors chosen for the experiment included temporal variables, such as utterance duration (measured in seconds and denoted as *tt*, as well as in words, *tw*) overall utterance rate (in words per second: *tr*), and distance of the potential boundary from start and end of utterance (in seconds and in words: *st, et, sw, ew*). Syntactic structural factors were also considered, by allowing the tree induction algorithm to examine part of speech in a window of four words surrounding each potential break (*j1, j2, j3, j4*: part of speech for w_{i-1}, w_i, w_j, w_{j+1}) and by taking syntactic constituency into account (*fs*: category of the smallest syntactic constituent dominating *wi* and *wj*; *fl*: category of the largest constituent dominating w_i and not w_j; *fr*: category of the largest constituent dominating w_j and not w_i).

The tree they obtained is given in Fig. 6.9, with a successful detection of 96 percent of the boundaries in the training data and of about 90 percent of boundaries from cross-validated data. When trying to make a distinction between minor and major phrases, the cross-validated success rate fell below 82 percent. These results have been implemented in the AT&T Bell Laboratories TTS system (Sproat *et al.,* 1992), although they still needed to be improved by enlarging the training corpus.

It seems clear that, with the increasing availability of large, labeled databases, these statistical, corpus-based analysis techniques have a bright future before them.

6.8. Sentential stress assignment

It was mentioned in Chapter One that phoneticians distinguish three stress levels—namely, *primary* and *secondary stress* (sometimes *tertiary stress* as well), as opposed to *unstressed* syllables. Although stress is an everyday notion, it is quite difficult to characterize it in acoustic terms (cf. Section 6.3). It is often stated that stressed syllables are perceptually related to some prominence, in the sense that they are typically produced louder or longer than unstressed ones, and that they are less amenable to reduction, although it happens not to be always the case.

Knowing in which terms stress assignment should be expressed is yet another matter. Stress indeed is subject to several linguistic factors, which contribute to classify it into two broad categories—namely, *lexical* and *sentential* stress. Lexical stress refers to the position of prominent syllables in words spoken in isolation. When words are grouped into sentences, however, they do not necessarily keep their lexical stress. It also happens that lexically unstressed syllables become stressed. Sentential stress accounts for such sentence-level phenomena.

One often distinguishes *grammatical* stress from *emphatic* and *contrastive* stress. The former is directly related to the syntactic-prosodic structure of the sentence. In French, for instance, grammatical stress always falls on the last pronounced syllable in a rhythmic group. In contrast, emphatic and contrastive stress originate in semantic and pragmatic considerations. They allow a speaker (or a reader) to draw special attention on some words by emphasizing them or by contrasting them.

Stress is also a language-dependent feature of speech. Lexical stress assignment, for instance, is an important problem for so-called *free stress languages*, while it is only a minor concern for so-called *fixed stress languages* in which lexical stress is assigned a fixed syllabic position. In French, for instance, it always falls on the last syllable of a word that does not contain a schwa [ə]. Free stress languages, as the name literally infers, allow stress to be *a priori* assigned to several syllabic positions, although a given word is generally associated with a unique stress pattern, occasionally two. The words *diplomat*, *diplomacy*, and *diplomatic*, for instance, have stress on the first, second, and third syllables respectively (as opposed to the French words *diplomate*, *diplomatie*, *diplomatique*, in which it always falls on the last pronounced syllable).

Notice that not all words receive word stress. Hence the distinction between *clitic* and *nonclitic* words: only the latter can be stressed.[31] Several criteria have been put forward to define the classes of clitic and nonclitic words (see Mertens, 1993a for a discussion). Some authors consider the [± clitic] feature of a word as part of its lexical identity.

It remains that the adequate assignment of stress on incoming words is of primary importance in a TTS system, for many reasons:

- An incorrect stress pattern is disruptive in itself. It makes a TTS system speak as a nonnative speaker, thereby reducing its naturalness.[32] Worse still, ambiguities arise when it falls on heterophonic homographs such as *recOrd* - *'rEcord*, in which case intelligibility suffers.
- In a number of TTS systems, stress is used as contextual information for MLRRs (remember the important stress contrast between *photograph* and *photography*). Vowel qualities (front/back, open/closed, short/long) are then assumed to be stress-dependent features. Failing to assign correct stress patterns therefore leads to phonemic errors, which themselves imply a loss of intelligibility.
- Last but not least, prosodic accents in continuous speech mostly coincide with characteristic fundamental frequency excursions related to the primary-stressed syllable of words (that is, not all primary-stressed syllables receive an accent, but most accents fall on primary stressed syllables). What is more, wrong lexical stress—that is, wrong determination of the possibly accented part of words—is perceptually more striking than an inappropriate selection of accented words themselves and can be clearly classified as an error (Böhm, 1992).

As can be expected, TTS systems most often focus on lexical and grammatical stress, leaving emphatic and contrastive stress for later considerations since it is hardly ever predictable on the basis of lexical, morphological, or syntactic information. The MITALK system, for instance, implements a modified set of the ordered rules developed by Halle and Keyser (1971) for English, the most well known of which are the *main* and *alternating stress rules* for predicting which syllable receives primary stress as a function of the weak/strong syllable pattern of the word. The notion of syllable weight has been used by Church (1986) as well and extended with part-of-speech, morphological, and etymological considerations. Monaghan (1990b) uses word-class information to assign primary, secondary, or no accent to lexical items (thereby obtaining an *overaccented* representation) and deletes some accent markers by applying a set of heuristics based on the principle of rhythmic alternation and incorporating phonological insights (an analysis of the sentence in terms of *prosodic domains*; see Section 6.7) where possible. Modern TTS systems make use of many knowledge sources simultaneously :

---

[31]Notice also, as already mentioned in Section 5.6, that word stress is not always maintained at the sentence level. The formation of *stress groups* depends on morpho-syntactic rules: clitic words are reorganized around the non-clitics on the basis of the syntactic relation between the clitics and the non-clitics.

[32]For an interesting description of prosodically related mispronunciation in nonnative students learning English, see Bagshaw (1994).

- Syllabification: syllables are marked as heavy or light as a function of their phonemic content and surrounding (such as vowel length and number of post-vocalic consonants), and stress assignment is submitted to contextual constraints such as the aforementioned main and alternating stress rules.
- Part-of-speech information is used not only to disambiguate heterophonic homographs (as *record*) but more broadly as a hint for stress assignment (i.e. for non-homographs as well). It has been noted, for instance, that over 90 percent of English bisyllabic nouns in running text have stress on the first syllable, while only about 15 percent of bisyllabic verbs are stressed on the first syllable (Klatt, 1987).
- Far beyond the simple part-of-speech category of isolated words, a complete phrase level parsing is sometimes necessary to assign stress, as it is the case for complex nominals in English (think of *computer communications network performance analysis primer*) (Sproat 90).
- Morphology can also contribute, by describing stress in terms of stressed roots that combine with affixes and/or other stressed roots, in combination with rules that account for stress shifts in the case of inflection and derivation, and for stress ordering (primary/secondary) in compounds, as mentioned in Section 5.3. This approach has been widely followed for German, as in (Böhm, 1992) in the form of MLRRs mainly addressing the graphemic, phonemic and morphological levels. Similarly, Martin (1990) describes how lexical stress in Italian can be located on one of the last four syllables in nouns, adjectives, pronouns, and adverbs and one of the last six syllables in verbal forms, on the basis of a prior morphological analysis combined with some morphophonemic rules.
- Finally, etymology can be made responsible for many irregularities in stress assignment, especially in proper names. Typically, English words borrowed from French (as for *AnnEtte*) will receive stress on the last syllable, and many Italian words (as *MarcOni*) will be stressed on the penultimate one.

6.9. From symbolic to acoustic representation of prosody

Assuming one has adequately analyzed a sentence and derived its syntactic-prosodic organization, in the form of hierarchically related groups of syllables marked with several levels of stress (at least two: stressed and unstressed), the last step toward providing the utterance with a consistent prosody lies in producing an acoustic representation of its intonation and of the duration of its phonemes and pauses.

6.9.1. Generating timing

Speech exhibits intricate temporal patterns, which must be mimiced by TTS systems in order to sound natural. More precisely, TTS systems should be able to produce segment and pause durations that do not significantly differ from natural ones. Although the *significance* of a duration is not easy to define (it is a linguistic concept that calls for an analysis of durations in term of their linguistic function rather than of their numerical value), it is often reduced to its perceptual component by defining *just noticeable differences* for segmental durations (often denoted as JNDs) as the smallest differences that can be discriminated (see Huggins, 1971; Klatt, 1976). Bartkova and Sorin (1987), for instance, accept a JND of 20 ms.

Deriving natural durations from text basically implies to answer two questions:

1. What type of target units will be considered? Should a TTS system try to account for the duration of subphonetic units (such as vowel onset, steady-state part, and offset), phonemes, syllables, feet,[33] words, or phrases?
2. How should the relationships between syntactic-prosodic information and these units be estiblished ?

6.9.1.1. Duration "units"

It is generally acknowledged that a perceptual study of duration should be based on a prior segmentation into duration "units." A number of authors, for example, have reported on the existence of invariants in the timing of sub-phonetic parts of speech as a function of speaking rate. Such information, however, is of little help for TTS systems based on the concatenation of segmental units (to which Part Two of this textbook is more particularly related), which generally do not incorporate information on the respective durations of their subsegmental components.

An old principle of *isochrony* is sometimes invoked in favor of an analysis in terms of syllable- or foot-sized units (see Lehiste, 1977), based on evidence on some interaction between speaker production and perception systems in order to produce segments perceived as isochronous. According to this purely rhythmic principle, speakers would unconsciously use an *internal clock* to schedule speech segments (typically synchronized with syllables for fixed stress languages and feet for free stress ones, although there is no general consent; see Wenk and Wioland, 1982). A change in speech rate, for instance, would simply imply a modification of the internal clock rate. Since, as just mentioned, many state-of-the-art TTS systems are based on phonetic elements as ultimate speech units, syllable-based duration systems incorporate an algorithm to derive segment durations from syllable ones, as proposed by Campbell and Isard (1991). According to an *elasticity principle*, Campbell reports that all segmental durations in a syllable frame are obtained by one and the same factor k by the formula

$$Dur_i = \exp(\mu_i + k\,\sigma_i) \tag{6.18}$$

in which Dur_i is the duration of phoneme i in a syllable and μ_i and σ_i are the mean and standard deviation of its log-transformed durations in a large corpus. Starting from syllable duration, taking into account the μ_is and σ_is of its segmental components, this formula easily provides the value of k, hence the actual durations of segments. In practice, however, it is easy to verify experimentally that a strict application of the isochrony principle leads to arbitrarily regular speech. It is therefore more adequate to say that languages exhibit a *tendency* to isochrony. Besides, recent work has introduced

---

[33] A *foot* is a rhythmic unit which contains one and only one stressed syllable.

the notion of *perceptual center* (of a vowel) and replaced syllables by *interperceptual center groups* to better account for this (see Barbosa and Bailly, 1994).

Most TTS systems actually focus on the duration of phonetic segments, although effects of syllable, word and phrase boundaries on segmental duration suggest that larger units should also be considered (van Santen, 1993a).

6.9.1.2. Duration models and parameter estimation

As van Santen (1993a) soundly observes, duration models have traditionally made use of four sources of information:

1. Verbal statement in the phonetics literature, based on articulatory or phonological principles.
2. Results from exploratory statistics to discover broad regularities in data, which result in information on which types of models are appropriate.
3. Parameter estimates based on experiments where a small number of factors is varied, as typically reported in the phonetics literature.
4. Parameter estimates based on experiments where many factors are varied, in large-scale corpora.

Two main trends can be distinguished for duration modeling. In the first and by far most common one, durations are computed by first assigning an *intrinsic duration* to segments (pauses being considered as particular segments), which is further modified by successively applying rules combining co-intrinsic and linguistic factors into additive or multiplicative factors (see Klatt, 1976; O'Shaughnessy, 1984; Bartkova and Sorin, 1987). Which factors are considered, however (and typically a small number of them), greatly depends the *prior* availability of information of type 1, 2 and 3. Bartkova and Sorin, for instance, analyzed several corpora to study speaker independent intrinsic durations and their modifications under the influence of a consonant upon the preceding vowel in a stressed syllable, of the type of word (content versus function word), of the location of the segment within a word, of its distance to major and minor phrase boundaries, and so on. as well as of the grouping of consonants into clusters at various positions of the word/prosodic phrase/sentence. This led to establish simple multiplicative rules:

$$\begin{aligned} &\textit{Vowel duration} = ID \quad V_i \quad m_c \\ &\textit{Consonant duration} = ID \quad C_{ij} \end{aligned} \tag{6.19}$$

in which ID are intrinsic durations, V_i and C_{ij} are shortening or lengthening coefficients dependent on the position i of a segment within a word and in a sentence, j refers to the phonetic class membership of consonants, and m_c reflects, in certain cases of closed syllables, the influence of a consonant or semivowel on the preceding vowel. All factor values (about sixty-three of them) were computed by averaging the data obtained by applying this model to corpora.

In a second and more recent approach, mainly facilitated by the availability of large speech corpora and of computational resources for generating and analyzing these corpora, duration models are restricted to type 4 information. A large number of possible control factors are simultaneously varied, a very general duration model is proposed (a general additive model in Kaiki *et al.*,1990; CARTs in Riley, 1992; neural networks in Campbell, 1992), and its parameters are automatically derived by the associated approximation algorithm (standard least squares methods, minimization of entropy, and error back-propagation, respectively). Riley's CARTs, for example (which slightly differ from those used by (Wang and Hirschberg, 1991) in that Riley used them in a *regression* mode, that is, to make *continuous* predictions), are based on an impressive set of factors: segment context (three segments to left, segment to predict, three segments to right[34]), stress (three levels), lexical position (segment count from start and end of word, vowel count from start and end of word), phrasing position (word count from start and end of phrase).

Van Santen (1993b) argues in favor of an intermediate position between these two approaches, combining information of types 2 and 4. After an analysis of the intrinsic complexity of duration modeling due to the *sparsity* of data (no corpus can claim to be exhaustive given the number of factors to consider) and to *paralinguistic (text-independent) variability,* he claims that statistical approaches cannot neglect linguistic knowledge as a means of judiciously restricting the search space and conditioning the form of models so as to ease the approximation of their parameters.

6.9.2. Generating fundamental frequency

How the F_0 curve is ultimately generated greatly depends on the type of prosodic model chosen for its representation (whether it is an acoustic, a perceptual or a linguistic one). Indeed, bridging the gap between a syntactic-prosodic description of the utterance in terms of hierarchically related groups of syllables and its actual melody involves two steps:

1. Establishing the relationship between the parameters of the model and the syntactic-prosodic description, and
2. Generating the final melodic curve from its parametric representation in the model.

Clearly, step two reduces to rather trivial mathematical processing in the case of acoustic and perceptual models (respectively addressed in Sections 6.8.2.1 and 6.8.2.2), which are intrinsically closer to the actual data than linguistic ones. This, however, is balanced by the complexity of step one. In contrast, linguistic models (Section 6.8.2.3) are closer to the linguistic-prosodic description and therefore facilitate step 1, while not increasing the complexity of step 2 too much. This advantage is gained as the result of preliminary analysis by the human expert who established the model.

Independently of the type of intonation model used, however, step one is by far the most intricate one, given the possible complexity of the syntactic-prosodic hierarchy which should be used as input. Besides, we have seen in Section 6.7 that currently available syntactic-prosodic parsers produce limited information. As a result, the aforementioned hierarchical description is often replaced by, or completed with, a collection of related parameters assumed to carry the same information in a more crude form (for example, parts of speech are used in place of a complete syntactic description, or the position of

---

[34]In order to keep things manageable, Riley codes segments in terms of four phonetic features related to manner and place of articulation. Such a distributed representation is commonplace when using statistical methods.

syllables or words within a syntactic-prosodic constituant is used in place of the complete rhythmic structure)

6.9.2.1. Generating *F0* with Fujisaki's model

Fujisaki's model has been extensively used in TTS systems (see Hirose *et al.*, 1986; Klatt, 1987; Bailly, 1989; Ljungqvist and Fujisaki, 1993; Möbius *et al.* 1993; Hirai *et al.*, 1994). If it virtually reduces step two to elementary filtering, it also makes step one more difficult. As a matter of fact, since the linguistic status of its accent and phrase components remains *a priori* unspecified, Fujisaki's model is theoretically compatible with different linguistic descriptions. In practice, however, it is hardly compatible with any. Indeed, since the timing and amplitudes of accent and phrase components are not constrained by the linguistic organization of the utterance (because Fujisaki's model is truly an acoustic model), it is very difficult to establish a relationship between the parameters obtained by the automatic analysis of a large number of naturally produced F_0 contours and their syntactic-prosodic organization. As a result, this relationship is often *a priori* imposed, by associating phrase commands with prosodic phrases and accent commands to accented syllables and by modifying the parameter approximation algorithm accordingly. Clearly, instead of obtaining timings and amplitudes (and durations in the case of accent commands) from scratch, the approximation algorithm is forced to account for the syntactic-prosodic organization of the utterance, which constrains the number and timing of phrase and accent components. Once such a modified, linguistically constrained approximation has been performed on a consistent corpus, the analysis of timing and amplitude in terms of linguistic features can be performed with statistical tools (see Möbius *et al.* 1993, Hirai *et al.* 1994). This results in a control model that produces parameter values (hence, melodic curves) when provided with syntactic-prosodic features at its input.

6.9.2.2. Generating *F0* as sequences of stylized contours

The F_0 representations provided by acoustic or perceptual stylization methods suffer from the same impairments as Fujisaki's model. They take the acoustic data for granted, which greatly complicates the design of control strategies.

Narrow stylization methods, which typically focus on segmentally related F_0 stylizations in the form of a limited number of target points to describe intonation on a syllable or phoneme basis (see Section 6.5.1.2), straightforwardly enable the constitution of prosodic databases in which segmental and suprasegmental features are put in correspondence. In Larreur *et al.* (1989), for instance, the duration of syllables, and phonemes and the initial, medial and final frequency of vowels are measured from a large speech corpus and associated with segmental information such as : the preceding, current and following phoneme, the type of the current syllable (open or closed), the type of the current word (function or content), the position of the current syllable within the word, the position of the current phoneme within the syllable, the broad phonetic class of surrounding phonemes, and some syntactic-prosodic code for the current word related to its position within the syntactic-prosodic hierarchy. TTS system developers are then faced with the problem of interpreting the data. Two strategies have basically been exploited to solve it.

In the first one, a top-down methodology is followed, in which a set of prosodically meaningful segmental features are *a priori* selected and hierarchically organized into a decision tree (or conversely into a set of classification rules). Melodic contours are associated with each leaf of this decision tree, by either averaging the melodic information provided by each of its representatives in the database (Aubergé, 1991), or by choosing arbitrarily one of them (or several as in Larreur *et al.,* 1989).

The second strategy consists in trying to design a direct transducer between the available syntactic-prosodic information and the observed stylized curves, without the need to introduce any intermediate and arbitrary model of the relationship between them. Traber (1992) has achieved such fundamental frequency control by using a *recurrent neural network* (a multilayer perceptron as the one of Fig. 2.4 with additional connections and weights between its output layer and its input one so as to account not only of contextual inputs but also of contextual outputs). The transducer was presented with two input streams: a stream for binary-coded accent, phrase, and word boundary symbols, and a stream for phonetic properties of each syllable. These streams were presented to the network through sliding contextual windows (one window for each stream), and the network was trained to associate a stylized melodic pattern with the symbol at the center of both windows. Recurrent connections ensure some continuity of the melodic curve among successive syllables. The concatenation of all stylized patterns yields the F_0 contour of the whole utterance, which was reported to sound very natural.[35] Clearly, F_0 generation based on narrow stylization methods has only started to be explored. No doubt it will benefit from the increased availability of large corpora and of associated statistical analysis techniques.

As opposed to the top-down approach followed with narrow stylization methods, broad stylization ones (which describe pitch curves in terms of a small number of pitch pattern points) often adopt a bottom-up strategy. The IPO approach, which is by far the most widely used stylization-based bottom-up prosody generation method (see Balestri *et al.*, 1993; Meyer *et al.*, 1993; Terken, 1993; Beaugendre, 1995), examines the relationships between the standardized patterns it obtains and the syntactic-prosodic hierarchy to which it is related (Collier, 1991). Declination lines, for example, are associated with intonational phrases and accents are produced with fall-rise patterns. How this is precisely organized, however, depends on the language analyzed. Notice also that there is not always a one-to-one relationship between patterns and syntactic-prosodic hierarchy (since patterns have been established on a perceptual rather than functional basis). The resulting control algorithm is generally composed of a small number of rules, starting from rules for isolated intonational phrases with no accent, one accent, two accents, and so on, then examining the case of embedded intonational phrases with a variable number of accents and similarly the case of final phrases in an interrogative sentence. Each rule assigns a pattern of a particular kind, its duration, its position relative to the onset of the vowel to which it is attached, and its amplitude within the current frequency range (between a top and a bottom declination line).

---

[35]The reader might argue that the descriptions provided in Section 6.8.2 suffer from a lack of precision. Most papers on implementing prosody generators provide rather vague descriptions about the nature of the investigation tools used (see also Chapter Eleven).

Top-down approaches using broad stylization have been proposed by Sagisaka (1990) and Sagisaka and Kaiki (1992), and by Taylor (1994). Sagisaka models the global F_0 shape of minor (intermediate) phrases with three F_0 values (corresponding to their starting, top, and end positions) as a function of higher-level syntactic-prosodic information, such as current minor phrase accentedness and length, total number preceding accented minor phrases in the current major (intonation) phrase, length of the preceding and following minor phrases, and local phrase structure. Control parameters are associated to F_0 values by training a three-layered perceptron (Sagisaka, 1990) (which is a similar approach as that of Traber (1992) except that no recurrent connections are needed in this case, since the broad stylization of a minor phrase can reasonably assumed to be independent of that of its neighbors) or by adjusting the parameters of a linear regression-based statistical model (Sagisaka and Kaiki, 1992).

6.9.2.3. Generating *F0* through sequences of tones

We have seen in Section 6.5.3.2 that, being based on a functional, phonological analysis of intonation curves, tone sequence theories explicitly restrict the allowed F_0 movements to be the acoustic manifestation of sequences of tones chosen in a very restricted set, allowed sequences being constrained by a grammar.

Although these models explicitly introduce hierarchical levels in the description of intonation (the intermediate and intonational phrases of Pierrehumbert, 1980, and the accent and intonation groups of Mertens, 1993), they do not explain when such or such allowed sequence is used. The use of tone-based models in TTS systems therefore requires to derive first a component dealing with the choice of tones rather than with their form in phonetic context. Its goal is to produce a description of the intonation of sentences in terms of tone sequences, as a function of their modality (declarative, interrogative, imperative), of the number of their syntactic-prosodic components, of their length, position, stress structure (which tone sequence should be used to account for emphatic stress or focus?), and so on. This corresponds to step one of the introduction of Section 6.8.2. Since it requires a deep insight into linguistic theories of segmental and supra-segmental aspects of speech, this analysis is always performed by a linguist (as opposed to being automatically achieved by machine). Pierrehumbert (1981), for instance, provides a rather crude but efficient description of neutral declarative intonation of English: pitch accents are all **H**, phrase-final tones are typically **L- L** for a terminal phrase and **L- H** for a non-terminal one. **H**s and **L**s are additionally associated with stress levels (from 1 to 5) established on the basis of a syntactic analysis of the sentence. Bruce and Grandström (1993) study a similar description of the intonation of Swedish, which additionally accounts for focus. They also make use of prominence levels (from 0 to 9). Mertens (1993) also establishes such relations (although much more elaborate) for his model of French intonation.

Deriving a sequence of tones for an utterance, however, is not equivalent to obtaining its F_0 curve. Hence the need for a second component, which transforms tones into numerical F_0 values (cf. step 2 of the introduction of Section 6.8.2): tones indeed are abstract units, as opposed to the stylized patterns examined in the previous sections. Clearly, many different approaches can be followed to achieve this goal, from the computation of average pitch contours corresponding to tones in context to the development of rule-based systems to produce F_0 targets and interpolate between them, through the use of statistical models or artificial neural networks. In Pierrehumbert's linguistic model, for instance, tones are only positioned either on stressed vowels (which may receive one or two such tones, with one singled out to fall right on the stress and the other leading or trailing) or a the margins of intermediate and intonational phrases. What is more, they do not correspond to absolute pitch values. Hence the use of the aforementioned stress levels. Pierrehumbert (1981) relates stress (or prominence) levels to positions of pitch targets within the local pitch range (as determined by declination lines and by the position of the pitch target), from the baseline to the topline. Targets are then combined into quadratic curves, whose precise shape depends on tones involved and of the associated prominence levels. Anderson and Pierrehumbert (1984) additionally introduce a smoothing filter, in the spirit of Fujisaki's model. Bruce and Grandström (1993) precisely relate their tones and prominence levels to the parameters of a Fujisaki-like model.

6.10. Summary

After a definition of prosody and a short introduction to its levels of representation, components and meanings, we have described how prosodic (and especially intonation) models can be classified as acoustic, perceptual, or linguistic, as a function of the degree of abstraction with which the data is examined. Acoustic models of intonation such as Fujisaki's model and those based on acoustic stylization methods merely aim to present melodic curves in a compact way. Perceptual models such as automatic perceptual stylization and the IPO model go one step further by only accounting for perceivable data. Eventually, linguistic models reduce prosodic facts to a limited number of functional prosodic units organized into phrase-like and clause-like units whose admitted structures are accounted for by an intonation grammar. We have briefly mentioned two main trends, known as *pitch contour theory* and *tone sequence theory*, which respectively describe intonation in terms of absolute, standardized pitch contours and of relative, elementary tonal units.

Although these models aim to account for the phonetic and phonological aspects of intonation independently of other aspects of speech (such as lexicon, syntax, semantics, and pragmatics) we have underlined that important but not yet completely understood relationships exist between the segmental and suprasegmental levels.

We then investigated the organization of the syntactic-prosodic parser, which finds the text structure (its organization into clause and phrase-like constituents) that is more closely related to its expected prosodic realization. We have shown that simple, *hand-derived heuristics* perform surprisingly well for phrase-level segmentation, which already solves an important part of the syntactic-prosodic parsing problem, since prosodic structures are known to be much flatter than their syntactic correlates. *Grammar-based systems* using advanced context-free grammars provide formal means of deriving full parses when possible. In combination with chart parsing and beam-search pruning, they are computationally efficient. The resulting syntactic structure is flattened to obtain a first prosodic structure that is then further simplified by collapsing the most strongly syntactically related prosodic phrases while taking constituent length conditions into account. Finally, *corpus-based methods* such as CARTs constitute an interesting alternative to formal grammars. Based on statistical considerations gathered from a training data set, they automatically derive contextual decision trees by choosing the most significant contextual factors among a large set of *a priori* useful ones. When provided with the right set of contextual factors (which requires some insight into the problem to be solved) and adequately cross-validated on test data, they lead to high success rates.

Stress assignment has been briefly examined. It has been recalled that TTS systems mainly focus on lexical and grammatical stress (as opposed to emphatic and contrastive stress), which are related to syllabification, part-of-speech information, phrase-level parsing, morphology and etymology in a highly language-dependent way.

Finally, we have addressed the issue of phoneme duration, pause duration and fundamental frequency assignment from the syntactic-prosodic description of a sentence. In particular, we have studied how the aforementioned models of intonation can be put to contribution in this goal. Acoustic and perceptual models constitute the engineering approach to the synthesis of intonation. They lend themselves to automatic and more or less accurate approximation techniques such as neural networks or statistical methods. Linguistic models proceed in two separate steps: they establish a connection between prosodic-syntactic information and their own symbolic description of intonation, and then propose methods to translate their abstract prosodic symbols into F_0 curves.

As can be seen, although prosody has been increasingly investigated these last years in the context of TTS synthesis, there is still a long way from text to automatically derived natural prosody. Even acceptably neutral intonation has not been satisfactorily achieved yet. It is our belief, however, that the linguistic information required to derive it is already in the range of our parsers (*it is our noun, adverbially, that the adjective noun verbed to verb it is adverbially in the noun of our nouns).* What mostly makes this task extremely difficult is the text-independent variability of the acoustic components of prosody.

References

ALLEN, J., 1992, "Overview of Text-to-Speech Systems", in *Advances in Speech Signal Processing,* S. Furui, M. Sondhi, eds., Dekker, New York, pp. 741-790.

ALLEN, J., S. HUNNICUT, and D. KLATT, (1987), *From Text to Speech, The MITALK System,* Cambridge University Press, Cambridge.

ANDERSON, M.D., and J. PIERREHUMBERT, (1984), "Synthesis by Rule of English Intonation Patterns", *Proceedings of the International Conference on Acoustics, Speech, and Signal Processing 84,* pp. 2.8.1-2.8.4.

AUBERGE, V., (1991), *La Synthèse de la Parole: des Règles aux Lexiques,* Ph.D. dissertation, Université Pierre Mendès France, Grenoble.

BACHENKO, J., and E. FITZPATRICK, (1990), "A Computational Grammar of Discourse-Neutral Prosodic Phrasing in English", *Computational Linguistics,* n°16, September, pp. 155-167.

BÄCKSTRÖM, M., K. CEDER, and B. LYBERG, "PROPHON - An Interactive Environment for Text-to-Speech Conversion", *Proceedings of Eurospeech 89,* Paris, vol. 1, pp. 144-148.

BAGSHAW, P.C., (1994), *Automatic Prosodic Analysis for Computer-Aided Pronunciation Teaching,* Ph.D. dissertation, University of Edinburgh.

BAHL, L.R., P.F. BROWN, P.V. de SOUZA, and R.L. MERCER, (1989), "A Tree-Based Statistical Language Model for Natural Language Speech Recognition", *IEEE Transactions on Acoustics, Speech, and Signal Processing,* vol. 37, n°7, pp. 1001-1008.

BAILLY, G., (1989), "Integration of Rhythmic and Syntactic Constraints in a Model of Generation of French Prosody", *Speech Communication,* vol. 8, pp. 137-146.

BALESTRI, M., S. LAZZARETTO, P.L. SALZA, and S. SANDRI, (1993), "The CSELT System for Italian Text-to-Speech Synthesis", *Proceedings of Eurospeech 93*, Berlin, pp. 2091-2094.

BARBOSA, P., and G. BAILLY, (1994), "Characterization of Rhythmic Patterns for Text-to-Speech Synthesis", *Speech Communication*, vol.15, n°1-2, pp. 127-139.

BARTKOVA, K., and C. SORIN, (1987), "A Model of Segmental Duration for Speech Synthesis in French", *Speech Communication*, vol. 6, pp. 245-260.

BEAUGENDRE, F., (1995), *Une Etude Perceptive de l'Intonation du Francais* Ph.D. dissertation, LIMSI, Paris.

BÖHM, A., (1992), *Maschinelle Sprachausgabe Deutschen und Englishe Textes*, Ph.D. dissertation, Ruhr-Universität Bochum.

BRIEMAN, L., J.H. FRIEDMAN, R.A. OLSHEN, and C.J. STONE, (1984), *Classification and Regression Trees*, Wadsworth & Brooks, Monterey, CA.

BRUCE, G., and B. GRANDSTRÖM, (1993), "Prosodic Modeling in Swedish Speech Synthesis", *Speech Communication*, n°13, pp. 63-73.

CAMPBELL, W.N., and D. ISARD, (1991), "Segment Durations in a Syllable Frame", *Journal of Phonetics*, 19, pp. 37-47.

CEDER, K., and B. LIBERG, (1992), "Yet Another Rule Compiler for Text-to-Speech Conversion?", *Proceedings of the International Conference on Spoken Language Processing*, Alberta, pp. 1151-1154.

CERICOLA, D., M. DANIELI, M.J. MOLLO, and D. VOLTOLINI, (1989), "Morpho-Syntactic Tools for Speech Processing", *Proceedings of Eurospeech 89*, Paris, vol. 1, pp. 386-389.

CHURCH, K. W., (1986), "Stress Assignment in Letter-to-Sound Rules for Speech Synthesis", *Proceedings of the International Conference on Acoustics, Speech, and Signal Processing 86*, Tokyo, pp. 2423-2427.

COLLIER, R., (1991), "Multi-Language Intonation Synthesis", *Journal of Phonetics*, vol. 19, pp. 61-73.

CRYSTAL, D., (1969), *Prosodic Systems and Intonation in English*, Cambridge University Press, Cambridge.

CRYSTAL, D., and P. QUIRK, (1964), *Systems of Prosodic and Paralinguistic Features in English*, Mouton, The Hague.

d'ALESSANDRO, C., and P. MERTENS, (1995), "Automatic Pitch Contour Stylization Using a Model of Tonal Perception", *Computer, Speech and Language*, 9, pp. 257-288.

DELATTRE, P., (1966), "Les Dix Intonations de Base du Français", *French Review*, n°15, pp. 1-14.

EJERHED, E., (1988), "Finding Clauses in Unrestricted Text by Stochastic and Finitary Methods", *Second ACL Conference on Applied Natural Language Processing*.

EMERARD, F., 1977, *"Synthèse par Diphones et Traitement de la Prosodie"*, Ph.D. dissertation, Université de Grenoble III.

FRENKENBERGER, S., B. SCHNABEL, M. ASSALI, and M. KOMMENDA, (1994), "Prosodic Parsing Based on Parsing of Minimal Syntactic Structures", *Proceedings of the second ESCA/IEEE workshop on Speech Synthesis*, New-Paltz, NY, pp. 143-146.

FUJISAKI, H., (1992), "The Role of Quantitative Modeling in the Study of Intonation", *Proceedings of the International Symposium on Japanese Prosody*, pp. 163-174.

GEOFFROIS, E., (1993), "A Pitch Contour Analysis Guided by Prosodic Event Detection", *Proceedings of Eurospeech 93*, Berlin, pp. 793-796.

GRICE, M., (1992), *The Intonation of Interrogation in Palermo Italian; Implications for Intonation Theory*, PhD dissertation, University College of London.

GUSSENHOVEN, C., (1984), *On the Grammar and Semantics of Sentence Accents*, Foris, Dordrecht.

HALLE, M., and S.J. KEYSER, (1971), *English Stress: Its Form, its Growth, and its Role in Verse*, Harper and Row, New York.

HALLIDAY, M.A.K., (1967), *Intonation and Grammar in British English*, Mouton, Berlin.

't HART, J., (1991), "*F0* Stylization in Speech: Straight Lines versus Parabolas", *Journal of the Acoustical Society of America*, vol. 90, n°6, pp. 3368-3370.

't HART, J., R. COLLIER, and A. COHEN, (1991), *A Perceptual Study of Intonation: an Experimental-Phonetic Approach to Speech Melody*, Cambridge University Press, Cambridge.

HESS, W.J., (1983), *Pitch Determination of Speech Signals : Algorithms and Devices*, Springer Verlag, Berlin.

HESS, W.J., (1992), "Pitch and Voicing Determination", in *Advances in Speech Signal Processing,* S. Furui, M. Sondhi, eds., Dekker, New York, pp. 3-48.

HIRAI, T., N. IWAHASHI, N. HIGUCHI, and Y. SAGISAKA, (1994), "Automatic Extraction of F_0 Control Parameters Using Statistical Analysis", *Proceedings of the Second ESCA/IEEE Workshop on Speech Synthesis,* New-Paltz, NY, pp. 57-60.

HIROSE, K., H. FUJISAKI, and H. KAWAI, (1986), "Generation of Prosodic Symbols for Rule-Synthesis of Connected Speech of Japanese", *Proceedings of the International Conference on Acoustics, Speech, and Signal Processing 86*, Tokyo, pp. 2415-2418.

HIRSCHBERG, J., (1991), "Using Text Analysis to Predict Intonational Boundaries", *Proceedings of Eurospeech 91,* Genova, pp. 1275-1278.

HIRST, D.J., N. IDE, and J. VERONIS, (1994), "Coding Fundamental Frequency Patterns for Multi-lingual Synthesis with INTSINT in the MULTEXT Project", *Proceedings of the Second ESCA/IEEE Workshop on Speech Synthesis,* New-Paltz, NY, pp. 77-80.

HIRST, D.J., P. NICOLAS, and R. ESPESSER, (1991), "Coding the F0 of a Continuous Text in French: An Experimental Approach", *Proceedings of the International Congress on Phonetic Sciences,* Aix en Provence, pp. 234-237.

HUGGINS, A.W.F., (1971), "Just Noticeable Differences for Segmental Duration in Natural Speech", *Journal of the Acoustical Society of America*, vol. 51, pp. 1270-1278.

KAIKI, N., K. TAKEDA, and Y. SAGISAKA, (1990), "Statistical Analysis for Segmentation Duration Rules in Japanese Text-to-Speech", *Proceedings of the International Conference on Spoken Language Processing 90,* pp. 17-20.

KARLSSON, K., (1990), "Constraint Grammars as a Framework for Parsing Running Text", *Proceedings of the Conference on Computational Linguistics*, Helsinki, vol. 3, pp. 168-173.

KELLER, E., B. ZELLNER, S. WERNER, and N. BLANCHOUD, (1993), "The Prediction of Prosodic Timing: Rules for Final Syllable Lengthening in French", *Proceedings of the ESCA Workshop on Prosody,* Lund, pp. 212-215.

KLATT, D. H., (1976), "Linguistic Use of Segmental Duration in English: Acoustic and Perceptual Evidence", *Journal of the Acoustical Society of America,* vol. 59, pp. 1208-1221.

KLATT, D. H., (1987), "Review of Text-to-Speech Conversion for English", *Journal of the Acoustical Society of America,* 82, 3, pp. 737-793.

KOHLER, K.J., (1991), "Prosody in Speech Synthesis: the Interplay Between Basic Research and TTS Application, *Journal of Phonetics*, n°19, pp. 121-138.

LADD, D.R., (1986), "Intonational Phrasing: the Case for Recursive Prosodic Structure", *Phonology Yearbook 3*, pp. 311-340.

LARREUR, D., F. EMERARD, and F. MARTY, (1989), "Linguistic and Prosodic Processing for a Text-to-Speech Synthesis System", *Proceedings of Eurospeech 89*, Paris, pp. 510-513.

LEHISTE, I., (1977), "Isochrony Reconsidered", *Journal of Phonetics*, vol. 5, pp. 253-263.

LIBERMAN, M.J., and K.W. CHURCH, (1992), "Text Analysis and Word Pronunciation in Text-to-Speech Synthesis", in *Advances in Speech Signal Processing*, S. Furui, M.M. Sondhi, eds., Dekker, New York, pp.791-831.

LINDSTRÖM, A., M. LJUNGQVIST, and K. GUSTAFSONN, (1993), "A Modular Architecture Suppoorting Multiple Hypotheses for Conversion of Text to Phonetic and Linguistic Entities", *Proceedings of Eurospeech 93*, Berlin, pp. 1463-1466.

LINDSTRÖM, A., and M. LJUNGQVIST, (1994), "Text Processing within a Speech Synthesis System", *Proceedings of the International Conference on Spoken Language Processing 94,* Yokohama.

LJUNGQVIST, M., and H. FUJISAKI, (1993), "Generating Intonation for Swedish Text-to-Speech Conversion Using a Quantitative Model for the *F0* Contour", *Proceedingd of Eurospeech 93*, Berlin, pp. 873-876.

LJUNGQVIST, M., A. LINDSTRÖM, and K. GUSTAVSON, (1994), "A New System for Text-to-Speech Conversion and its Application to Swedish", *Proceedings of the International Conference on Spoken Language Processing 94*, Yokohama.

MARTIN, P., (1982), "Phonetic Realization of Prosodic Contours in French", *Speech Communication,* n°1, pp. 283-294.

MARTIN, P., (1987), "Prosodic and Rhythmic Structures in French", *Linguistics*, n°25, pp. 925-949.

MARTIN, P., (1990), "Automatic Assignment of Lexical Stress in Italian", *Proceedings of the ESCA Workshop on Speech Synthesis*, Autrans, pp. 149-152.

MERTENS, P., (1987), *L'intonation du Français: de la Description Linguistique à la Reconnaissance Automatique*, Ph.D. dissertation, Katholieke Universiteit Leuven.

MERTENS, P., (1989), "Automatic Recognition of Intonation in French and Dutch", *Proceedings of Eurospeech 89,* Paris, pp. 46-49.

MERTENS, P., (1990), "Intonation", in *Le français parlé*, Blanche-Benveniste, Cl. *et al.*, eds., éditions du CNRS, Paris.

MERTENS, P., (1993), "Accentuation, Intonation et Morphosyntaxe", *Travaux de Linguistique*, Katholieke Universiteit Leuven , n°26, pp. 21-69.

MERTENS, P., (1993) "Intonational Grouping, Boundaries, and Syntactic Structure in French", *Proceedings of the ESCA Workshop on Prosody,* Lund, pp. 156-159.

MEYER, P., H.W. RÜHL, R. KRÜGER, M. KUGLER, L.L.M. VOGTEN, A. DIRKSEN, AND K. BELHOULA, (1993), "PHRITTS - A Text-To-Speech Synthesizer for the German Language", *Proceedings of Eurospeech 93*, Berlin, pp. 877-890.

MÖBIUS, B., M. PÄTZOLD, and W. HESS, (1993), "Analysis and Synthesis of German F_0 Contours by Means of Fujisaki's Model", *Speech Communication,* vol. 13, pp. 53-61.

MOL, H., and E.M. UHLENBECK, (1956), "The Linguistic Relevance of Intensity in Stress", *Lingua* 5, pp. 205-213.

MONAGHAN, A.I.C., (1989), "Phonological Domains for Intonation in Speech Synthesis", *Proceedings of Eurospeech 89* Paris, pp. 502-506.

MONAGHAN, A.I.C., (1990), "A Multi-Phrase Parsing Strategy for Unrestricted Text", *Proceedings of the ESCA Workshop on Speech Synthesis*, Autrans, pp. 109-112.

MONAGHAN, A.I.C., (1990), "Rhythm and Stress Shift", *Computer Speech and Language*, n°4, pp. 71-78.

OHMAN, S., (196), "Word and Sentence Intonation: a Quantitative Model, *K.T.H. Quarterly Progress and Status Report*, vol. 2, pp. 25-54.

O'Connor, J.D., and G.F. ARNOLD, (1973), *Intonation of Colloquial English*, Longman, New-York.

O'SHAUGHNESSY, D., (1984), "A Multispeaker Analysis of Durations in Read French Paragraphs", *Journal of the Acoustical Society of America,* vol. 76, pp. 1664-1672.

O'SHAUGHNESSY, D., (1987), "Specifying Intonation in a Text-to-Speech System Using Only a Small Dictionary", *Proceedings of the International Conference on Acoustics, Speech, and Signal Processing 87,* pp. 1430-1433.

O'SHAUGHNESSY, D., (1990), "Relationships Between Syntax and Prosody for Speech Synthesis", *Proceedings of the ESCA tutorial day on speech synthesis,* Autrans, pp. 39-42.

PIERREHUMBERT, J., (1980), *The Phonology and Phonetics of English Intonation*, PhD dissertation, MIT, Indiana University Linguistics Club.

PIERREHUMBERT, J., (1981), "Synthesizing Intonation", *Journal of the Acoustical Society of America,* 70 (4), pp. 985-995.

PIERREHUMBERT, J., and M. BECKMAN, (1988), *Japanese Tone Structure*, MIT Press, Cambridge, MA.

QUENE, H., and R. KAGER, (1989), "Automatic Accentuation and Prosodic Phrasing for Dutch Text-to-Speech Conversion", *Proceedings of Eurospeech 89*, Paris, pp. 214-217.

RILEY, M.D., (1992), "Tree-Based Modeling for Speech Synthesis", in *Talking Machines: Theories , Models, and Designs*, G. Bailly and C. Benoît, eds., North Holland, pp. 265-273.

RIVEST, R.L., (1987), "Learning Decision Lists", *Machine Learning*, 2, pp. 229-246.

ROSSI, M., A. DI CRISTO, D. HIRST, P. MARTIN, and Y. NISHINUMA, (1981), *L'Intonation, de l'Acoustique à la Sémantique*, Klincksieck, Paris.

ROSSI, M., (1993), "A Model for Predicting the Prosody of Spontaneous Speech (PPSS Model)", *Speech Communication*, vol. 13, n°1-2, pp. 87-107.

RUSSI, T., (1992), "A Framework for Morphological and Syntactic Analysis and its Application in a Text-to-Speech Ssystem for German", in *Talking Machines: Theories , Models, and Designs*, G. Bailly and C. Benoît, eds., North Holland, pp. 163-182.

SAGISAKA, Y., (1990), "On the Prediction of Global *F0* Shape for Japanese Text-to-Speech", *Proceedings of the International Conference on Acoustics, Speech, and Signal Processing 90*, pp. 325-328.

SAGISAKA, Y., and N. KAIKI, (1992), "Optimization of Intonation Control Using Statistical *F0* Resetting Characteristics", *Proceedings of the International Conference on Acoustics, Speech, and Signal Processing 92*, vol. 2, pp. 49-52.

SCHEFFERS, M.T.M., (1988), "Automatic Stylization of *F0* Contours", *Proceedings of the Seventh FASE Symposium*, Edinburgh, vol. 3, pp. 981-987.

SELKIRK, E.O., (1984), *Phonology and Syntax: the Relation Between Sound and Structure*, MIT Press, Cambridge, MA.

SILVERMAN, K., M. BECKMAN, J. PITRELLI, M. OSTENDORF, C. WHIGHTMAN, P. PRICE, J. PIERREHUMBERT, and J. HIRSCHBERG, (1992), "ToBI: a standard for labeling English prosody", *Proceedings of the International Conference on Spoken Language Processing*, Alberta, pp. 867-870.

SLUIJTER, A.M.C., S. SHATTUCK-HUFNAGEL, K.N. STEVENS, and V. VAN HEUVEN, (1995), "Supralaryngeal Resonance and Glottal Pulse Shape as Correlates of Stress and Accent in English", *Proceedings of the XIIIth International Congress on Phonetic Sciences*, vol. 2, pp. 630-633.

SORIN, C., D. LARREUR, and R. LLORCA, (1987), "A Rhythm-Based Prosodic Parser for Text-to-Speech Systems in French", *Proceedings of the XIth International Congress on Phonetic Sciences*, Tallin, vol.1, 125-128.

SPROAT, R., (1990), "Stress Assignment in Complex Nominals for English Text-To-Speech", *Processing of the ESCA Workshop on Speech Synthesis*, Autrans, pp. 129-132.

SPROAT, R., J. HIRSHBERG, D. YAROWSKY, (1992), "A Corpus-Based Synthesizer", *Proceedings of the International Conference on Spoken Language Processing 92*, Alberta, pp. 563-566.

STEEDMAN, M., (1991), "Structure and Intonation" *Language*, n° 67, pp. 260-286.

TERKEN, J., (1993), "Synthesizing Natural-Sounding Intonation for Dutch: Rules and Perceptual Evaluation", *Computer Speech and Language*, vol. 7, pp. 27-48.

TAYLOR, P., (1994), "The Rise-Fall-Connection Model of Intonation", *Speech Communication.*

TOMITA, M., (1986), *Efficient Parsing for Natural Language. A Fast Algorithm for Practical Ssystems*, Kluwer, Boston, MA.

TOMITA, M., (1988), "Linguistic Sentences and Real Sentences", *Proceedings on Coling 88*, Budapest, pp. 453-456.

TRABER, C., (1992), "*F0* Generation with a Database of Natural *F0* Patterns and with a Neural Network", in *Talking Machines: Theories , Models, and Designs*, G. Bailly and C. Benoît, eds., North Holland, pp. 287-304.

TRABER, C., (1993), "Syntactic Processing and Prosody Control in the SVOX TTS System for German", *Proceedings of Eurospeech 93*, Berlin, vol. 3, pp. 2099-2102.

VAISSIERE, J., (1983), *Prosody: Models and Measurements*, A. Cutler and D.R. Ladd, eds., Springer-Verlag, Berlin, pp. 53-66.

van SANTEN, J.P.H., (1993), "Timing in Text-to-Speech Systems", *Proceedings Eurospeech 93*, Berlin, 1993, pp. 1397-1404.

van SANTEN, J.P.H., (1993), "Quantitative Modeling of Segmental Duration", *Proceedings of the ARPA Human Language Workshop.*

WANG, M. Q., and J. HIRSCHBERG, (1991), "Predicting Intonational Boundaries Automatically from Text: The ATIS Domain", *Proceedings of the DARPA Speech and Natural Language Workshop*, pp. 378-383.

WENK, B. J., and F. WIOLAND, (1982), "Is French Really Syllable-Timed ?", *Journal of Phonetics*, 10, pp. 193-216.

WITHGOTT M.M., and F.R. CHEN, (1993), *Computational Models of American Speech*, CSLI Lecture notes n° 32, 143 pp.

PART TWO

FROM NARROW PHONETIC TRANSCRIPTION TO SPEECH

I'll try and say "How doth the little –" and she crossed her hands on her lap as if she were saying lessons, and began to repeat it, but her voice sounded hoarse and strange, and the words did not come the same as they used to do...

Lewis Carroll, *Alice's Adventures in Wonderland*

Once the NLP module of a TTS system has completed its task (and we will assume that it has *successfully* worked out the incoming text, although we have mentioned in Part One that much work remains to be done to approach human performance in that respect), the reading machine is in about the same state as a person knowing *what* to say but still wondering *how* to say it. Bridging the gap between abstract, *symbolic* commands and actual, *numerical* speech waveforms is the subject of the second part of this textbook, with which TTS synthesis enters the realm of digital signal processing.

Chapter Seven gives an overview of the main synthesis strategies currently available in genuine TTS systems (namely *synthesis by rule* and *synthesis by concatenation*), introduces the reader with the related system designs, and emphasizes in each case the operations to which speech quality is found to be more sensitive. Synthesis by concatenation is presented in more detail, and it is shown that the segmental quality of the resulting TTS system strongly (but not solely) depends on the speech model it embraces, together with the algorithms adopted to support it. These algorithms are used in speech analysis, coding, prosody matching, concatenation, and synthesis.

Three types of models (respectively, production models, spectral models and temporal algorithms) are then successively examined in Chapter Eight, Nine, and Ten, with special emphasis on their increasingly positive effect on delivering natural sounding speech. The very classical and extensively used autoregressive (AR) model is taken as the obvious reference in Chapter Eight. Among the possible enhancements to this production-based approach, a glottal auto-regressive model (GAR) is introduced, which, in theory, strongly reduces modeling errors, but heavily increases the analysis complexity. Spectral models are then investigated in Chapter Nine, with the help of the very powerful but computationally expensive multiband excited (MBE) model. Its application to synthesis by concatenation is shown to yield high quality, although it seems that some buzziness cannot be avoided, due to a loss of phase coherency at the junction of segments. As a logical extension of our quest for segmental quality, the special case where no model at all is assumed, is presented in Chapter Ten. Its implications on prosody matching and its restrictions on database memory requirements and concatenation, as developed in the highly efficient time domain pitch synchronous overlap-add (TD-PSOLA) approach, are comprehensively reviewed. We also introduce variants of TD-PSOLA, which make use of previous models—namely, the multiband resynthesis pitch synchronous overlap-add (MBR-PSOLA) and the linear prediction pitch synchronous overlap-add (LP-PSOLA), based on the MBE and the AR models, respectively. While maintaining the very high segmental quality of PSOLA, the latter models reduce its memory requirements and exhibit the powerful concatenative capabilities of the AR and MBE models.

Chapter Seven
SYNTHESIS STRATEGIES

It was not a very difficult question to answer, as there was only one road through the wood, and the two finger-posts both pointed along it. "I'll settle it," Alice said to herself, "when the road divides and they point different ways." But this did not seem likely to happen. She went on and on, a long way, but wherever the road divided there were sure to be two finger-posts pointing the same way, one marked "TO TWEEDLEDUM'S HOUSE" and the other "TO THE HOUSE OF TWEEDLEDEE."

Lewis Carroll, *Through the Looking Glass*

The block diagram introduced in Fig. 1.7 and the definition of a general interface between the NLP and DSP blocks presents the advantage of allowing a separate study of both processes, whether for their requirements, for their functionalities, for their architectural descriptions, or for their results. Consequently, we shall now assume that top-quality information is delivered at the input of the digital signal processing module, like what would be directly copied from a human reading.

Intuitively, the operations involved here are the computer analogue of dynamically controlling the articulatory muscles and the vibratory frequency of the vocal folds so that the output signal matches the input requirements. In order to do it properly, the DSP module should obviously, in some way, take articulatory constraints into account,[1] since it has been known for a long time that phonetic transitions are more important for the understanding of speech than stationary portions of speech (Liberman, 1959). This, in turn, can be basically achieved in two ways:

- Explicitly, in the form of a series of rules that formally describe the influence of phonemes on one another;
- Implicitly, by storing examples of phonetic transitions and coarticulations into a speech segment database, and using them just as they are, as ultimate acoustic units (in place of phonemes).

---

[1]This is desirable, even if the actual synthesis technique describes speech in terms of time-varying parameters that generally have no close relationship with articulatory ones: as already mentioned, planes do not flap their wings.

Two main classes of TTS systems have emerged from these two alternatives, which quickly turned into synthesis philosophies given the divergence they present in their means and objectives: *synthesis by rule* and *synthesis by concatenation.*

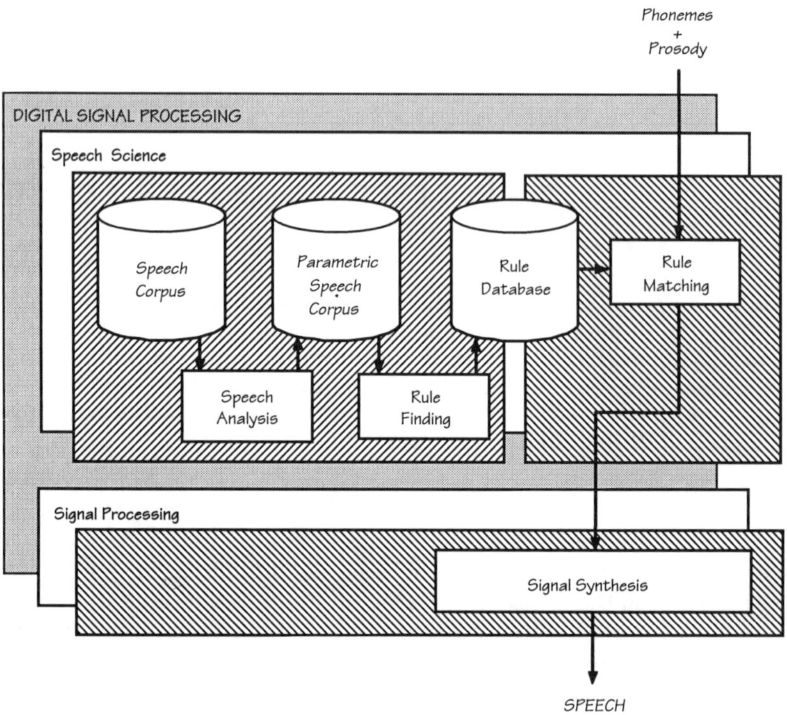

Fig. 7.1 A typical rule-based synthesizer. Most of the work is effectively concentrated in the speech science field.

7.1. Rule-based synthesizers

Rule-based synthesizers (Fig. 7.1) are mostly in favor with phoneticians and phonologists, as they constitute a cognitive, generative approach of the phonation mechanism (Pols, 1990). The widespread use of the Klatt synthesizer (Klatt, 1980), for instance, is principally due to its invaluable assistance in the study of the characteristics of natural speech, by analytic listening of rule-synthesized speech (Klatt and Klatt, 1990). What is more, the existence of relationships between articulatory parameters and the inputs of the Klatt model make it a practical tool for investigating physiological constraints (Stevens, 1990). They are organized as in Fig. 7.1.

In a preliminary phase, a large number of words, generally consonant-vowel-consonant (CVC) sequences, are read by a professional speaker, digitally recorded and stored. They are chosen so as to constitute a speech corpus, representative of the many transitions and coarticulations to study. Thanks to a speech analyzer, the resulting digital data are then given a parametric form that typically separates the respective contributions of the glottal folds and of the vocal tract, and presents the latter in a compact way, more suitable for further study. A rule-finding stage is then applied on the data, which is generally performed by human experts. A first inspection of the speech data from all the speakers is done to get the rule form. The actual values of their parameters (duration or formant frequency values, for example) are then adjusted for one speaker, since rough interspeaker means have little significance at this level.[2] Rules from already existing synthesizers—extracted from other speakers' voices and by different persons—may advantageously help as far as they model general articulatory features rather than speaker peculiarities and as they do not correct specific system deficiencies. Finally, a long trial-and-error work is undertaken to optimize the synthetic quality.

When a sufficient number of rules is gathered, synthesis can start. Rules are matched to the phonetic inputs, and a parametric speech signal is produced. It is finally transformed back into digital speech by a synthesizer that implements the model chosen in the analysis stage.

Consequently, the final *segmental quality* of the synthetic speech delivered by a rule synthesizer depends

- On the efficiency of its internal rules—their capacity to describe the basic parametric corpus with few audible errors;
- On the quality of the corpus—that is:
    - On the actual choice and recording quality of the utterances stored into it;
    - On the accuracy of the speech *model* used in the analysis block to effectively describe high-quality speech: even if no rule finding or rule matching were applied, synthetic utterances could differ from the original ones due to *intrinsic* modeling errors. As a matter of fact, a model typically constrains the type of signal to be analyzed. If a model is too simplistic, it may not be able to account for the richness of speech, and even a simple analysis-resynthesis (as opposed to the more complex text-to-speech synthesis) would therefore lead to distortions;
    - On the actual algorithm that is used to fix up the (time-varying) values of the parameters of a given model, which can be responsible for *extrinsic* modeling errors, denoted as such because they result from the inability of the analysis *algorithm* to produce a correct parametric representation, rather than from an incapacity of the model itself, even though this effect is often related to the peculiar model chosen;
- On the improvements gained by the trial and error stage, which can eventually compensate for recording or algorithmic shortages.

For historical and practical reasons (mainly the need for a physical interpretability of the model), rule synthesizers always appear in the form of *formant synthesizers*. These describe speech as the dynamic evolution of up to 60 parameters (Stevens, 1990), mostly related to formant and antiformant frequencies and bandwidths together with glottal

---

[2]The existence of such an inter-speaker inconsistency is also encountered at the prosody generation level.

waveforms.[3] Consequently, they can be made almost free of intrinsic modeling errors, as shown by the recent production of high-quality speech based on a formant analysis/resynthesis (Karlsson and Noevius, 1994). On the contrary, the large number of (coupled) parameters complicates the analysis stage and tends to produce extrinsic errors. What is more, formant frequencies and bandwidths are inherently difficult to estimate from speech data. The need for intensive trials and errors in order to cope with extrinsic errors, makes them time-consuming systems to develop (several years are commonplace). Yet the synthesis quality achieved up to now reveals typical buzziness problems, which originate from the rules themselves: introducing a high degree of naturalness is theoretically possible, but the rules to do so are still to be discovered.[4]

Rule-based synthesizers remain, however, a potentially powerful approach to speech synthesis. They make it possible, for instance, to study speaker-dependent voice features so that switching from one synthetic voice to another can be achieved with the help of specialized rules in the rule database. Following the same idea, synthesis-by-rule seems to be a natural way of handling the articulatory aspects of changes in speaking styles (as opposed to their prosodic counterpart, which can be accounted for by concatenation-based synthesizers as well). It can account for the modification of formant trajectories correlated with pitch movements, for instance, a feature that is hardly achievable with concatenation-based synthesis (see Pierrehumbert and Frisch, 1994, for instance). No wonder then that it has been widely integrated into TTS systems—MITALK (Allen *et al.*, 1987) and the JSRU synthesizer (Holmes *et al.*, 1964) for English, Santos and Nombela (1982) for Spanish, the multilingual INFOVOX system (Carlson *et al.*, 1982), and the I.N.R.S system (O'Shaughnessy, 1984), or (Bailly *et al.,* 1988) for French.

7.2. Concatenation-based synthesizers

As opposed to rule-based synthesizers, *concatenation-based synthesizers* possess a very limited knowledge of the data they handle: most of it is embedded in the segments to be chained up. This clearly shows in Fig. 7.2, where all the operations that could indifferently be used in the context of a music synthesizer (without any explicit reference to the inner nature of the sounds to be processed) have been grouped into a *signal processing* block, as opposed to the upper *speech science* block whose design requires at least some understanding of phonetics.

7.2.1. Database preparation

As before, a series of preliminary stages have to be fulfilled before the synthesizer can produce its first utterance. At first, segments are chosen so as to minimize future concatenation problems. A combination of diphones (units that begin in the middle of the

---

[3]Since rule-based synthesizers are not the object of this textbook, we invite interested readers to refer to (Holmes, 1983) and (Allen *et al.*, 1987) for detailed descriptions of formant synthesizers.

[4]The lack of naturalness is not peculiar to rule-based synthesizers. Even with segment concatenation methods, increasing the naturalness of a given synthesizer is far from being simple. Ideas such as incorporating stochastic components into speech models or rules, for instance, most often merely result in additional noises.

stable state of a phone and end in the middle of the following one)[5], half-syllables, and triphones (which differ from diphones in that they include a complete central phone) are chosen as speech units, since they involve most of the transitions and coarticulations while requiring an affordable amount of memory.

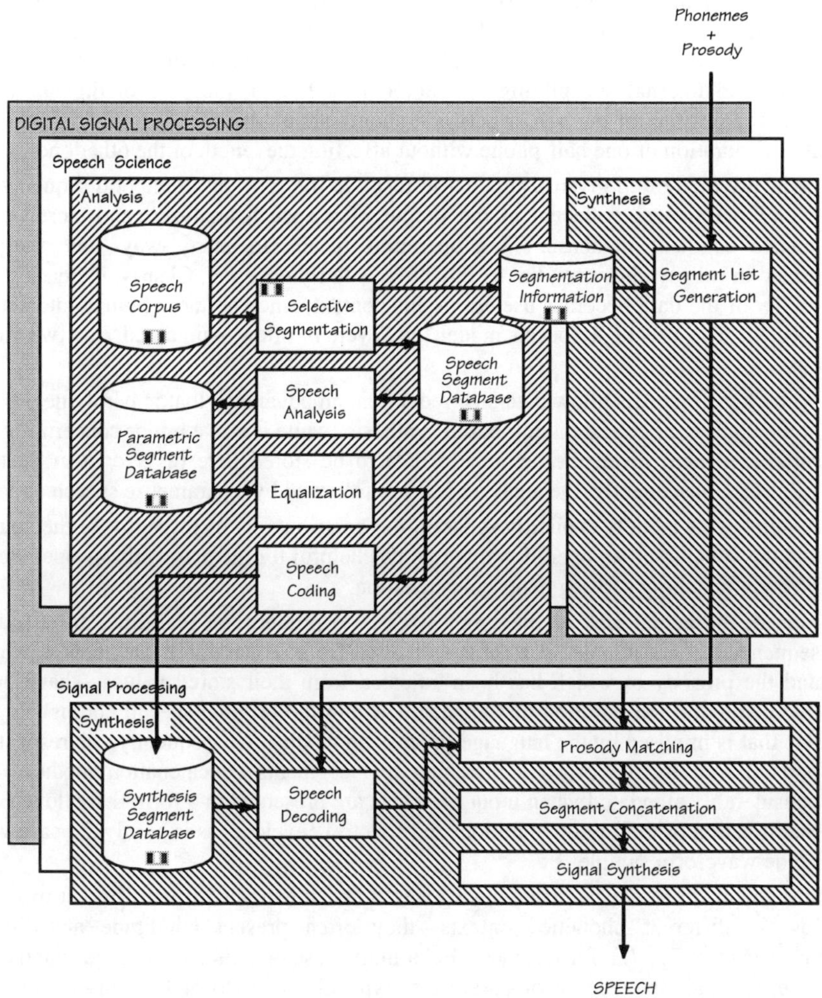

Fig. 7.2 A general concatenation-based synthesizer. Language-dependent blocks and data are indicated by a flag.

---

[5]A consequence of this very imprecise definition being that diphones mostly remain obscure units when highly transient sounds are involved.

When a complete list of segments has emerged, a corresponding list of words is carefully completed, in such a way that each segment appears at least once (twice is better, for security). Unfavorable positions, like inside stressed syllables or in strongly reduced (overcoarticulated) contexts, are excluded. A corpus is then digitally recorded and stored, and the elected segments are spotted, either manually with the help of signal visualization tools, or automatically thanks to segmentation algorithms, the decisions of which are checked and corrected interactively. Segment waveforms are then collected in a speech segment database[6] and the corresponding segmental information (segment names, durations, and internal subsplittings) is stored in a file. In the case of diphones, for example, the position of the border between phones should be stored, so as to be able to modify the duration of one half-phone without affecting the length of the other one.

Segments are then often given a parametric form, in the form of a temporal sequence of vectors of parameters collected at the output of a speech analyzer, and stored in a *parametric* segment database. This operation recalls in many ways the analysis performed in rule synthesizers, but its objective is fairly different. Clearly, increasing the readability of the data becomes useless in this context, since no more human interaction is expected. The use of a speech model, however, is often maintained[7] for two main reasons:

- Well-chosen models allow data-size reduction, an advantage that is hardly negligible in the context of concatenation-based synthesis (while it was a minor concern for the rule-based one) given the amount of data to be stored here (as shown in Section 7.2.3.1). Consequently, the analyzer is often followed by a parametric speech coder.
- A number of models explicitly separate the contributions of respectively the source and the vocal tract, an operation that remains helpful for the pre-synthesis operations: prosody matching and segments concatenation.

  Indeed, the actual task of the synthesizer is to produce, in real-time, an adequate sequence of concatenated segments, extracted from its parametric segment database and the prosody of which has been adjusted from their stored values—that is, the intonation and the duration that appeared within the original speech corpus—to the one that is imposed by the language-processing module. Consequently, the respective parts played by the prosody matching and segments concatenation modules are considerably alleviated when input segments are presented in a form that allows easy modification of their pitch, duration, and spectral envelope, as is hardly the case with crude waveform samples.[8]

Since segments to be concatenated have generally been extracted from different words—that is, in different phonetic contexts—they often present amplitude and timbre mismatches (see Fig. 7.3, for instance). Even in the case of stationary vocalic sounds, for instance, a rough sequencing of parameters typically leads to audible discontinuities.

---

[6] Segments are usually stored with their left and right context (typically 50 to 100 ms), so as to avoid transients when running the following analysis block.

[7] Even though it is generally not a formant model, as it is the case in rule synthesis.

[8] Although, as we shall see in Chapter Ten, there now exist simple algorithms for the prosodic modification of speech waveforms.

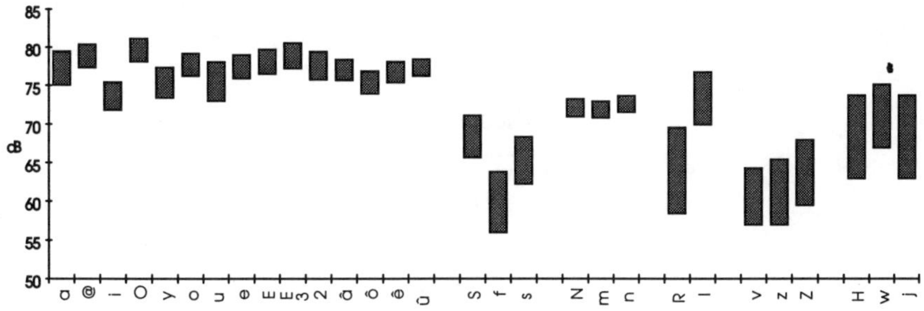

Fig. 7.3. Average instantaneous power per phoneme in a nonequalized segment database for French (after Dutoit, 1993), plotted as (μ-σ,μ+σ) bars. Power excursions are characteristic of phoneme families. Vowels and nasal consonants have a small standard deviation ($\sigma \approx 1.5$ dB), while semivowels, liquids, and fricatives are subject to wider energetic variations ($\sigma \approx 4$ dB)

These can be coped with during the constitution of the synthesis segments database, thanks to an *equalization* in which related endings of segments are given similar amplitude spectra, the difference being distributed over their neighborhood. In practice, however, this operation is restricted to amplitude parameters: the equalization stage smoothly modifies the energy levels at the beginning and at the end of segments, in such a way as to eliminate amplitude mismatches (by setting the energy of all the phones of a given phoneme to their average value). In contrast, timbre conflicts are better tackled at run-time, by *smoothing* individual couples of segments when necessary rather than equalizing them once for all, so that some of the phonetic variability naturally introduced by co-articulation is still maintained. In practice, amplitude equalization can be performed either before or after speech analysis (on crude samples or on speech parameters).

Once the parametric segment database has been completed, synthesis itself can begin.

7.2.2. Speech synthesis

A sequence of segments is first deduced from the phonemic input of the synthesizer, in a block termed as *segment list generation* in Fig. 7.2, which interfaces the NLP and DSP modules. It queries the speech segment database for global information on the units it contains. Segment durations are chosen accordingly, given the desired lengths of the embedded phonemes. In the simplest case, boundary phonemes are distributed proportionally to the initial respective contributions of the segments concerned. For example, let us suppose that the word *see* is to be synthesized. After some language processing, the command '_ #120 s #70 ɪ #150 _ #100' (in which *#nnn* stands for "with a duration of *nnn* ms") is presented at the input of the synthesizer.[9] Successive accesses to

[9]For the sake of clarity, prosodic events were not introduced here. They are conveyed from phonemes to synthesis segments in a similar manner.

the speech segment database inform the segment list generator that the following units are available:

Segment names	Duration of subsegments	
...	...	...
_s	#100	#50
...	...	...
sɪ	#30	#90
...	...	...
ɪ_	#90	#200
...	...	...

Clearly, durations have to be adapted. Assuming the *a priori* equal importance of the left and right parts[10] of any diphone, a constant shortening (or lengthening) ratio can be applied throughout a given phoneme, which results in the final synthesis command: _s #120 #44 sɪ #26 #75 ɪ_ #75 #100 (where #*nnn* now denotes the desired duration of subsegments), in which, for instance, 26 is to 44 what 30 is to 50. Notice that such a strategy implies that the explosion of plosives gets expanded in the same way as their preceding closure, which is obviously not very natural. In practice, however, this causes little degradation, if any, provided the expansion ratio is kept reasonably small.

Once prosodic events have been correctly assigned to individual segments, the *speech decoder* queries the synthesis segment database for the actual parameters of the elementary sounds to be used and decodes them. The *prosody matching* module then adapts them one by one to the required prosody. A convenient way to do this is to make use of two *prosody transformation functions* that completely describe the prosodic changes to be made. The first one is a monotonous time alignment function $t'(t)$ that associates a synthesis instant t' to any analysis instant t. It is easily computed (see Fig. 7.4) by interpolating the (t,t') pairs corresponding to the limits of the subsegments of the current segment, as imposed by the segment database (for t) and by the synthesis command string (for t'). In a like manner, the second function $\omega_0(t')$ provides frequency values at any synthesis instant t'. The way it is computed actually depends on the intonation description model implemented in the prosody generator module (see Section 6.5). In the case of the IPO approach, for instance, ω_0 is described as a piecewise linear function of time.

The segment concatenation block is in charge of dynamically matching segments to one another, by smoothing discontinuities. Here again, an adequate model of speech is highly profitable, provided simple interpolation schemes performed on its parameters approximately correspond to smooth acoustical transitions between sounds. Since such modifications are most often relevant to both ends of a segment, the concatenation block

---

[10]It is preferable to speak in terms of *parts*, rather than *halves*, since respective contributions are obviously not equal in general.

is inherently non-causal, as shown in Fig. 7.5, in which a one-segment delay[11] has been introduced in order to force causality.

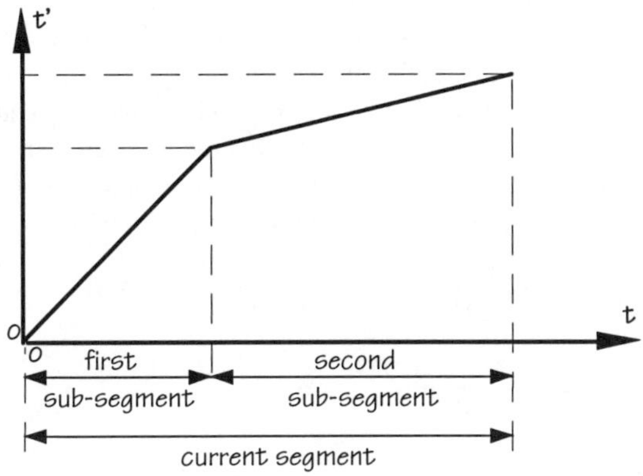

Fig. 7.4. Example of a linear time alignment function for a segment with two subsegments (e.g., a diphone).

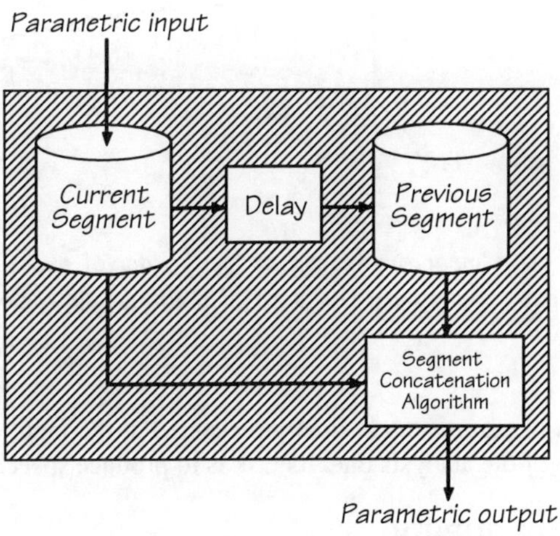

Fig. 7.5. Inside the segment concatenation block.

[11]It should be emphasized that, as opposed to the transitions computed by rule-driven systems, the spectral matching performed by the concatenation block is independent of the type of segment processed. Segments are considered as sounds rather than as individual speech units. Consequently, there is no reason, or even means, of adapting a given segment according to any of its nonnearest neighbors.

Given the natural acoustical closeness of the sounds to be chained, simple linear smoothing is often applied, as shown on Fig. 7.6. Let us denote the left and right segments to be concatenated by L and R. If we consider a set $\mathbf{p}$ of parameters $\{p_1, p_2, ..., p_N\}$, the values of which are $\mathbf{p_L}^0$ at the end of L and $\mathbf{p_R}^0$ at the beginning of R, linear smoothing consists of distributing the difference $(\mathbf{p_R}^0\text{-}\mathbf{p_L}^0)$ among a number M_L of vectors $\{\mathbf{p_L}^{-(M_L-1)}, \mathbf{p_L}^{-1}, \mathbf{p_L}^0\}$ before and including $\mathbf{p_L}^0$, and a number M_R of vectors $\{\mathbf{p_R}^0, \mathbf{p_R}^1, ..., \mathbf{p_R}^{(M_R-1)}\}$ after and including $\mathbf{p_R}^0$. The values M_L and M_R are different, in all generality, so that they can be imposed by the segments themselves, in order to restrain the interpolation process to their steady state parts only. If we denote as $\mathbf{p'}$ the parameters after smoothing, the interpolation laws are given by

$$\mathbf{p'}_L^{-i} = \mathbf{p}_L^{-i} + (\mathbf{p}_R^0 - \mathbf{p}_L^0)\frac{(M_L - i)}{2M_L} \tag{7.1}$$

$$\mathbf{p'}_R^{j} = \mathbf{p}_R^{j} + (\mathbf{p}_L^0 - \mathbf{p}_R^0)\frac{(M_R - j)}{2M_R} \tag{7.2}$$

for i=0. . . M_L-1 and j=0. . . M_R-1

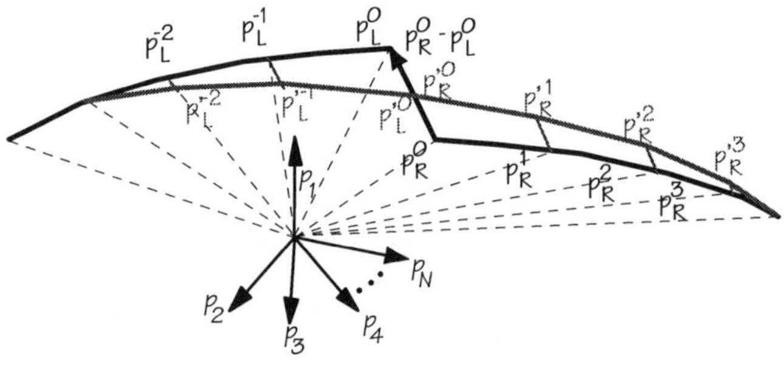

Fig. 7.6. Parametric linear smoothing at the border of successive segments (M_L=2; M_R=3; $p_1, p_2, . . . , p_N$ are the axes of the multidimensional parametric space).

The resulting stream of parameters is finally presented at the input of a synthesis block, the exact counterpart of the analysis one. Its task is to produce speech.

7.2.3. Segmental quality

Figure 7.2 leads to a description of concatenation-based synthesizers in terms of segmental quality. Once the sampling frequency of the synthesizer has been chosen,[12] its efficiency to produce high quality speech is subordinated to the

---

[12]Telephone quality voices require 8 kHz, while 10 kHz is a good choice for medium-quality, general purpose systems (the difference with 8 kHz being quite noticeable). However, 16 kHz or higher is a must for high-quality speech synthesizers, although the normalized frequencies 11.25 and 22.5 kHz are increasingly used in TTS synthesis, at least for multimedia applications.

- Type of segments chosen
- Corpus they were extracted from
- Segmentation quality
- Speech signal model, to which the analysis and synthesis algorithms refer
- Amount of degradation introduced by the speech coding phase
- Prosody matching efficiency, which is strongly related to the model
- Capabilities of the concatenation algorithm

Each bullet item will be reviewed in the following.

7.2.3.1. The choice of segments

Segments should obviously exhibit some basic properties:

- They should account for as many coarticulatory effects as possible.
- Given the restricted smoothing capabilities of the concatenation block, they should be easily connectable.
- Their number and length should be kept as small as possible.

On the other hand, longer units decrease the density of concatenation points, therefore providing better speech quality. Similarly, an obvious way of accounting for articulatory phenomena is to provide many variants for each phoneme. This is clearly in contradiction with the limited memory constraint. Some tradeoff is necessary.

Diphones (or *dyads*) are not too numerous (about 1,200 for French, including lots of phoneme sequences that are only encountered at word boundaries—that is, 3 minutes of speech or approximately 5 Mbytes of 16 bits samples at 16 kHz), and they do incorporate most phonetic transitions. No wonder then that they have been extensively used. They imply, however, a high density of concatenation points (one per phoneme), which reinforces the importance of an efficient concatenation algorithm. Besides, they can only partially account for the many coarticulatory effects of a spoken language, since these often affect a whole phone rather than just its right or left halves independently: formant targets themselves tend to be changed, or a formant starts rising (or falling) during the realization of one phoneme and continues that movement smoothly throughout the ensuing one without any steady state (for a deeper insight, see O'Shaughnessy, 1990, in which contextual effects are presented in terms of formant movements). Such effects are especially patent when somewhat transient phones, such as liquids and (worst of all) semivowels, are interconnected.

Assimilations[13] are still a bigger source of concern. In French, for instance, sonorant consonants (r, l, m, n, j, ɥ, w) become partially or totally unvoiced when they are immediately preceded or followed by a nonsonorant consonant (like in *carte, plein*). Partial assimilation cases are approximately covered by diphones, while total ones would require the introduction of several realizations per phoneme in the database (as many as there are allophones) and a corresponding number of diphones. In the case of sonorant consonants for example, this would necessitate the recording, segmentation, and storage of voiced and unvoiced versions of each phone (remember the voiced and unvoiced

---

[13] At least the ones that do not simply change a phoneme into something that is sufficiently close to another one for a simple substitution to yield acceptable results.

versions of the French semiconsonant [ɥ] in Fig. 1.6). Such a strategy is sometimes denoted as *allo-diphone synthesis.*

Disyllables (units that would start and end in the middle of vowels) could precisely obviate these difficulties, as only vowels would be concatenated.[14] Their number, however, is highly problematic, and their size is naturally larger than diphones (more than 10,000 theoretically possible units for French: about 30 minutes of speech and 50 Mbytes). Another option, namely, restricting concatenation points to low-energy sounds (consonants), could also lead, or not, to a higher quality (this is still an open question). If this case, syllables (units of the form *Ci - V - Cf*, where *Ci* and *Cf* are consonant clusters and *V* is one vowel) could prevail. For the time being, neither solution has been adopted as such. It makes more sense to use them in combination with diphones.

Besides, triphones constitute a variant to the use of syllabic units as a complement to diphones. They can embody the strong dynamic effects quoted above, providing these only affect one phone—excluding sequences of sonorant consonants. To a larger extent, associating diphones with tri- and tetraphones to cope with a limited number of very specific contextual effects is often denoted as a *polyphone* approach (Aubergé, 1991). Bigorne *et al.* (1993), for instance, report on an enhanced French polyphone database composed of 1290 diphones, 153 triphones including [j], 54 triphones including [w], 544 diphones including either [l] or [r], and 288 quadriphones including [j] and either [l] or [r]. Preliminary intelligibility tests have shown that the use of triphones reduce the intelligibility error rate by 20 percent.

Another way of keeping the database size small while still somewhat taking syllables into account is to split them in two, at their vocalic nucleus, which acts as a co-articulation barrier in most cases. Such demisyllables (of the form *Ci - V* or *V - Cf*) have been fully adopted for the synthesis of German, in which long consonant clusters are manifold. A complete demisyllable database, containing about 2000 units, plus some high frequency functional words in order to decrease the concatenation density, is described in (Kraft and Andrews, 1992). Similarly, a combination of HAlf-syllables, DIphones, and sufFIXes is the basis of the HADIFIX system of Portele *et al.* (1990).

Notice, finally, that all the types of speech units proposed above are based on some pre-established phonetic knowledge. In contrast, nothing truly impedes them from being directly automatically derived on the basis of connectivity and representativity requirements. This has led Nakajima and Hamada (1988) to propose a corpus-based, automatic unit set design algorithm termed as *context-oriented clustering* (COC). They start with a speech database with phonetic labels and constitute initial clusters composed of all the speech segments with the same phonetic label. A context tree is then automatically derived by a greedy algorithm, in a way that recalls the CART technique of Section 6.1.2.3. At each step, the cluster with the highest variance (and with more than a minimum number of segments) is split into two classes on the basis of a partition of a context factor, and provided the resulting split leads to an average sub-cluster variance

[14]Syllable protagonists also often argue that children start to babble syllables before they actually begin to talk. The hypothetical syllable-timed nature of languages is still largely debated (Wenk and Wioland, 1982).

that is significantly lower than the initial one. Intra-cluster variances are computed on segments after a normalization of their duration, based on a linear warping technique. Possible context factors are the phonetic labels of neighboring phones up to a given distance from the central one. Broad phonetic classes can also be used as contextual factors, so as to reduce the size of the decision tree and to cope with the problem of data sparsity (see Wang *et al.*, 1993). The best context factor is chosen at each step, as the one that yields the most compact clusters that are most widely acoustically separated. As a result, a context tree is derived for each phoneme, which automatically accounts for acoustic variations due to neighboring phones—that is, for *coarticulation.*[15] This idea has been extended so as to account for phonetic variations related to stress and/or syntax as early as in the design phase of the unit set. Nakajima (1994) incorporates stress and syntactic boundaries in the training database (by describing its content in the form of a multi-level data structure: one layer for phonemic symbols, one for stress, and one for syntactic-prosodic group boundaries), and admits them as potential cluster splitting contextual factors. The resulting *multilayered COC* (or ML-COC) algorithm is still closer to the CART tree-based data clustering technique of Section 6.1.2.3 than the initial single-layered COC (SL-COC) approach (see Fig. 7.7 for an example). It has been tested for the synthesis of English and found to reflect the phonological structure of the target language more adequately than the SL-COC units.

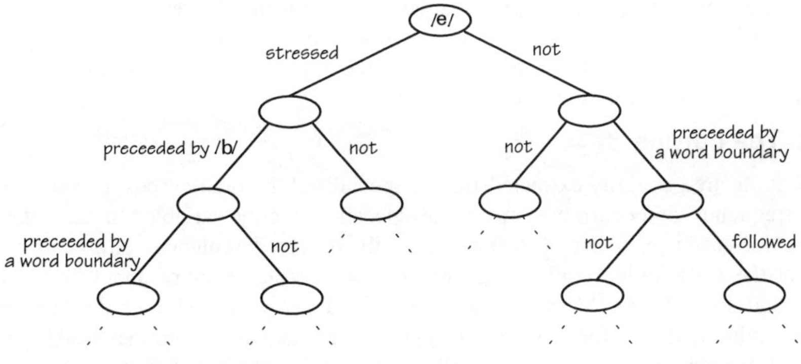

Fig. 7.7. An example of the ML-COC cluster splitting process for occurrences of the phoneme /e/ (adapted from Nakajima, 1994).

By further extending this principle to prosodic features (pitch, duration,and power), Campbell and Black (1995) and Hunt and Black (1996) have drastically reduced the amount of signal processing required to correct the prosodic characteristics of units

[15]Notice that, as the algorithm automatically accounts for coarticulation, a broad phonetic transcription of the initial speech database suffices. Hence, COC can be seen as a numerical, speech-based equivalent to the symbolic, text-based postphonetic processing of Section 5.5. Besides, COC is very similar to the tree-based decomposition of CART techniques, also mentioned in Section 5.5, the major difference being that CARTs are trained in a *supervised* way (they account for preestablished decisions), while COC methods are *unsupervised.* Hence the use of an acoustic distance to obtain the best contextual factor, in place of the maximum mutual information principle of CARTs.

(thereby strongly increasing the segmental quality of their synthesizer). At synthesis time, choosing a list of segments for a given phonetic sequence is no longer trivial. Several instances of each unit are considered as potential synthesis segment, each with a different phonetic and prosodic context in the database. Segment selection is based on using a Viterbi search minimizing two cost functions : a *target cost*, which is an estimate of the difference between a database unit and the target unit it is supposed to represent, and a *concatenation cost,* which is an estimate of the quality of the joint between consecutive units.

It should be clear that such an approach could speed up the constitution of the segment database—for multilingual synthesis for instance. It has proved to lead to very high quality synthesis, at the expense of large databases (from 10 to 150 minutes of speech) and high computational cost (typically about real time on a Sun SparcStation 20). It deserves great consideration, especially given the increasing availability of large labeled corpora.

To come to a conclusion, finding an optimal set of segments for a given memory capacity is still an open problem. The impossibility of associating scores to sets of units, so as to compare them, does not help matters. Even if quantitative intelligibility measurements (such as those obtained by the diagnostic rhyme test (DRT) or the consonant-vowel-consonant (CVC) test; see Section 7.3) were collected for similar synthesizers addressing different segment databases for a given language, contrasting the results would still be highly debatable, since segmental quality has also a lot to do with the peculiar speech corpus from which segments originate. Hence the interest of corpus-based methods, which are likely to result in reproducible sets of units.

7.2.3.2. The corpus

Acoustic units are generally extracted from a specialized, tailored corpus, because their relative frequency of occurrence in natural speech is highly variable. In the case of diphones, for instance, a list of 100 phonetically balanced sentences covers only 43 percent of the 1,200 units required for French, with a redundancy of about 80 percent (Tubach and Boë, 1986). Moreover, as emphasized above, the shorter the segments (whether explicitly due to the type of units chosen, or implicitly, given the speaking rate adopted when reading), the more their realizations depend on their phonological context. The design and recording of the corpus should therefore be given special attention.

Just think for instance of the synthesis of [iʃɔ̃] with diphones [iʃ] and [ʃɔ̃], respectively extracted from the French words *dénicher* and *baluchon*. Clearly, both acoustic realizations of [ʃ] are very different from one another, given their respective vocalic contexts, so that [iʃ] and [ʃɔ̃] will generally not match exactly (this is illustrated in Fig. 7.8). As far as possible, such coarticulation effects should be avoided during the design of the synthesis corpus. At least, they should be kept small enough for the segments concatenation block in figure 7.2 to be able to smooth the resulting discontinuities.

Yet it is not clear whether segments should be extracted from nonsense syllabic sequences (also called *logatoms*), existing isolated words, or meaningful sentences. Bigorne *et al.* (1993) have reported on the advantage of using logatoms instead of meaningful words when using automatic segmentation methods. This does not imply,

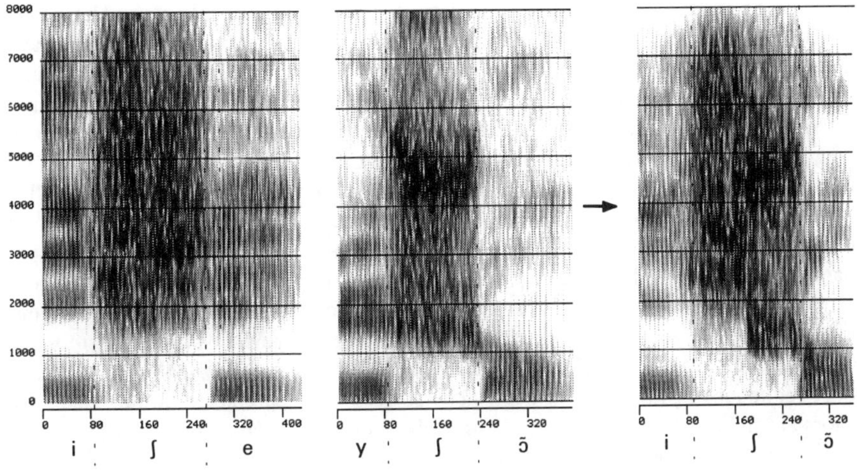

Fig. 7.8. Coarticulation affects the realization of the [ʃ] in [iʃe] and [uʃö]. As a result, a spectral discontinuity appears when synthesizing [iʃö].

however, that high quality synthetic speech cannot be achieved with word-based units (Dutoit, 1993). Even the question of best positioning units in the corpus is still widely debated. Stressed syllables are longer, thus less submitted to coarticulation, which results in easily chainable units; while unstressed ones are more numerous in natural speech, so that producing them well could both increase segmental quality and reduce memory requirements. Likewise, coarticulations (and to a lesser extent assimilations) are also strongly subordinate to speaker 's fluency, so that imposing a slow speaking rate results in more intelligible, but highly overarticulated units. To some extent, these issues can be seen as part of a regrettable, but still necessary, tradeoff between intelligibility and naturalness.

Notice in this connection that the corpus-based approaches of Nakajima and Hamada (1988), Nakajima (1994), Campbell and Black (1995), and Hunt and Black (1996) precisely avoid this tradeoff, by allowing segments to automatically account for different phonetic and prosodic contexts.

7.2.3.3. Segmentation

Except for approaches in the spirit of Nakajima and Hamada (1988), which explicitly banishes human intervention in the segments generation process, a fully automatic segmentation of corpora is hardly conceivable in the context of concatenation-based synthesis. Indeed, most units presented in past discussions originate in phonological considerations rather than in acoustic notions. Isolating them therefore requires a deep prior knowledge of their specific features, so that fully unsupervised segmentation, that is, segmentation on acoustic principles only, without any reference to the phonetic transcription of the material to segment (which Van Hemert, 1991, calls *implicit segmentation*) can merely aspire the outline of a solution, the refinement of which has to be performed by human experts. Segment and subsegment limits indeed are typically misplaced, or just missing, while undefined ones appear.

This does by no means imply that automatic segmentation methods are useless. On the contrary, they are very interesting for speeding up the segmentation process, especially in a multilingual perspective or for the creation of new voices. Efficient algorithms are obtained by enabling the computer to get additional information about the problem it is supposed to solve (at least by providing it with the phonetic transcription of the material to segment, and possibly by allowing some initial training of the segmentation algorithm on a speech database, whether preliminarily segmented or not). In Boeffard *et al.* (1992) and Boeffard *et al.* (1993), phoneme HMMs (one model per phoneme) have been automatically trained with the Viterbi algorithm on French, Spanish, and German speech logatom corpora, with the only help of their phonetic transcription (with no prior manual segmentation of any other corpus). Indeed, instead of modeling sequences of symbols chosen in a given vocabulary, as it was the case for the HMMs introduced in Chapter Two as stochastic versions of nondeterministic regular grammars, HMMs can be used to model sequences of numbers, or sequences of vectors of parameters in this case. These vectors are the result of some parametric analysis of the speech corpus, similar to the one that was introduced on Fig. 7.2 but applied to the whole speech corpus. At the end of the training phase, the emission and transition probabilities of each HMM account for the various acoustic realizations of the related phoneme (some "numerical" grammar has been inferred). The automatic segmentation of a given speech material can then be achieved by "parsing" it with the grammar constituted as a sequence of HMMs corresponding to the phonemes it is composed of. Compared with a manual segmentation used as a reference, Boeffard *et al.* achieved 85 percent of success for the three languages (i.e. 85% of the time, the difference between automatically and manually positioned marks was less than 30 ms). A similar experiment has been performed by Ljolje and Riley (1993) on a training corpus composed of 50 phonetically balanced sentences and a test data composed of 1,158 read sentences. More than 80 percent of the automatically assigned boundaries were within 11.5ms of the boundary selected by a human segmenter. Clearly, one cannot spare further fine tuning by a human transcriber.

It should be noted, finally, that, whatever method is adopted for the segmentation (incorporating *a priori* information by designing an expert system, or allowing automatic learning with statistical or neural models), the final segmentation system should clearly exhibit a speaker-dependent behavior—that is, it should establish its segmentation strategy with reference to a single voice. Indeed, it is rather illusive to try to design a system whose knowledge would not be adapted to individual speakers, since the many speaker peculiarities will always introduce segmentation inconsistencies. One cannot simply ignore them, given the important contribution of segments to the final synthetic speech quality.

7.2.3.4. The model

As in the case of rule synthesizers, describing speech in terms of mathematical equations, the effect of which is revealed by the synthesis algorithm, generally introduces *intrinsic* modeling errors. Yet *extrinsic* ones (analysis biases) can also occur if the analysis algorithm does not deliver the best achievable parameters—the ones that would allow the synthesizer (hence the model) to produce its best-quality speech.

Both types of modeling errors still have a major contribution in the somewhat limited naturalness provided by most actual TTS systems. They can be isolated by combining the analysis and synthesis algorithms alone into a *copy synthesizer* (Fig 7.9), the aim of which is to generate the synthetic signal that best mimics the natural speech presented at its input. They can even be somewhat separated from one another by presenting at the input of the related copy synthesizer some synthetic speech that has precisely been produced by the model, and contrasting its input and output signals: extrinsic errors are responsible for the difference, since the model is in intrinsically capable of accounting for the input speech (after all, it has produced it).

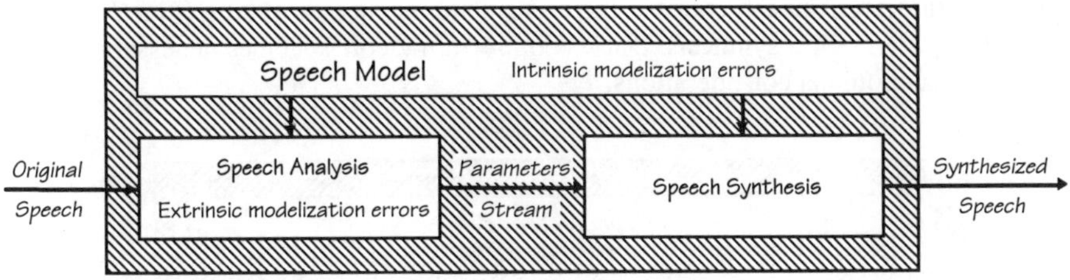

Fig. 7.9. Copy synthesis.

Models can be roughly divided into two classes.

Physiological ones are the foundation of *articulatory synthesis*—from the early two-mass vocal folds model of Ishizaka and Flanagan (1972) to the recent region and mode theory of Mrayati *et al.* (1988), through the very popular Maeda 's model (Maeda 79) and the time-frequency domain articulatory synthesizer of Sondhi and Schroeter (1987). Given the high complexity of the mathematical equations underlying these models (they are based on nonlinear partial differential equations), both the analysis and synthesis operations remain open problems. Assessing articulatory parameters from rough speech data, often denoted as the *acoustic-articulatory inversion*, cannot be performed by some simple signal processing algorithms. Neural networks, for instance, have been proposed as a black-box type solution (Soquet *et al.,* 1990). For a comprehensive review of the problem, see Schroeter and Sondhi (1994). As a result of their complexity, the interest of these models in the context of TTS synthesis is somewhat reduced for the time being. So far, they have mostly been used for cognitive research purposes.

As opposed to articulatory models, *terminal-analogue* models originate, as their name literally infers, in the fact that the actual segmental quality of a synthesizer is related to its effectiveness to produce natural sounding speech, no matter what the means to do it. They are again roughly classified into two groups, depending on their relationship with the actual phonation process. *Production models* provide mathematical substitutes for the part respectively played by vocal folds, nasal and vocal tracts, and by the radiation at the lips. Their most representative members are linear prediction coding (LPC) synthesizers, and the formant synthesizers we mentioned in Section 7.1. On the contrary, *phenomenological models* intentionally discard any reference to the human production mechanism. Among these pure digital signal processing tools, spectral and time-domain approaches are increasingly encountered in TTS systems.

7.2.3.5. The parametric speech coder

It is well known by speech processing specialists that the strong inertia of the articulatory muscles is responsible for a large amount of redundancy in speech signals.[16] Data reduction techniques try to take this into account. However, whatever the form in which speech is passed to a speech coder, whether as a sequence of samples or as a list of parameters referring to a given model, as respectively performed by *waveform* and *parametric coding algorithms*, compression brings distortion. The maximum compression ratio for which distortions remain inaudible, which is generally obtained by optimizing the given speech coder in the corresponding *vocoder* setting (Fig. 7.10), is an important figure in TTS synthesis. If it is too low, a tradeoff becomes necessary between quality and economical considerations.

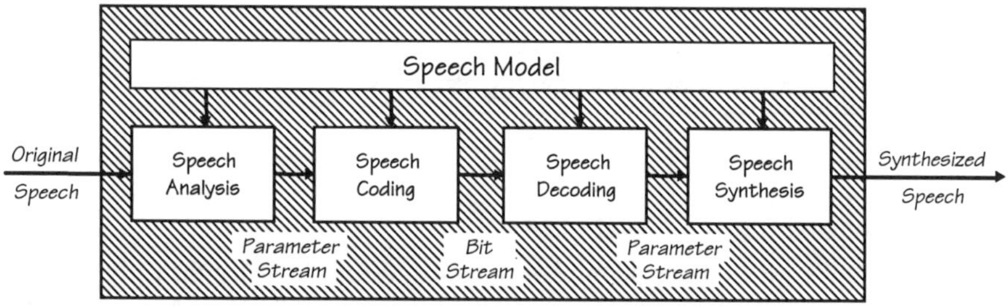

Fig. 7.10. A general vocoder.

7.2.3.6. Prosody matching

Being able to change the pitch and/or the duration of segments while maintaining naturalness is of primary importance in TTS synthesis. It definitely is one of the main criteria for the choice of a model. Given the natural range of the laryngeal frequency in reading (yet with no emphasis), and assuming segments have been recorded with an average frequency lying in the middle of this range, the prosody matching module should typically make it possible to apply a factor of 0.5 to 2 on the pitch of segments, without affecting the position and bandwidth of formants.[17] [18] Similarly, in order to account for the different lengths of stressed and unstressed vowels, a temporal contraction/dilation of 0.5 to 2 should be possible. Most of the aforementioned models do it quite well.

---

[16] Yet only the signal-level redundancy is taken into account here. Additional superfluity is encountered in all languages at the linguistic level, as a means naturally adopted by humans to protect their message against misunderstandings.

[17] As opposed to the effect obtained by varying the pitch of a cassette player, which results in well known bull-dog or mouse voices, since the whole spectrum is contracted/dilated in frequency.

[18] We have already mentioned, however, that given the physical connection between the glottis and the laryngeal tract (the latter playing the role of a finite impedance on the former), pitch movements are somewhat related to formant variations. Such variations are not accounted for in the context of concatenative synthesis.

7.2.3.7. Concatenation

Transitions between speech sounds are theoretically defined in the articulatory domain. Whatever speech model chosen, they are in all generality expressed in terms of nonlinear relations (even in the case of a formant-based description). They are consequently best accounted for by rule (as in the synthesis-by-rule approach), statistical, or neural models. However, when sounds to be concatenated are sufficiently close to one another, linear relations give good results, provided they are performed in an adequately chosen representation space. In other words, the efficiency of linear laws to describe simple transitions in such a way that they are still perceived as natural, depends on the particular representation space in which they are expressed.[19]

The importance of the concatenation operation for synthesis-by-concatenation has been previously underlined in the context of the natural mismatches between segments in a given database. Moreover, the availability of an efficient concatenation algorithm actually interferes with the database preparation, in the sense that over-articulation can be avoided during the recording phase, which results in faster and more natural synthetic speech. Until recently, however, concatenation had not received the attention it deserves. A reason for this can be found in an experiment by Kraft (1994). He synthesized German speech by concatenating either segments taken from a genuine German TTS synthesizer or segments that had precisely been extracted from the sentences to be synthesized, pronounced by the speaker whose voice had been used for the TTS synthesizer. In each case, he varied prosody between machine- and human-like and asked twenty-two subjects to choose the segment inventory they preferred. He found no difference between the two segments inventories for stimuli with poor prosodic quality, but a highly significant preference for well-articulated, smooth units when natural prosody was applied. Consequently, as soon as prosody generation reaches a certain quality level, concatenation becomes a determining issue. This level of quality, however, is only being approached now by state-of-the-art TTS systems.

7.3. Quality assessment

The possible commercial breakthrough of a TTS system is inevitably (but not solely) conditioned by its success to some quality tests. In that respect, quality assessment methods play an important part in the evaluation of the performances of both the digital signal and natural language processing blocks. In this section, however, we shall restrict ourselves to the assessment of the former, in the purpose of contrasting concatenative synthesis algorithms regarding their *intelligibility* and *naturalness*, respectively. Both features, indeed, are not as strongly related as one might think at first. It is generally admitted that intelligibility depends on the efficiency achieved in producing phoneme-to-phoneme transitions, especially VC and CV ones, while naturalness is the result of the congruence of segmental and supra-segmental features.

Given the intrinsic complexity of the speech assessment problem and the rather vague acceptance of the notions of intelligibility and naturalness, no clear standardization has

---

[19]This will clearly appear in Chapter Eight (see Fig. 8.10).

emerged yet.[20] Still, a number of methods are in favor with speech scientists, for the accuracy and reliability of their results, the fairly wide range of phenomena they account for, and the comparisons allowed by the number of tests in which they have already been used. They are, for phoneme-level intelligibility: the *diagnostic rhyme test* (DRT), the *modified rhyme test* (MRT), the *cluster identification tests (consonant-vowel-consonant* (CVC) or *vowel-consonant-vowel* (VCV) tests), for sentence-level intelligibility, the *semantically unpredictable sentences* (SUS) *test*, and for naturalness: the *paired comparison*, and the *mean opinion score* (MOS) test.[21]

The DRT (Voiers, 1983) and the less popular MRT (House *et al.*, 1965) protocols were originally proposed as phonemic intelligibility tests for American English, but they have been adapted to many other languages (a French version of the DRT is described in (Peckels and Rossi, 1973)). Listeners are presented with one isolated word, and are asked to identify it in a list of two or six words, respectively for the DRT or the MRT. Words in the lists only differ by their initial consonant in the DRT and by their initial or final consonants in the MRT. After a number of words have been presented (192 for the DRT, 50 for the MRT), a diagnostic is established about the phonemic attributes that are badly produced by the tested system. The advantages of these methods are their proved reliability, the reduced number of listeners they require, and the availability of a lot of previous test results, mostly for the DRT. Their weakness consists in the fact that only initial and final phonemic transitions are tested, and that their limited number of alternatives may propel listeners to modify their perception (Steeneken, 1992).

In that respect, tests based on cluster identification are more inclusive, as they insert nonsense words with a given syllable structure (often CVC, VCV) in short carrier phrases and only provide listeners with the overall list of words to be played. They are more reproducible than DRT ones and require a limited number of trained listeners (typically, four is sufficient). CVC tests form the subject of a recommendation of the CCITT. On the other hand, VCV tests have been proposed as a standard segmental test for European languages (SAM, 1992).

Figure 7.11 presents a comparison of several phoneme-level intelligibility tests regarding their discriminative power. Scores in each test are associated with the results of a MOS intelligibility test. The CVC test clearly has the most linear scale—that is, the best resolution— which makes it a good candidate for testing high quality systems.

Semantically unpredictable sentences aim to minimize the contextual semantic information given in sentences by using raw sentence structures filled by random selection of words out of predefined lists of possible candidates (Benoit, 1992). The grammatical structures are kept simple:

- Subject - verb - adverbial: *The table walked through the blue truth*
- Subject - verb - direct object: *The strong way drank the day*
- Adverbial- verb - direct object: *Never draw the house and the fact*

[20]For a more comprehensive review of speech assessment methods and a discussion of their validity and reliability, the reader may advantageously refer to Calliope (1989), van Santen (1993), and Jekosch (1993).

[21]These tests can also be used for intelligibility evaluation.

- Q-word - transit. verb - subject - direct object: *How does the day love the bright word?*
- Subject - verb - complex direct object: *The plane closed the fish that lived*

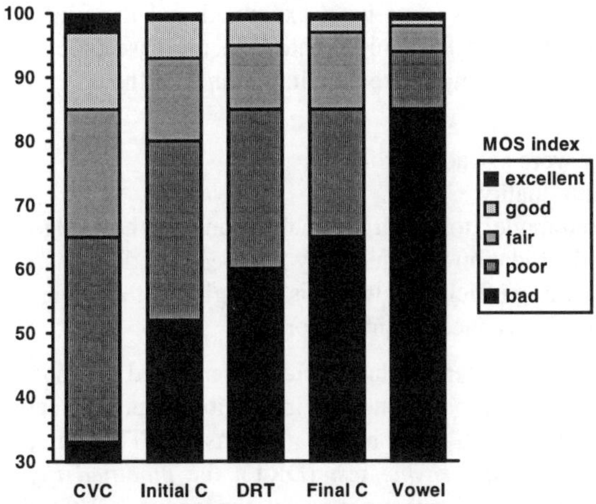

Fig. 7.11. Comparing the discriminative power of intelligibility tests (Deketelaere 94, after Steeneken, 1992). Initial C, Vowel, and Final C are obtained from CVC tests in which the contributions of each component are separately examined.

Each test involves fifty sentences (five for each structure, in a random order), and listeners are simply asked to write down what they hear.

In paired comparisons, listeners are presented two stimuli and asked to express their preference. After a series of comparisons, in which identical stimuli are sometimes proposed in order to verify the reliability of the listener, a preference matrix is obtained. Conclusions, however, may be difficult to draw if transitivity is not respected, as it sometimes happen (A is better than B, which is better than C, which is itself better than A).

Finally, MOS tests rate systems on an absolute scale, usually from 1 to 5:

5. Excellent   4. Good   3. Fair   2. Poor   1. Bad

In general, MOS results change from one listener to another. Hence, MOS tests are essentially used as relative comparison tools. In particular, MOS and CVC results have recently been reported for seven speech coders (Steeneken and Agterhuis, 1992).

7.4. Summary

In this chapter, we have gone one step further in our description of the DSP module of TTS synthesizers. The architectural organization of the two main synthesis strategies currently available in state-of-the-art TTS systems—namely, rule-based and concatenation-based synthesis—has been examined and their influence on the overall synthetic speech quality has been highlighted. In the case of concatenative synthesis, more specifically, we have emphasized the importance of the

- Type of segments chosen
- Corpus they were extracted from
- Segmentation quality
- Speech signal model, to which the analysis and synthesis algorithms refer
- Amount of degradation introduced by the speech coding phase
- Prosody matching efficiency, which is strongly related to the model
- Capabilities of the concatenation algorithm

Before concluding this chapter, we have briefly mentioned the most widely used speech assessment methods, aimed to estimate the intelligibility and the naturalness of synthetic speech (whether rule-based or concatenation-based). They are, for phoneme-level intelligibility: the *diagnostic rhyme test* (DRT), the *modified rhyme test* (MRT), the *cluster identification tests (consonant-vowel-consonant* (CVC) or *vowel-consonant-vowel* (VCV) tests), for sentence-level intelligibility, the *semantically unpredictable sentences* (SUS) *test*, and for naturalness: the *paired comparison*, and the *mean opinion score* (MOS) test.

References

ALLEN, J., S. HUNNICUT, and D. KLATT, (1987), *From Text to Speech, The MITALK System*, Cambridge: Cambridge University Press.

AUBERGE, V., (1991), *La synthèse de la parole: des règles aux lexiques*, Ph.D. dissertation, Institut de la Communication Parlée, Grenoble.

BAILLY, G., G. MURILLO, O. AL DAKKAK, and B. GUERIN, (1988), "A Text-to-Speech System for French Using Formant Synthesis', *Proceedings of SPEECH '88, 7th FASE Symposium*, Edinburgh, pp. 255-260.

BENOIT, C., (1992), "Semantically Unpredictable Sentences: Theoretical Basis for Generation", in *Specification of listener dimensions*, part of the SAM ESPRIT PROJECT final report, 21pp.

BIGORNE, D., O. BOEFFARD, B. CHERBONNEL, F. EMERARD, D. LARREUR, J.L. LE SAINT-MILON, I. METAYER, C. SORIN, and S. WHITE, (1991), "Multilingual PSOLA Text-to-Speech System", *Proceedings of the International Conference on Acoustics, Speech, and Signal Processing 93,* Minneapolis, vol. 2, pp. 187-190.

BOEFFARD, O., B. CHERBONNEL, F. EMERARD, S. WHITE, (1993), "Automatic Generation and Quality Evaluation of Speech Unit Inventories for Concatenation-Based, Multilingual PSOLA Text-to-Speech Systems", *Proceedings of Eurospeech 93,* Berlin, pp. 1449-1452.

BOEFFARD, O., L. MICLET, and S. WHITE, (1992), "Automatic Generation of Optimized Unit Dictionaries for Text-to-Speech Synthesis', *Proceedings of the International Conference on Spoken Language Processing,* Alberta, pp. 1211-1214.

CALLIOPE, (1989), *La Parole et son Traitement Automatique,* CNET-ENST, Collection technique et scientitfique des télécommunications, Paris: Masson, pp 410-414.

CAMPBELL, N., and A. BLACK, (1995), "Prosody and the Selection of Source Units for Concatenative Synthesis", in *Progress in Speech Synthesis*, J. van Santen, R. Sproat, J. Olive, and J. Hirshberg, eds., Springer Verlag.

CARLSON, R., B. GRANSTRÖM, and S. HUNNICUT, (1982), "A Multi-Language Text-to-Speech Module", *Proceedings of the International Conference on Acoustics, Speech, and Signal Processing 82*, Paris, vol. 3, pp. 1604-1607.

DEKETELAERE, S., (1994), *Application de la Quantification Vectorielle au Codage Bas-Débit d'un Signal de Parole*, Ph.D. dissertation, Faculté Polytechnique de Mons.

DUTOIT, T., (1993), *High Quality Text-to-Speech Synthesis of the French Language*, Ph.D. dissertation, Faculté Polytechnique de Mons.

HOLMES, J., (1983), "Formant Synthesizers - Cascade or Parallel?", *Speech Communication*, vol. 2, pp. 251-273.

HOLMES, J., I. MATTINGLY, and J. SHEARME, (1964), "Speech Synthesis by Rule", *Language and Speech*, vol. 7, pp.127-143

HUNT, A.J., and A. BLACK, (1996), "Unit Selection in a Concatenative Speech Synthesis System Using a Large Speech Database", *Proceedings of the International Conference on Acoustics, Speech, and Signal Processing 96*, Atlanta.

ISHIZAKA, K., and J.L. FLANAGAN, (1972), "Synthesis of Voiced Sounds from a Two-Mass Model of the Vocal Cords", *Bell Systems Technical Journal*, n°51, pp. 1233-1268.

JEKOSCH, U., (1993), "Speech Quality Assessment and Evaluation", *Proceedings of Eurospeech 93*, Berlin, pp. 1387-1394.

KARLSSON, I., and L. NOEVIUS, (1994), "Rule-Based Female Speech Synthesis - Segmental Level Improvements", *Proceedings of the Second ESCA/IEEE Workshop on Speech Synthesis*, New Paltz, NY.

KLATT, D.H., (1980), "Software for a Cascade /Parallel Formant Synthesizer", *Journal of the Acoustical Society of America*, vol. 67, pp. 971-995.

KLATT, D.H., and L.C. KLATT, (1990), 'Analysis, Synthesis, and Perception of Voice Quality Variations among Female and Male Talkers', *Journal of the Acoustical Society of America*, vol. 87, pp. 820-857.

KRAFT, V., (1994), "Does the Resulting Speech Quality Improvement Make a Sophisticated Concatenation of Time-Domain Synthesis Units Worthwhile?", *Proceedings of the Second ESCA/IEEE Workshop on Speech Synthesis*, New Paltz, NY, pp. 65-68.

KRAFT, V., and J.R. ANDREWS, (1992), "Design, Evaluation, and Acquisition of a Speech Database for German Synthesis-by-Concatenation', *Proceedings of SST 92*, Brisbane, pp. 724-729.

LIBERMAN , A.L., (1959), 'Minimal Rules for Synthesizing Speech', *Journal of the Acoustical Society of America*, pp. 1490-1499.

LJOLJE, A., and M.D. RILEY, (1993), "Automatic Segmentation of Speech for TTS", *Proceeding of Eurospeech 93*, Berlin, pp. 1445-1448.

MAEDA, S., (1979), "An Articulatory Model of the Tongue Based on a Statistical Analysis', *Journal of the Acoustical Society of America*, vol. 65, S-22.

MRAYATI, M., R. CARRE, and B. GUERIN, (1988), "Distinctive Regions and Modes: a New Theory of Speech Production', *Speech Communication*, vol. 7, pp. 257-286.

NAKAJIMA, S., (1994), "Automatic Synthesis Unit Generation for English Speech Synthesis Based on Multi-Layered Context Oriented Clustering", *Speech Communication*, n°14, p. 313-324.

NAKAJIMA, S., and H. NAMADA, (1988), "Automatic Generation of Synthesis Units Based on Context-Oriented Clustering", *Proceedings of the International Conference on Acoustics, Speech, and Signal Processing 88*, 51.S14.2, pp. 659-662.

O' SHAUGHNESSY, D., (1984), "Design of a Real-Time French Text-to-Speech System', *Speech Communication*, vol. 3, pp. 233-243.

O' SHAUGHNESSY, D., (1990), "Spectral Transitions in Rule-Based and Diphone Synthesis", *Proceedings of the First ESCA Workshop on Speech Synthesis,* Autrans, pp. 21-25.

PECKELS, J.P., and M. ROSSI, (1973), "Le Test de Diagnostic par Paires Minimales, Adaptation au Français du "Diagnostic Rhyme Test" de Voiers", *Revue d'Acoustique*, n° 27, pp. 245-262.

PIERREHUMBERT, J., and S. FRISCH, (1994), "Source Allophony and Speech Synthesis", *Proceedings of the Second ESCA/IEEE Workshop on Speech Synthesis*, New Paltz, NY, pp. 1-4.

POLS, L., (1990), "Does Improved Performance of a Rule Synthesizer also Contribute to More Phonetic Knowledge?', *Proceedings of the ESCA Tutorial Day on Speech Ssynthesis,* Autrans, pp. 50-54.

PORTELE, T., W. SENDLMEIER, and W. HESS, (1990), "HADIFIX, a System for German Speech Synthesis Based on Demisyllables, Diphones, and Suffixes", *Proceedings of the First ESCA Workshop on Speech Synthesis,* Autrans, pp. 161-164.

SAM (1992), *Multi-Lingual Speech Input/Output Assessment, Methodology, and Standardisation - Final Report*, SAM-UCL-G0004, ESPRIT Project 2589.

SANTOS, J.M., and J.R. NOMBELA (1982), "Text-to-Speech Conversion in Spanish: a Complete Rule-Based System', *Proceedings of the International Conference on Acoustics, Speech, and Signal Processing 82*, Paris, pp. 1593-1596.

SCHROETER, J., and M.M. SONDHI, "Techniques for Estimating Vocal-Tract Shapes from the Speech Signal", *IEEE Transactions on Speech and Audio Processing,* vol. 2, n°1, pp. 133-150.

SOQUET, A., M. SAERENS, and P. JOSPA, (1990), "Acoustic-Articulatory Inversion Based on a Neural Controller of a Vocal-Tract Model', *Proceedings of the First ESCA Workshop on Speech Synthesis,* Autrans, pp. 71-74.

SONDHI, M.M., and J. SCHROETER, "A Hybrid Time-Frequency Domain Articulatory Speech Synthesizer", *IEEE Transactions on Acoustics, Speech, and Signal Processing,* vol. 35, n°7, pp. 955-967.

STEENEKEN, H.J.M., (1992), *On Measuring and Predicting Speech Intelligibility*, Ph.D. dissertation, 1992, TNO Institute for Perception, Eindhoven.

STEENEKEN, H.J.M., and E. AGTERHUIS, (1992), *Speech Intelligibility and Speech Quality of Seven Speech Coders for the HERMES Project*, report IZF 1992 C-41, TNO Institute for Perception, Eindhoven.

STEVENS, K.N., (1990), "Control Parameters for Synthesis by Rule", *Proceedings of the ESCA Tutorial Day on Speech Synthesis,* Autrans, pp. 27-37.

TAKEDA, K., K. ABE, Y. SAGISAKA, and H. KUWABARA, (1989), "Adaptive Manipulation of Non-Uniform Synthesis Units", *Proceedings of Eurospeech 89*, vol. 2, pp. 195-198.

TUBACH, J.P., and L.J. BOE, (1986), "Quantitative Knowledge on Word Structure From a Phonetic Corpus, with Application to Large Vocabularies Recognition Systems", *Proceedings of the International Conference on Acoustics, Speech, and Signal Processing 86*, Tokyo, pp. 61-64.

VAN HEMERT, J.P., (1991), "Automatic Segmentation of Speech", *IEEE Transactions on Speech Processing,* vol. 39, n°4, pp. 1008-1012.

VAN SANTEN, J.P.H, (1993), "Perceptual Experiments for Diagnostic Testing of Text-to-Speech Systems", *Computer Speech and Language*, n°7, pp. 49-100.

VOIERS, W.D., (1983) "Evaluating Processed Speech Using the Diagnostic Rhyme Test", *Speech Technology*, January-February, pp. 30-39.

WANG, W.J., W.N. CAMPBELL, N. IWAHASHI, and Y. SAGISAKA, (1993), "Tree-Based Unit Selection for English Speech Synthesis", *Proceedings of the International Conference on Acoustics, Speech, and Signal Processing 93*, vol. 2, pp.-191-194.

WENK, B.J., and F. WIOLAND, (1982), 'Is French Really Syllable-Timed?', *Journal of Phonetics*, vol. 10, pp. 193-216.

Chapter Eight

LINEAR PREDICTION SYNTHESIS

It is a poor sort of memory that only works backwards.
Lewis Carroll, *Through the Looking Glass*

The most basic speech production model used in speech processing is, undoubtedly, the source-filter model. Since 1960, the year of its first appearance in Fant (1960) and its simplification into the auto-regressive model, it has given birth to dozens of profitable interpretations (see Markel and Gray, 1976 for a review), from maximum likelihood to polynomial approaches, through the extensively used spectral interpretation. All of them share the same mathematical reality observed in various lights. Instead of providing the reader with yet another overview of these developments, we have chosen to present a somewhat original geometrical interpretation of the linear prediction (LP) framework, an approach that is seldom investigated in the literature (it is introduced in Kroon, 1985, and Alexander, 1986), even though it does yield a significant insight into auto-regressive models[1]. Particular care has been taken in its presentation, in order to avoid abusively simplistic interpretations, which often result from ill-considered generalization of three- to *N*-dimension concepts. A clear understanding of the notions involved in the first few sections of this chapter will also be of some use in Section 8.9, which is employed to examine to what extent the glottal autoregressive model (GAR) constitutes an important refinement to the classical AR model.

8.1. The autoregressive (AR) model

Let us first recall that speech approximately originates in the passage of some excitation signal through the filter constituted by our vocal, nasal, and pharyngeal tract. This signal could be the glottal pulses produced by the periodic opening and closing of our vocal folds, or just some air flow pushed by our lungs, or even a combination of both at the same time.

[1]After all, a rough sketch is often more profitable than lengthy equations.

In the late 1950s, Fant proposed a linear model of speech production (Fant, 1960), based on the hypothesis that the glottis and the vocal tract are fully discoupled (glottal source impedance much larger than vocal tract input impedance). In the case of voiced sounds, the glottal volume velocity waveform is supposed to be a sequence of glottal pulses, modeled as the impulse response of a low-pass filter with two real poles, the bandwidth of which is set to about 100 Hz:

$$G(z) = \frac{1}{(1-e^{-\alpha}z^{-1})(1-e^{-\beta}z^{-1})} \tag{8.1}$$

In contrast, white noise is used as input for the filter to produce unvoiced sounds.[2] The glottal waveform is presented at the input of a cascade of second order resonators, *V(z)*, and finally transformed into an acoustic pressure waveform at the output of the lips filter *L(z)*:

$$S(z) = E(z)\, G(z)\, V(z)\, L(z) \tag{8.2}$$

where *E(z)* stands for the excitation of the glottal filter. Assuming the latter is approximately proportional to the volume velocity waveform derivative and making some further simplifications, it is easy to show (Boite and Kunt, 1987) that:

$$S(z) \cong E(z)\frac{1}{\sum_{i=0}^{p} a_i z^{-i}} = E(z)\frac{1}{A_p(z)} \qquad (a_0 = 1) \tag{8.3}$$

Consequently, the Fant proposal results in an all-pole model of speech production (see Fig. 8.1), in which $1/A_p(z)$ and $A_p(z)$ respectively define the *synthesis filter* and the *analysis* (or *inverse*) *filter*. The parameters of the model are the excitation pitch period (a value of zero being associated to unvoiced sounds, so that the pitch value includes a voiced/unvoiced (V/UV) decision), the amplitude factor σ that controls the amplitude of the excitation signal, and the filter coefficients.

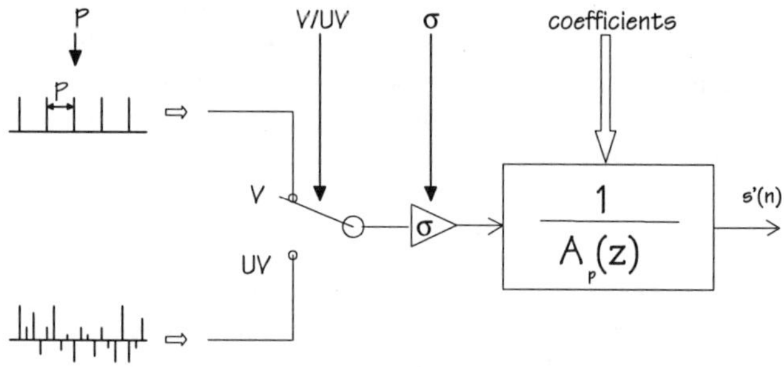

Fig. 8.1 The classical autoregressive model of speech production. The excitation signals are restricted to either an impulse train with unitary amplitude or white Gaussian noise with unitary variance and zero mean.

---

[2]Mixed sounds, like voiced fricatives, were not adequately taken into account in the original model.

It should be noticed that no provision is made in the Fant model for phase distortions, which nevertheless can hardly be avoided when no special attention is paid to the recording system (microphone, preamplifier, tape recorder, or sampler). It rather focuses on the evolution of the amplitude spectrum of speech. Consequently, the AR model is incapable of accounting for actual speech waveforms as such. This, however, was done on purpose, since constant or slowly varying phase distortions have no audible effect.

Equation (8.3) introduces in turn the concept of linear predictability of speech, which states that speech samples can be approximately expressed as a weighted sum of the p previous ones:

$$s(n) = e(n) + \sum_{i=1}^{p} - a_i s(n-i) \tag{8.4}$$

It is assumed in (8.4) that the a_i coefficients vary with time, so as to account for the nonstationarity of speech.

8.2. A mathematical framework for linear prediction analysis

From a given signal, a practical problem is to find the best set of AR parameters—that is, the set that minimizes modeling errors—by trying to restrain audible differences between the original signal and the one that is produced by the synthesizer of Fig 8.1. We shall now try to answer this question, since it is determines the quality of the LPC segment database to be used in any LPC-based TTS synthesis system.

In the context of TTS synthesis, pitch is imposed by the language-processing module, so that the computation and storage of the pitch of the synthesis segments *seem* useless. This is not true, as will be seen in Section 8.4: it is necessary to know the pitch of the each segment to modify it properly. Pitch computation, however, is not examined here.[3] We rather focus on the optimization of σ and of the filter coefficients, by so-called *pitch-asynchronous linear prediction* techniques. Parameters are estimated every 10 ms on speech frames of about 30 ms, during which they are assumed to be constant, given the inertia of the articulatory muscles.

8.2.1. The linear prediction problem

We now define a linear prediction of order p on a sequence $\{s(0), s(1),...s(N-1)\}$ of N speech samples as the search for a set $\{a_1,a_2,...a_p\}$ of *prediction coefficients* that verify

$$s(n) \overset{\text{LS}}{=} \sum_{i=1}^{p} - a_i s(n-i) \qquad \text{for } n = p, p+1, ..., N-1 \tag{8.5}$$

Since N is generally greater than p, the equality is understood in the sense of least squares approximation. Implicitly introducing the *prediction residual* (or *error*) $f(n)$ into (8.5) leads to

[3]Readers interested in pitch determination algorithms can advantageously resort to a tutorial by Hess (1992).

$$s(n) = \sum_{i=1}^{p} -a_i s(n-i) + f(n) \qquad \text{for } n = p, p+1, \ldots, N-1$$
$$\text{with } a_1, \ldots, a_p \text{ chosen so as to minimize } \sum_{n=p}^{N-1} f(n)^2 \tag{8.6}$$

This can be written in vector notation

$$\mathbf{s_p} = \sum_{i=1}^{p} -a_i \mathbf{s_{p-i}} + \mathbf{f_p} \tag{8.7}$$

as the search for the linear combination of the prediction vectors $\mathbf{s_0}, \ldots, \mathbf{s_{p-1}}$ that is closest to $\mathbf{s_p}$, with $\mathbf{s_i} = [s(i), s(i+1), \ldots, s(N-p+i-1)]^T$ (Fig. 8.2). As $\mathbf{s_p}$ is not, in general, included in the subspace $\{\mathbf{s_0}, \ldots, \mathbf{s_{p-1}}\}$, this decomposition is performed under the constraint of minimizing the norm of $\mathbf{f_p} = [f(p), f(p+1), \ldots, f(N-1)]^T$, noted α_p and termed as the *(forward) prediction error vector.*

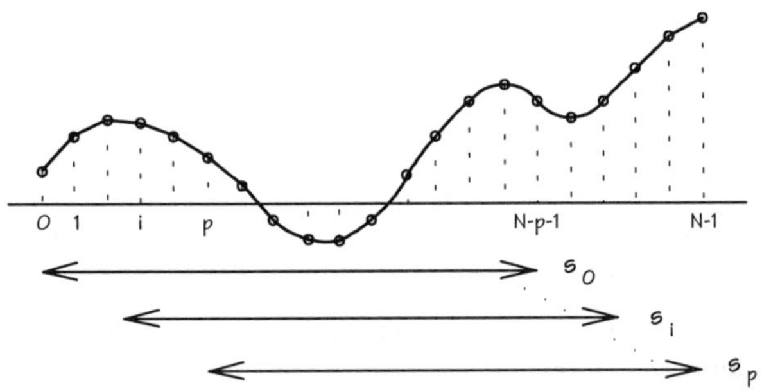

Fig. 8.2. Prediction vectors.

Equation (8.7) leads to a simple matrix form

$$\mathbf{s_p} = \mathbf{\Phi a} + \mathbf{f_p} \tag{8.8}$$

in which $\mathbf{\Phi}$ is the matrix of column vectors $\mathbf{s_0}, \ldots, \mathbf{s_{p-1}}$ and $\mathbf{a}$ is the vector $[-a_1, -a_2, \ldots, -a_p]^T$.

8.2.2. The Yule-Walker equations

The linear combination that minimizes the norm α_p of the prediction-error vector $\mathbf{f_p}$ is simply obtained by projecting $\mathbf{s_p}$ orthogonally to the prediction subspace $\{\mathbf{s_0}, \ldots, \mathbf{s_{p-1}}\}$ (Fig 8.3).

The resulting projected vector $\mathbf{s'_p}$ verifies the so-called Yule-Walker equations:

$$\mathbf{f_p} \perp \mathbf{s_j} \qquad \text{for } j = 0, 1, \ldots, p-1 \tag{8.9}$$

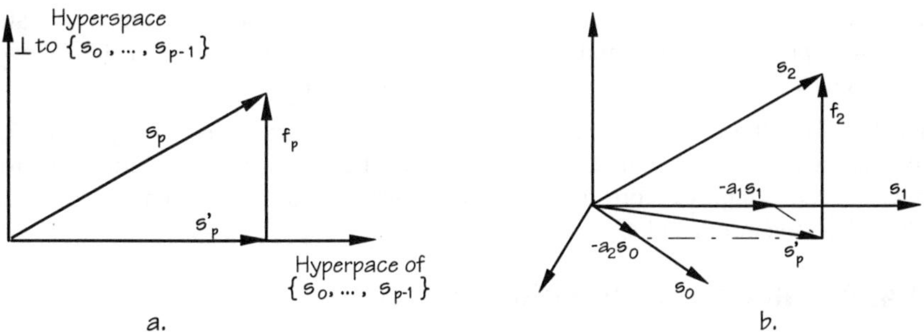

Fig. 8.3. The forward prediction error vector $\mathbf{f_p}$ is the component of $\mathbf{s_p}$ that is orthogonal to the prediction subspace (a: general order *p*; b: *N* = 3, *p* = 2).

that is,

$$\begin{aligned} \mathbf{s_p}\,\mathbf{s_j} &= \mathbf{s'_p}\,\mathbf{s_j} \\ &= \sum_{i=1}^{p} -a_i \mathbf{s_{p-i}}\,\mathbf{s_j} \qquad \text{for } j = 0,1,\ldots,p-1 \end{aligned} \tag{8.10}$$

or, in matrix form,

$$\begin{aligned} \mathbf{\Phi^T \Phi\, a} &= \mathbf{\Phi^T s_p} \\ \mathbf{a} &= (\mathbf{\Phi^T \Phi})^{-1} \mathbf{\Phi^T s_p} \end{aligned} \tag{8.11}$$

in which $(\mathbf{\Phi^T\Phi})^{-1}$ is the pseudo-inverse of $\mathbf{\Phi}$, well known in least squares theory (Golub and Van Loan, 1989).

8.2.3. Covariance versus autocorrelation

So far, we have assumed that $\mathbf{s_p}$ is composed of *N* speech samples. This leads to the so-called *covariance* formalism, for which $\mathbf{\Phi^T\Phi}$ is symmetrical positive definite and is often denoted as the covariance matrix[4] **C**. Equations (8.10) can be solved with the simple LDL^T decomposition provided by the Cholesky algorithm or, as shown in Section 8.2.5, with the *lattice* algorithm.

In practical cases, given the fact that *N>>p*, **C** is more than symmetrical: it is almost Toeplitz[5] symmetric—$\mathbf{s_i}\,\mathbf{s_j}$ almost only depends on |i-j|. As a matter of fact, most of the terms included in $\mathbf{s_i}\,\mathbf{s_j}$ also contribute to $\mathbf{s_{i+k}}\,\mathbf{s_{j+k}}$, for small *k*. Moreover, **C** can be imposed a true Toeplitz symmetric form by imposing $\mathbf{s_i} = [x(i), x(i+1),\ldots, x(N-p+i-1)]^T$, with

$$\{x(0),x(1),\ldots,x(p-1),x(p),\ldots,x(N-p-1),x(N-p),\ldots,x(N-1)\} = \{\underbrace{0,0,\ldots,0}_{p},s(0),\ldots,s(N-2p-1),\underbrace{0,0,\ldots,0}_{p}\}$$

[4]Mostly for historical reasons, since true covariance computation includes a preliminary mean extraction stage.

[5]A Toeplitz matrix is a square matrix in which diagonals parallel to the main diagonal contain identical elements.

This corresponds to the *autocorrelation* formalism, the main interest of which is to allow a faster computation of the prediction coefficients, as will be shown in the following sections, respectively devoted to the *Levinson* and *Schur* algorithms.

Since the artifact in the construction of $\mathbf{s}_i$ introduces some bias, owing to artificial prediction errors at the beginning and at the end of the frame, a weighting window is applied on the samples before the analysis, so that the covariance and autocorrelation methods lead to closer results.

8.2.4. A criterion for the choice of σ

It makes sense to choose σ so as to equal the energy of the analysis and synthesis frames:

$$\sigma_s^2 = \sigma_a^2 \tag{8.12}$$

But what does σ_S depend on? Obviously, the samples produced by the synthesis filter are a function of the excitation, of the filter coefficients, and of the signal history, as stored in the internal memory of the filter. What is more, given the overlap of the analysis frames and the duration changes imposed by prosody matching, synthesis is generally performed on a number *M* of samples, with *M* different from the analysis vector space dimension *(N-p)*. So that if the synthesized frame is separated into its free and forced response components (the forced response being assumed for σ equal to 1)

$$\tilde{s}(n) = \tilde{s}_{Free}(n) + \sigma\, \tilde{s}_{Forced}(n) \tag{8.13}$$

the value of σ_S is given by

$$\sigma_s = \sum_{n=0}^{M-1} \left(\tilde{s}_{Free}^2(n) + \sigma^2 \tilde{s}_{Forced}^2(n) + 2\sigma\, \tilde{s}_{Free}(n)\, \tilde{s}_{Forced}(n) \right) \tag{8.14}$$

The theoretical way to verify (8.12) is thus to compute the free and forced responses separately and solve (8.14) for σ (Markel and Gray, 1976).

In practice, however, σ is approximated from steady-state considerations. The forward prediction error vector is the optimal excitation, in the sense that its filtering by $1/A_p(z)$ would result in the analysis frame itself, according to equation (8.7) and provided the filter memory is filled with the $\{s(0), s(1), .., s(p-1)\}$ samples. Consequently, the optimal excitation energy is best given by

$$\sigma_e^2 = \frac{M}{N_e}\alpha_p^2 \tag{8.15}$$

where N_e is the effective number of samples predicted:

$$N_e = N - p \quad \text{or} \quad N - 2p, \tag{8.16}$$

for the covariance and autocorrelation methods, respectively.

For synthesis, an artificial excitation $e(n)$ is substituted, which is either an impulse train with unitary amplitude and period *P*, or white noise with zero mean and unit variance, both multiplied by σ. In the case of voiced sounds, the excitation energy depends on the position of the pulses within the frame. On the average, it is given by:

$$\sigma_e^2 = \sum_1^M e(n)^2 = \sigma^2 \frac{M}{P} \tag{8.17}$$

Otherwise, assuming white noise is generated as a sequence of pseudo-random numbers uniformly distributed in $[-q,+q]$, the variance of which is known to be $q^2/3$:

$$\sigma_e^2 = \sigma^2 M \frac{q^2}{3} \tag{8.18}$$

Combining the four last equations gives the optimal value of σ for voiced and unvoiced frames.

8.2.5. The covariance method, with the Grahm-Schmidt algorithm

In the general case of the covariance method, an alternative to the Cholesky algorithm was introduced by Itakura and Saito (1969). It is reviewed here since it lends itself to a very simple geometrical interpretation. What is more, it is a first step to the highly efficient Levinson and Schur methods, which will allow us to derive, with geometrical arguments only, a lattice structure for the LPC synthesis filter in Section 8.2.7.

Let us first consider, for a given speech vector $\mathbf{s_p}$ (defined by equation 8.7), the successive forward prediction of order m, for m = 1 to p—that is, the orthogonal components of $\mathbf{s_p}$ on $\{\mathbf{s_{p-m}}, \mathbf{s_{p-m-1}}, .., \mathbf{s_{p-1}}\}$. The corresponding forward prediction error vectors are $\mathbf{f_1}$, $\mathbf{f_2}$, .., $\mathbf{f_p}$. We shall generalize them by introducing $\mathbf{f_0} = \mathbf{s_p}$: $\mathbf{f_0}$ is the orthogonal component of $\mathbf{s_p}$ on nothing. Figure 8.4 gives an idea of the evolution of $\mathbf{f_m}$ for $m = 0,1,2$ in a three-dimensional space. It is clear that forward prediction error vectors are generally not orthogonal to one another.

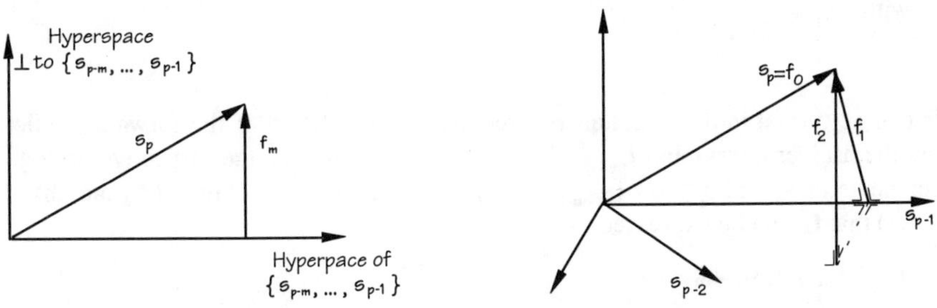

Fig. 8.4. Forward prediction error vectors $\mathbf{f_m}$.

Just like forward predictions of order m, which apply $\mathbf{s_p}$ on the vector subspace of the m previous vectors $\{\mathbf{s_{p-1}}, .., \mathbf{s_{p-m}}\}$, it is possible to define *backward predictions* of order m, as the linear decomposition of $\mathbf{s_{p-m-1}}$ on the vector subspace of the m following vectors $\{\mathbf{s_{p-m}}, .., \mathbf{s_{p-1}}\}$. The associated backward prediction error vectors $\mathbf{g_m}$, the norms of which are the square roots of the *backward prediction error energies* β_m, are thus the orthogonal components of $\mathbf{s_{p-m-1}}$ on $\{\mathbf{s_{p-m}}, .., \mathbf{s_{p-1}}\}$.They are shown in Fig. 8.5 for $m = 0,1,2$, where $\mathbf{g_0}$ was set to $\mathbf{s_{p-1}}$.

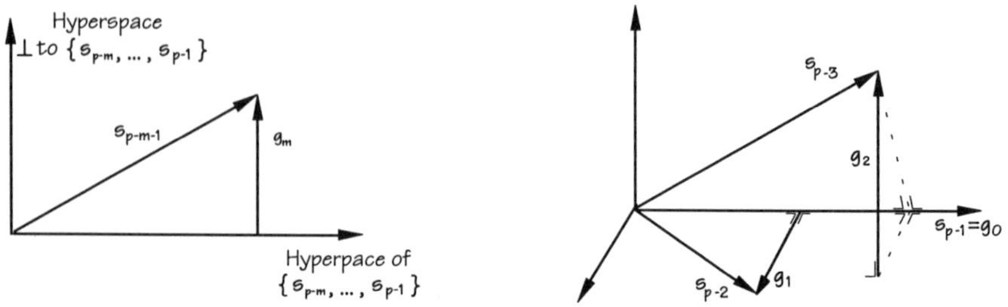

Fig. 8.5. Backward prediction error vectors g_m.

This time, $\{\mathbf{g_0}, \mathbf{g_1}, ..., \mathbf{g_m}\}$ are orthogonal to one another, by construction. What is more, they span the same subspace as $\{\mathbf{s_{p-m}}, .., \mathbf{s_{p-1}}\}$, so that $\{\mathbf{g_0}, \mathbf{g_1}, ..., \mathbf{g_m}\}$ is an orthogonal basis of $\{\mathbf{s_{p-m}}, .., \mathbf{s_{p-1}}\}$. The corresponding matrix, **G**, is orthogonal.

This feature results in a considerable simplification in their computation: $\mathbf{g_m}$ is also the orthogonal component of $\mathbf{s_{p-m-1}}$ on $\{\mathbf{g_0}, \mathbf{g_1}, ..., \mathbf{g_{m-1}}\}$, so that successive backward prediction error vectors can be computed recursively, according to the so-called *Grahm-Schmidt orthogonalization process*:

$$\mathbf{g_0} = \mathbf{s_{p-1}}$$

$$\mathbf{g_m} = \mathbf{s_{p-m-1}} - \sum_{i=0}^{m-1} g_i\, \mathbf{g_i} \qquad \text{for } m = 1...p-1 \tag{8.19}$$

$$\text{with} \quad g_i = \frac{\mathbf{s_{p-m-1}}\ \mathbf{g_i}}{\|\mathbf{g_i}\|^2}$$

This can be turned into a simultaneous recursive computation of the forward prediction errors themselves. Assuming $\mathbf{f_{m-1}}$ and $\mathbf{g_{m-1}}$ are known—that is, the respective orthogonal components of $\mathbf{s_p}$ and $\mathbf{s_{p-m}}$ on $\{\mathbf{s_{p-m-1}}, .., \mathbf{s_{p-1}}\}$—it is easy to see (Fig. 8.6, generalized in Fig. 8.7) that $\mathbf{f_m}$ can be expressed as[6]

$$\mathbf{f_m} = \mathbf{f_{m-1}} + k_m\, \mathbf{g_{m-1}} \tag{8.20}$$

in which k_m is the opposite of the projection of $\mathbf{f_{m-1}}$ on $\mathbf{g_{m-1}}$:

---

[6]Such efficient recursive backward-forward decompositions are not particular to linear prediction. It is striking to notice that the idea of computing the successive orthogonal components of a given signal on increasingly complex subspaces V_j (with $V_{j+1} \subset V_j$) by using the previous decomposition on V_{j-1} and the orthogonal complement of V_j in V_{j-1} is precisely the key point of the wavelet-based multiresolution approximations of Mallat (1989).

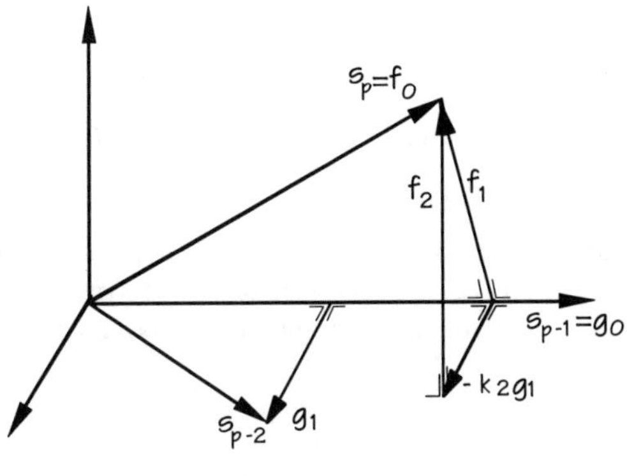

Fig. 8.6. Building f_2 from f_1 and g_1.

$$k_m = -\frac{\mathbf{f_{m-1}}\,\mathbf{g_{m-1}}}{\|\mathbf{g_{m-1}}\|^2} \tag{8.21}$$

$\{k_0, k_1,.., k_p\}$ are the well-known PARCOR coefficients. They actually have a wider geometrical interpretation than (8.21). Since $\mathbf{f_{m-1}} = \mathbf{f_{m-2}} + k_{m-1}\,\mathbf{g_{m-2}}$, according to (8.21), and $\mathbf{g_{m-2}} \perp \mathbf{g_{m-1}}$, equation (8.21) can be rewritten as:

$$k_m = -\frac{\mathbf{f_{m-2}}\,\mathbf{g_{m-1}}}{\|\mathbf{g_{m-1}}\|^2}$$

and so on, down to

$$k_m = -\frac{\mathbf{f_0}\,\mathbf{g_{m-1}}}{\|\mathbf{g_{m-1}}\|^2} = -\frac{\mathbf{s_p}\,\mathbf{g_{m-1}}}{\|\mathbf{g_{m-1}}\|^2} \tag{8.22}$$

In other words, PARCOR coefficients are the negative of the projections of $\mathbf{s_p}$ on the backward prediction error vectors[7,8]. As a result, $\{k_1, k_2, ..., k_p\}$ carry the same information as $\{a_1, a_2, ..., a_p\}$; they can be interchanged by a simple change of co-ordinates in the $\{\mathbf{s_{p-m}}, .., \mathbf{s_{p-1}}\}$ subspace[9].

---

[7]Since $\mathbf{f_m}$, $\mathbf{f_{m-1}}$ and $\mathbf{g_{m-1}}$ form a rectangular triangle, expression (8.20) also leads to the well-known energy interpretation $\alpha_m^2 = \alpha_{m-1}^2 - k_m^2\,\beta_{m-1}^2$.

[8]Deriving spectral interpretations of k_i is more problematic, except for k_1: since $\mathbf{g_0}=\mathbf{s_{p-1}}$, k_1 looks like the negative and normalized version of the first autocorrelation coefficient of the analysis frame. This is why it is often referred to as an indicator of the spectral tilt: the more k_1 is negative, the more important the high-frequency components are.

[9]Equation (8.24) reveals a linear relationship between **a** and **k**. Linearity, however, is only ensured as long as $\mathbf{\Phi}$ (and thus **G**) is left untouched. Since the $\mathbf{\Phi}$ matrix originates from the *N-1* first samples of the analysis frame, only one degree of freedom remains, corresponding to a change of *s(N)*. As a consequence, linear operations on **a** vectors, such as interpolations, do not generally result in corresponding linear operations on PARCOR vectors.

$$\mathbf{s'_p} = \sum_{i=1}^{p} -a_i \mathbf{s_{p-i}} = \sum_{i=1}^{p} -k_i \mathbf{g_i} \tag{8.23}$$

$$\mathbf{\Phi a} = \mathbf{Gk}, \qquad \text{where } \mathbf{k} \text{ is the vector } [-k_1, -k_2, ..., -k_p] \tag{8.24}$$

Equation (8.23) is illustrated in Fig. 8.7.[10]

Such a k_i to a_i transformation is hardly useful in practice (as far as synthesis is concerned), since, as we shall see, a lattice version of the synthesis filter $1/A_p(z)$ can be derived in the particular case of the autocorrelation method; its coefficients are precisely the k_i.

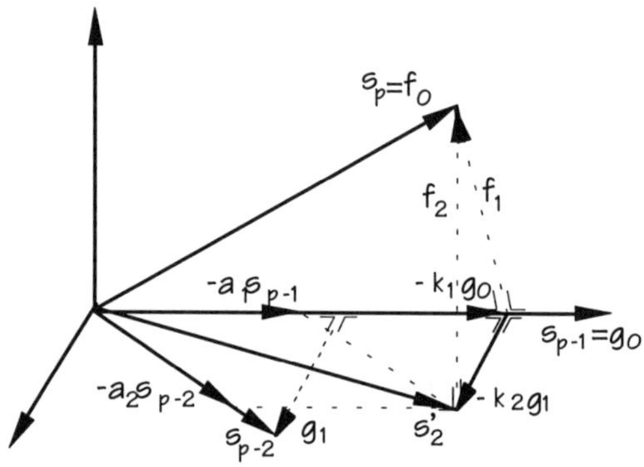

Fig. 8.7. Prediction and PARCOR coefficients are different expressions of the same decomposition process.

8.2.6. The autocorrelation method, with the lattice algorithm

Up to now, the very peculiar relationship among $\mathbf{s_i}$ vectors, which originates in the way they have been extracted from the analysis frame, has not been taken into account. It does, however, lead to important simplifications, provided the covariance matrix $\mathbf{\Phi}$ is further assumed to be Toeplitz symmetric.

Let us first define a *Toeplitz sequence of vectors* $\{\mathbf{s_0}, \mathbf{s_1}, ..., \mathbf{s_p}\}$ as any sequence of vectors constructed as shown in Fig. 8.3, and the covariance matrix of which is Toeplitz symmetric. Clearly, such vectors have identical norms, and the scalar product $\mathbf{s_i}\ \mathbf{s_j}$ depends only on *i-j*. They are shifted versions from one another, in the sense that any Toeplitz sequence is entirely defined by its last vector $\mathbf{s_p}$ and its order *p*, from which any other vector $\mathbf{s_i}$ is deduced by a circular right shift of order *p-i* of its coordinates.[11]

[10]The identity $k_m=a_m$ also clearly appears: the ratio of $-a_m\ \mathbf{s_{p-m}}$ on $\mathbf{s_{p-m}}$ is the same as the ratio of $-k_m\ \mathbf{g_{m-1}}$ on $\mathbf{g_{m-1}}$.

[11]More exactly, two Toeplitz sequences can be derived from any vector $\mathbf{s_p}$, a direct and a reverse one, as in {(1 2 3 4),(4 1 2 3),(3 4 1 2),(2 3 4 1)} and {(1 2 3 4),(-4 1 2 3),(-3 -4 1 2),(-2 -3 -4 1)}.

Obviously, the Toeplitz sequence introduced in Section 8.2.3 is generated by $\mathbf{s_p}$ = $(s(0),...,s(N\text{-}2p\text{-}1),0,..., 0)$.

Because PARCOR coefficients are computed on the basis of the angles between vectors in a Toeplitz sequence and of their norm, they are insensitive to orthogonal transformations applied on the initial Toeplitz sequence. They are therefore related to *Toeplitz figures* (in an *N*-dimension space) rather than to Toeplitz sequences themselves. As such, subsequences of vectors in a Toeplitz sequence (which are themselves Toeplitz sequences) have identical PARCOR coefficients, provided their order is the same. Furthermore, reversing a Toeplitz sequence has no effect on the related PARCORs.

These properties are turned into account in the Levinson algorithm to speed up the computation of the backward prediction error vectors. As a matter of fact, it suffices to notice that

1. Computing $\mathbf{g_m}$—the orthogonal component of $\mathbf{s_{p\text{-}m\text{-}1}}$ on $\{\mathbf{s_{p\text{-}m\text{-}2}}, .., \mathbf{s_{p\text{-}1}}\}$—is equivalent to computing the orthogonal component of $\mathbf{s_{p\text{-}m}}$ on $\{\mathbf{s_{p\text{-}m\text{-}1}}, .., \mathbf{s_p}\}$, which we shall denote as $\mathbf{g'_m}$, and circularly shifting the coordinates of the result one element to the right. This coordinate shift is an orthogonal transformation described by a *circulant* matrix $\mathbf{S}$.

2. Computing $\mathbf{g'_m}$ can itself be achieved with a similar strategy as in equation (8.20), that is by using the orthogonal component of $\mathbf{s_{p\text{-}m}}$ on $\{\mathbf{s_{p\text{-}m\text{-}1}}, .., \mathbf{s_{p\text{-}1}}\}$, i.e. $\mathbf{g_{m\text{-}1}}$, and adding it a correction vector parallel to the orthogonal component of $\mathbf{s_p}$ on $\{\mathbf{s_{p\text{-}m\text{-}1}}, .., \mathbf{s_{p\text{-}1}}\}$, i.e. to $\mathbf{f_{m\text{-}1}}$. In other words, $\mathbf{g'_m}$ can be considered as the *m*th *forward* prediction error vector of the Toeplitz sequence $\{\mathbf{s_p}, \mathbf{s_{p\text{-}1}}, .., \mathbf{s_{p\text{-}m}}\}$. This sequence has the same PARCORs as its reversed sequence, $\{\mathbf{s_{p\text{-}m}}, \mathbf{s_{p\text{-}m+1}}, .., \mathbf{s_p}\}$—the k_i computed with (8.21).

Remarks 1 and 2 lead to the very important relation:

$$\mathbf{g_m} = \mathbf{S}\,\mathbf{g'_m} = \mathbf{S}\,(\mathbf{g_{m-1}} + k_m\,\mathbf{f_{m-1}}) \tag{8.25}$$

which, when used in combination with (8.20) and (8.21), makes it possible to solve the linear prediction problem with a single recursion over the prediction order. This method is known as the *lattice algorithm*, since it is completely defined by the lattice inverse filter of Fig. 8.8, which should be understood as a sequence of blockwise operations that progressively produce forward and backward prediction error vectors from a given analysis frame.

This filter will be used in Section 8.6 to derive a lattice structure for the LPC synthesis filter. It can be shown (see Boite and Kunt, 1987) that its stability is preserved if and only if the modulus of every k_m coefficient is lower than one. This is always the case in the autocorrelation method, since, as has just been said, $\mathbf{f_{m\text{-}1}}$ and $\mathbf{g_{m\text{-}1}}$ in equation (8.21) have identical norms.

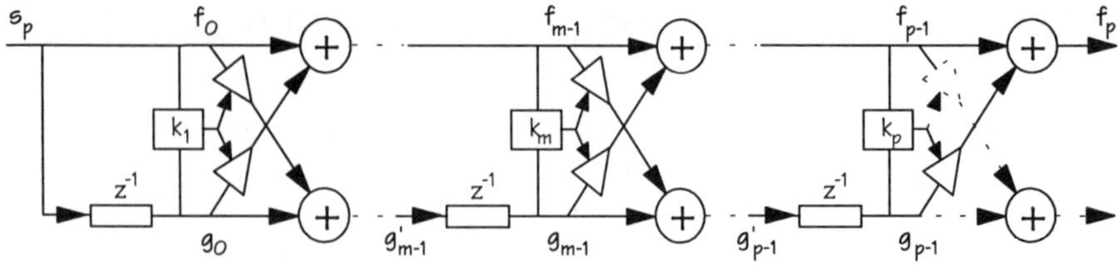

Fig. 8.8. The lattice inverse filter.

8.2.7. The autocorrelation method, with the Levinson, Schur, and split algorithms

A number of algorithms can be derived from the previous one, to further speed up computations when prediction error vectors are not needed. Since these algorithms do not yield significant insight into synthesis problems, they will only be outlined here.

In the *Levinson algorithm*, $\mathbf{f}_{m-1}$ and $\mathbf{g}_{m-1}$ in equation (8.21) are evaluated on $\{\mathbf{s}_{p-m},\mathbf{s}_{p-m+1},..,\mathbf{s}_p\}$, so that k_m is expressed as

$$k_m = -\frac{\sum_{i=0}^{m-1} a_i^{m-1}\mathbf{s}_{p-m}\,\mathbf{s}_{p-i}}{\|\mathbf{g}_{m-1}\|^2} \tag{8.26}$$

in which a_i^{m-1} is the *i*th prediction coefficient of order *m-1*. Scalar products in the numerator of (8.26) are directly found in the covariance matrix $\mathbf{\Phi}$. Its denominator is computed recursively by

$$\|\mathbf{g}_m\|^2 = \|\mathbf{g}_{m-1}\|^2 (1-k_m{}^2) \tag{8.27}$$

Equations (8.20) and (8.25) are themselves turned into a compact recursion on the prediction coefficients:

$$a_i^m = a_i^{m-1} + k_m a_{m-i}^{m-1}, \quad \text{for i = 1..}m\text{ - 1,} \quad \text{and } a_m^m = k_m \tag{8.28}$$

In contrast, the *Schur algorithm* does not require the computation of prediction coefficients. Equation (8.21) is expressed as

$$k_m = -\frac{\mathbf{f}_{m-1}\,\mathbf{s}_{p-m}}{\mathbf{g}_{m-1}\,\mathbf{s}_{p-m}} \tag{8.29}$$

in which the numerator and denominator are respectively obtained from (8.20) and (8.25) as

$$\begin{aligned} \mathbf{f}_{m-1}\,\mathbf{s}_{p-m} &= \mathbf{f}_{m-2}\,\mathbf{s}_{p-m} + k_{m-1}\mathbf{g}_{m-2}\,\mathbf{s}_{p-m} \\ \mathbf{g}_{m-1}\,\mathbf{s}_{p-m} &= \mathbf{g}'_{m-1}\,\mathbf{s}_{p-m+1} = \mathbf{g}_{m-2}\,\mathbf{s}_{p-m+1} + k_{m-1}\mathbf{f}_{m-2}\,\mathbf{s}_{p-m+1} \end{aligned} \tag{8.30}$$

More recently, it has been noticed by Delsarte and Genin (1987) that the Levinson recursion could be further simplified, by expressing it in terms of $\mathbf{p_i} = \mathbf{f_i}+\mathbf{g_i}$ *or* $\mathbf{q_i} = \mathbf{f_i}-\mathbf{g_i}$ instead of $\mathbf{f_i}$ *and* $\mathbf{g_i}$. The resulting *split Levinson algorithm* saves about half the operations of the original Levinson method. The idea has also been used to derive *split lattice* and *split Schur* algorithms.

8.2.8. Line spectrum pairs

Line spectrum pairs (LSP) have been first introduced by Itakura (1975) as interesting alternatives to PARCOR coefficients for speech coding. They are related to PARCORs and prediction coefficients by a non-linear relationship which, as we shall see, also provides them with interesting synthesis properties.

Let us denote by $P(z)$ and $Q(z)$ the filters that respectively convert $\mathbf{s_p}$ into $\mathbf{f_p}+\mathbf{g_p}$ and $\mathbf{f_p}-\mathbf{g_p}$, respectively. It is easy to show that

$$P(z) = A_p(z) + z^{-(p+1)} A_p(z^{-1}) \quad \text{and} \quad Q(z) = A_p(z) - z^{-(p+1)} A_p(z^{-1}) \tag{8.31}$$

so that $P(z)$ and $Q(z)$ are simply the reciprocal and antireciprocal parts of $A_p(z)$. As such, they possess the noticeable property that their complex conjugate zeros[12] lie on the unit circle (hence their denomination as *spectral lines*). Let us respectively denote them as $\left\{e^{j\phi_1}, e^{j\phi_2}, ..., e^{j\phi_m}, e^{-j\phi_1}, e^{-j\phi_2}, ..., e^{-j\phi_m}\right\}$ and $\left\{e^{j\theta_1}, e^{j\theta_2}, ..., e^{j\theta_m}, e^{-j\theta_1}, e^{-j\theta_2}, ..., e^{-j\theta_m}\right\}$. What is more, $A_p(z)$ can be shown to be stable if and only if the zeros of $P(z)$ and $Q(z)$ alternate—that is, if $0 < \phi_1 < \theta_1 < \phi_2 < \theta_2 < ... < \pi$.

It is found experimentally that spectral lines can often be associated by *pairs* (ϕ_i, θ_i), the closeness of which tends to indicate the existence of a formant at frequency $(\phi_1+\theta_1)/2$. This property can be taken into account in the context of speech coding, and allows a bit rate reduction of about 30 percent in comparison with PARCORs.

Efficient a_i-to-LSP and LSP-to-a_i computation methods can be found in (Kabal and Ramachandran, 1986). They approximately double the computational complexity of LPC analysis.

8.3. Database compression

PARCOR coefficients are extensively used for low-bit rate speech coding, in the form of log area ratios (LARs). It has been shown indeed that the sensitivity of PARCORs to coding is non-uniform: the acoustic result of a given error on the value of a k_i depends on value of this k_i. LARs are defined as

$$LAR_i = \log_{10} \frac{1+k_i}{1-k_i} \tag{8.32}$$

Their sensitivity to coding is more uniform. They also exhibit more interesting interpolation properties (see Section 8.5).

---

[12] Real zeros can also be encountered at $z=+1$ or -1, depending p is even or odd.

A nearly transparent[13] 2400 bits per second (bps) coder has been standardized by the U.S. Department of Defense (Tremain, 1982), for a sampling frequency f_s= 8000 Hz and p=10. Yet error correction codes are included. Linear predictive coding has seldom been explored for f_s= 16000 Hz (wide-band conditions). In Fuldseth *et al.* (1991), 20 LPC coefficients are coded in the form of LARs with 4250 bps. Keeping the same bit allocation for P, σ, and the V/UV decision as in LPC-10 (respectively, 6, 5, and 1 bits every 22.5 ms) leads to an overall rate of about 4400 bps for p=18.

LSPs allow a slightly higher compression ratio, since the aforementioned 2400 bps can easily be reduced to about 2000 bps (Crosmer and Barnwell, 1985). For f_s= 16000 Hz, Roy and Kabal (1991) announce 2400 bps for coding 16 LSPs, so that a nearly transparent coding for p=18 could reasonably be achieved with about 3800 bps. White (1990) reports on an average rate of 4500 bps for 18 LSP coefficients in the context of a variable rate CELP coder.

8.4. Prosody matching with the AR model

The AR model lends itself to very simple prosody matching operations.

Imposing the pitch is straightforward, since it explicitly appears in the parameters of the model. A simultaneous modification of the amplitude σ is performed, following (8.17), so as to keep the synthesis energy equal to its original value σ_0:

$$\sigma = \sigma_0 \sqrt{\frac{P_0}{P}} \tag{8.33}$$

Changing the duration of segments from their original values to the desired ones is performed by applying a simple duration factor to each frame—by modifying the temporal coverage of each set of filter coefficients. An additional parameter is introduced in the synthesis stream: the synthesis frame length N_s (in samples).

8.5. Segment concatenation

Segment concatenation is performed as in equations (7.1) and (7.2) on vectors $[LAR_1, LAR_2, ..., LAR_p, \sigma_p]$: transitions are smoothed by linearly distributing spectral envelope mismatches among several frames before and after the segment limits (typically 4 or 5 on the whole, which results in a concatenation area of about 50 ms).

As far as TTS synthesis is concerned, the main advantage of the lattice structure in comparison with the canonical form of $1/A_p(z)$ lies in the good interpolation properties of the LAR coefficients. A striking example is given in Fig. 8.9, which compares the spectral results of linear interpolations of a_i, k_i, LAR_i and LSP_i. LAR, PARCOR, and LSP coefficients allow smooth and stable transitions between sounds, while prediction coefficients interpolations often result in unnatural, not to say unstable, formant

[13] It is transparent in the sense that speech synthesized with coded PARCORs can hardly be distinguished from speech synthesized with uncoded ones. In both cases, however, synthetic speech sounds buzzy in comparison with natural speech.

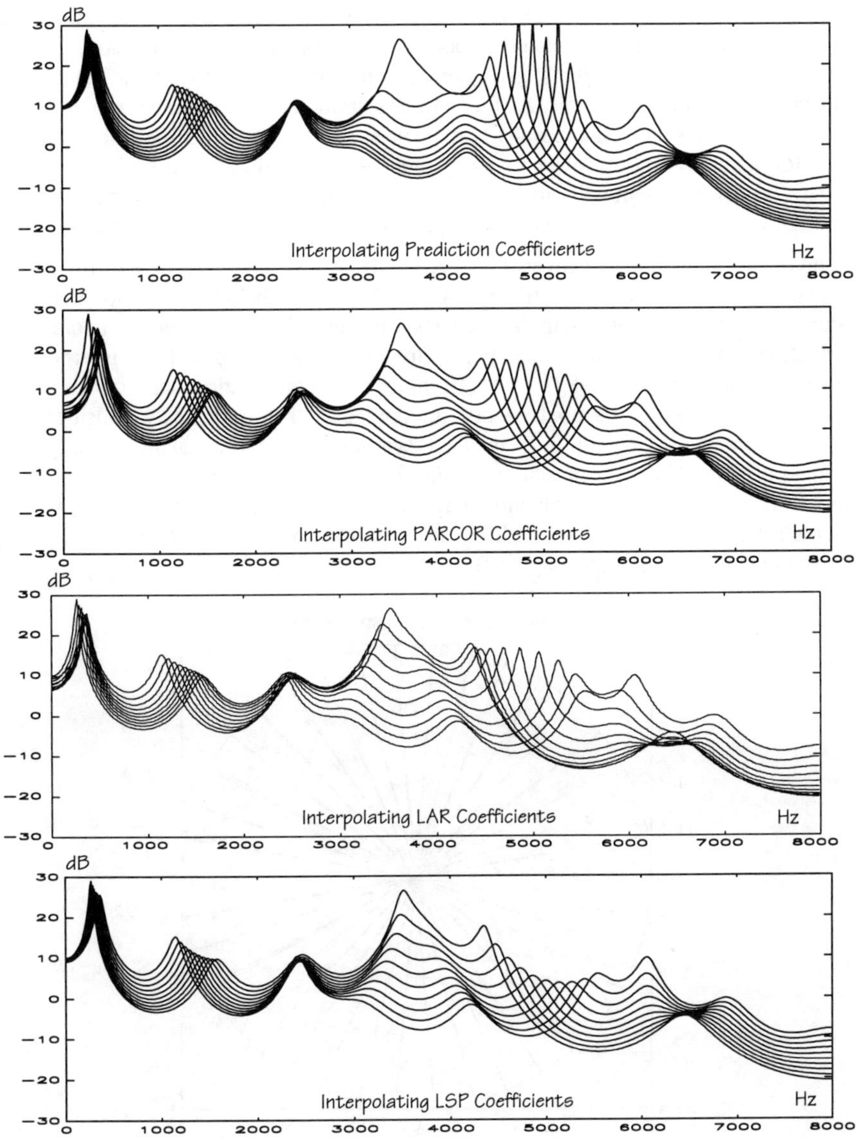

Fig. 8.9. Linear interpolation of filter parameters reveals that even in the case of small acoustic differences, linear concatenation laws may lead to nonoptimal results. Interpolation is performed between /l/ phones, respectively encountered in diphones /al/ and /lo/ and the resulting filter transfer functions are plotted (p=18).

transitions, since nothing keeps the zeros of $A_p(z)$ from moving out of the unit circle. LSPs are generally found to produce smoother transitions than LARs or PARCORs. In practice, all sets of coefficients give perceptually similar results in the context of segments concatenation synthesis. As far as filter stability is concerned, *interpolating* two sets of stable LARs or LSPs always results in stable filters, since the corresponding PARCORs maintain a norm lower than one and LSP tracks do not cross. This property, however, is not automatically maintained when linear *smoothing* is considered, as described by equations (7.1) and (7.2), though it happens to be the case in practice.

One should not conclude, however, that PARCOR or LSP coefficients lead to optimal interpolations. Transitions are still encountered, in which formant amplitudes are unnaturally affected when their frequencies are modified. In order to avoid such imperfections, the prediction filter poles are sometimes proposed as interpolation parameters (Papamichalis, 1987). [14] This requires the computation and storage of the roots of $A(z)$ instead of its coefficients, an operation that is scarcely considered in vocoders while it is easy to face in TTS synthesis since the database analysis is performed off-line. However, as shown in Fig. 8.10, grouping filter poles by pairs is not always a straightforward operation and interpolating them can be disastrous. Such pairing problems are not encountered with LSP's (see Fig. 8.10, on which LSPs for the left and right segments are almost superimposed).

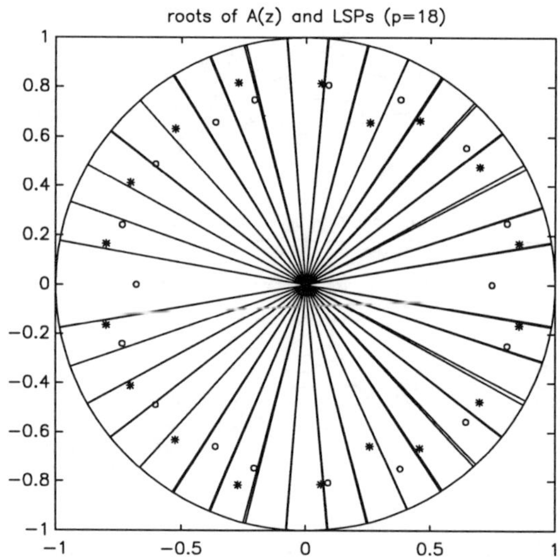

Fig. 8.10. Roots of A(z) for diphones /ɔv/ (o) and /vɛ/ (*), for p=18. LSP sets are presented in the form of radial lines, for they are very close to one another.

[14]They also obviously make it possible to control the filter stability.

Some authors have also proposed to interpret prediction filter poles in terms of the amplitudes, frequencies, and bandwidths of spectral amplitude peaks, and to smooth their trajectories over diphones (see Depalle *et al.*, 1990, for example). However, since the reverse transformation (from peaks to prediction coefficients) is not easy to perform, they implement a frequency domain synthesizer (see Chapter Nine).

8.6. Speech synthesis

Synthesis is performed as in Fig. 8.1, in which parameters are received from the segment concatenation module. Given the bad interpolation properties of the prediction coefficients, the IIR synthesis filter $1/A_p(z)$ is never implemented, either in its direct or transposed form. PARCOR coefficients can be used directly in a lattice synthesis filter derived from the lattice inverse filter (Boite and Kunt, 1987) (Fig. 8.11), which is easily found to require ($4p$-2) elementary operations (additions and multiplications) per sample—70 operations for order 18. On the other hand, an LSP filter has been proposed by Sugamura and Itakura (1986). Its computational load is about 10 percent higher than for the lattice filter.

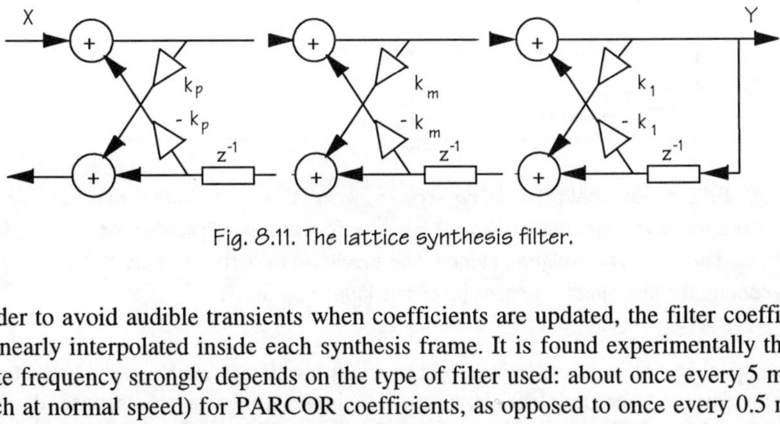

Fig. 8.11. The lattice synthesis filter.

In order to avoid audible transients when coefficients are updated, the filter coefficients are linearly interpolated inside each synthesis frame. It is found experimentally that the update frequency strongly depends on the type of filter used: about once every 5 ms (for speech at normal speed) for PARCOR coefficients, as opposed to once every 0.5 ms for the LSP parameters. Thus, since LSP parameters have no significant advantage over PARCOR coefficients, except their slightly higher compression ratio, and given the additional computational load they imply, they have never been used so far for TTS synthesis.

8.7. Segmental quality

The selection of an analysis method with respect to optimal quality is not simple. The covariance method theoretically is a better spectral estimator than the autocorrelation method (Kay, 1988); it is generally admitted that it provides more naturalness, as far as pitch asynchronous analysis is concerned (Hunt *et al.*, 1989). Differences, however, tend to vanish when $N>>p$.

Even though, it should be noticed that the resulting prediction coefficients are not free of extrinsic modeling errors. This is easy to show in the case of periodic signals. Let us imagine that a truly autoregressive periodic signal—that is, a signal that has specifically been created with the synthesizer of Fig. 8.1 with an impulse train as excitation—is analyzed by the covariance method. Clearly, as shown in Fig. 8.12, the excitation sequence is generally not orthogonal to the prediction subspace.[15] This is not astonishing, since the algorithm minimizes the norm of $\mathbf{f_p}$, with no constraint on its actual waveform.[16] In the case of unvoiced autoregressive signals, errors still appear due to the approximation of **C** in (8.10).

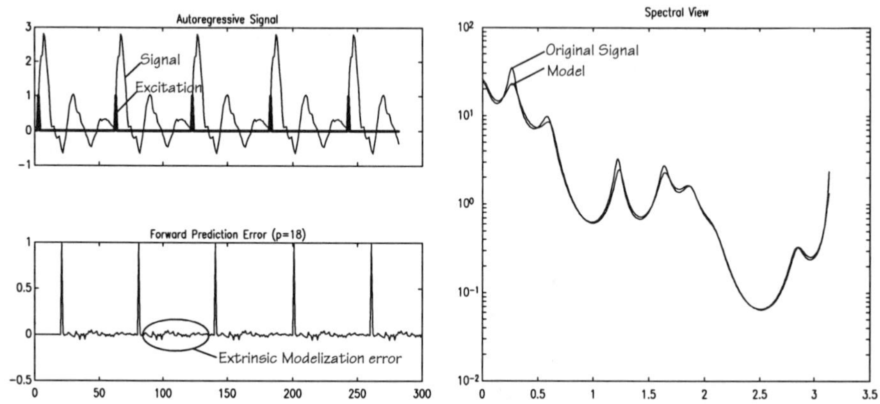

Fig. 8.12. Extrinsic modeling errors with the covariance method. An autoregressive signal is analyzed, but the forward prediction error does not equal the excitation signal. Hence, the prediction coefficients do not exactly account for the spectral envelope of the input signal.

Finally, specific extrinsic errors may occur when nasals are analyzed. These introduce zeros (antiformants) around 1000 Hz in the vocal tract transmission response, due to the existence of two parallel radiation sources (Flanagan, 1972, p. 79). The inability of the AR model to take them into account (which itself constitutes an intrinsic modeling error) often prevents the actual nasal formants from being correctly estimated (Fig. 8.13).

---

[15]More precisely, as can be deduced from the geometrical interpretation we have introduced, extrinsic errors on the filter parameters can be avoided when frames are chosen *between* pulses. *Pitch synchronous* linear prediction algorithms have been derived to take this feature into account. The rigorous pitch extraction they require prevents them from being widely used. Extrinsic errors may also vanish when frames include several pitch periods, provided impulse responses do not significantly overlap. This is hardly the case in speech, particularly for female utterances, which exhibit a high pitch frequency.

[16]Besides, as mentioned in Section 8.1, imposing the temporal evolution of the prediction error would be nonsense, unless an additional all-pass filter is introduced in the processing for phase equalization. This point is further discussed in Section 8.9.

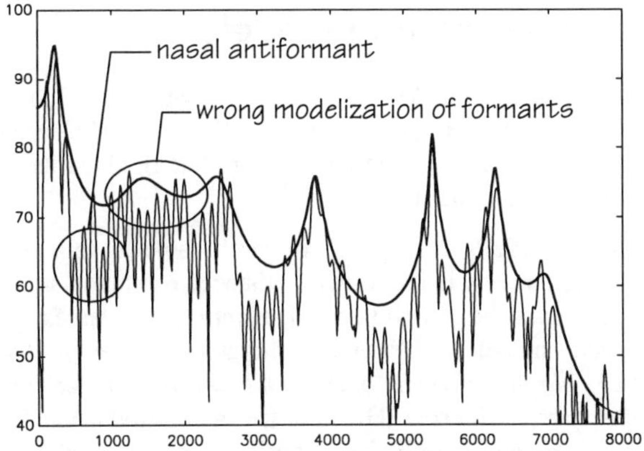

Fig. 8.13 Wrong detection of formant frequencies and bandwidths when analyzing consonant [n] because of the impossibility to account for the nasal anti-formant.

Intrinsic errors are far more troublesome. It is generally admitted that AR models, as presented in Fig. 8.1 are incapable of producing high-quality speech (whatever analysis algorithm is employed), for many reasons:

- It was first shown by Atal and David (1979), and further studied by Kang and Everett (1985), that the pulse-like nature of the LPC excitation introduces buzziness into synthetic speech, due to its resulting high peak factor. A significantly higher quality was obtained by providing the AR filter input with synthetic excitation, the amplitude spectrum of which was kept constant with time and frequency, while its phase was given a frequency-dependent (but time-constant) value.
- A rough V/UV decision often cannot efficiently describe speech sounds. The concept of open or closed glottis itself seems unrealistic, since substantial air flow remains almost at any time (Hunt *et al.*, 1978). Experiments with a frequency-selective mixed-source model were first performed by Makhoul *et al.* (1978) and revealed a certain "fullness" that was absent in the binary V/UV synthesis.
- The glottal waveform of the Fant model is clearly oversimplified. What is more, the absence of coupling between the vocal and nasal tracts and the sub-glottal cavities is a rough approximation. Some inter-dependency actually appears periodically, which seems to be responsible for sub-glottal formants and antiformants (Ananthapadmanabha *et al.*, 1982; Deller, 1982).
- The vocal tract configuration sometimes evolves very quickly, as is the case for the burst of plosives. Such highly transient sounds are unrealistically modeled with analysis frames of about 30 ms.
- Finally, as mentioned above, the AR model is inherently incapable of producing high-quality nasals.

8.8. Advanced production models

Despite all these impediments, the AR model has been widely used in narrow band TTS systems—that is, for f_S= 8000 Hz (see the SPARTE system of Courbon and Emerard, 1982 for French, and Van Coile and Martens 89 for Dutch), given its simplicity and the exceptional data compression ratio it allows.

It was not until the mid-1980s that speech research came to further developments in linear prediction models, mainly in the form of enhanced excitation signals. The resulting advanced approaches can be roughly divided into two classes: linear prediction techniques that directly model the LPC residual signal (such as *multipulse excited LP*, *codebook excited LP*, *harmonic coding of the residual waveform*, or *residue excited LP*), and so-called *glottal inverse filtering* (GIF) ones, which introduce additional production constraints in the LP equations (8.6) by imposing a glottal model for the excitation waveform. Some of these developments have been applied to TTS systems.

The *multipulse linear prediction* technique was introduced by Atal and Remde (1982). It depicts the LP residual signal as a sequence of pulses, the positions and amplitudes of which are typically adjusted by an analysis-by-synthesis algorithm, so as to minimize (in the least-squares sense) the perceptual difference between the MP-LPC synthesis signal and the original one. Even though only suboptimal analysis algorithms can de derived, the segmental quality of MP-LPC speech is very high. However, given its rather indirect connection to pitch and duration, the MP excitation signal does not easily lend itself to prosody modifications. Waveform editing techniques have been proposed for that purpose (Caspers and Atal, 1983; Varga and Fallside, 1987), and further enhanced by the use of the TD-PSOLA algorithm in the MP-LPC-PSOLA approach of Moulines and Charpentier (1988), which is to our knowledge the only published implementation of an MP model in a TTS system (although Joe Olive had also tested it in the early eighties at the Bell Labs).

In the case of *harmonic coding of the residual waveform*, which is simply an extension of the sinusoidal coding of Mac Aulay and Quatieri (1986) to the LP residual signal, excitation is modeled as a sum of sinusoids, the frequencies of which are harmonically related to produce voiced sounds. There again, amplitudes, phases and frequencies are estimated with an analysis-by-synthesis algorithm. Its successful application to TTS synthesis has been reported by Del Carpio (1989).

In both cases, however, it is striking to notice that it is possible to get rid of the intrinsic segmental quality limitations due to the use of an auto-regressive model by applying similar formalisms directly on speech, rather than on LP residuals. Indeed, time-domain algorithms are now being widely used *per se* (see Chapter Ten), and hybrid harmonic/stochastic methods provide a basis for high-quality TTS systems (see Chapter Nine).

Glottal inverse filtering techniques are more extensively discussed in the next section.

8.9. Glottal inverse filtering

The computation of the glottal excitation waveform directly from the speech signal has drawn considerable attention over the last twenty years, i.e. since linear prediction of speech has been widely used. The numerous papers related to it show two main trends.

For some applications, the main interest of the glottal wave resides in its most significant parameter: the instant of glottal closure (see Strube, 1974; Wong *et al.*, 1979; Cheng and O'Shaughnessy, 1989, for a review). Being the only physical reference instant available in any voiced speech waveform, its precise estimation has become a fundamental requirement for some speech coding and synthesis techniques, such as the recently developed and very promising TD-PSOLA.

For other analysis-synthesis schemes, known as glottal inverse filtering (GIF) algorithms, the complete glottal waveform is under consideration (see Hedelin, 1984; Milekovic, 1986; Isaksson and Millnert, 1989, for a review). These schemes are derived from classical linear prediction, from which they may be seen as a step forward, based on a more realistic modeling of the glottal velocity waveform. As we shall see, their algorithmic complexity is enormous, in comparison with classical linear prediction. However, this computational load is not a major drawback in speech synthesis from unrestricted text, where the analysis is performed once for all for the constitution of a speech segment database.

Successful synthesis experiments have been reported (Fujisaki and Ljungqvist, 1987; Lobo and Ainsworth, 1989), using a complex GARMA model (glottal autoregressive mean adjusted). In the sequel, we shall rather focus on the more tractable (yet very complex) glottal autoregressive (GAR) model.

8.9.1. The glottal autoregressive (GAR) model

The GIF equations used here are a somewhat modified version of equation (8.7):

$$\mathbf{s_p} = \sum_{i=1}^{p} -a_i \mathbf{s_{p-i}} + \mathbf{u_p}(\theta) + \mathbf{f_p} \tag{8.34}$$

or in matrix form:

$$\mathbf{s_p} = \Phi \mathbf{a} + \mathbf{u_p}(\theta) + \mathbf{f_p} \tag{8.35}$$

where $\mathbf{u_p}(\theta)$, which accounts for the glottal excitation, is a non-linear function of θ, and $\theta = [\theta_1, \theta_2, ..., \theta_r]^T$ is the vector of r glottal waveform parameters.

Equations (8.35) implicitly introduce a glottal autoregressive (GAR) speech production model, in which speech is created by filtering an excitation signal $x(n)$, composed of a glottal pulse component $u(n)$ and a noise component $f(n)$, with an all pole filter (Fig. 8.14). As a matter of fact, coming back the Fant model in the case of voiced sounds and posing

$$E(z)G(z)L(z)=U(z) \tag{8.36}$$

in equation (8.2) results in

$$S(z) = U(z)\frac{1}{A_p(z)} \tag{8.37}$$

which differs from (8.3) in that no strict assumption has been made on the glottal volume velocity waveform.

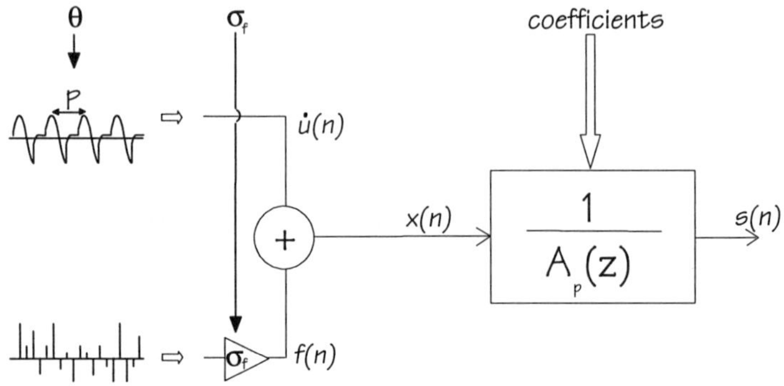

Fig. 8.14. The glottal auto-regressive (GAR) speech production model.

It should be clear from equation (8.36) that the components of $\mathbf{u_p}$ are supposed to be the time derivatives of the glottis volume velocity *u(n),* denoted as $\dot{u}(n)$[17]. They are defined as a piecewise non-linear function of several parameters θ_i, which define the vector $\boldsymbol{\theta}$ of glottal parameters. Several models have been proposed for the glottal waveform (see Fujisaki and Ljungqvist, 1986, for a review). Fujisaki's model, which happens to cover all the other previous ones, describes the glottal waveform as a piecewise polynomial function of time (composed of four chunks), which depends on seven parameters (Fig. 8.15).

As in the linear prediction framework, system (8.35) has to be solved, on $\mathbf{a}$ and $\boldsymbol{\theta}$ simultaneously, under least-squares constraints, by minimizing the norm of the new error $\mathbf{f_p}$. This can be understood as a quest for the point of the hyper-surface $\mathbf{u_p}(\boldsymbol{\theta})$ which verifies that its orthogonal component on the prediction vectors $\{\mathbf{s_0},...,\mathbf{s_{p-1}}\}$ is minimized in norm (Fig. 8.16).

This generally leads to optimal prediction coefficients different from those obtained through classical linear prediction of $\mathbf{s_p}$: the prediction coefficients of the GAR model determine the vocal tract transfer function rather than the global short-term spectrum envelope of the signal produced, as it is the case for the classical prediction coefficients. On the other hand, if the hyper-surface $\mathbf{u_p}(\boldsymbol{\theta})$ is well chosen—that is, if the glottal waveform model is adequate—it is clear that the new error norm could be smaller than the classical one.

---

[17] Notice also that A_p in (8.37) is not the same as the one of (8.3), since A_p in (8.37) is obtained from pre-emphasized speech.

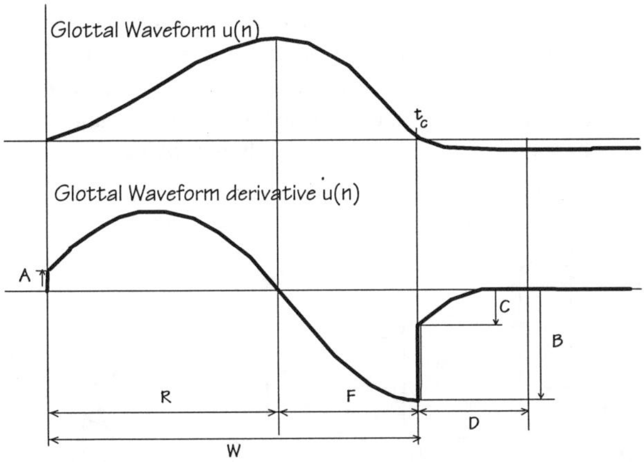

Fig. 8.15. The glottal volume velocity waveform proposed by Fujisaki and Ljungqvist (after Fujisaki and Ljungqvist , 1986) involves three amplitude parameters (*A*, *B*, *C*), and four temporal ones (a rise time *R*, fall time *F*, delay *D* + the glottal closure instant t_c). It is characterized by two slope discontinuities, the main one being positioned at the glottal closure instant (copyright ©1986, IEEE.)

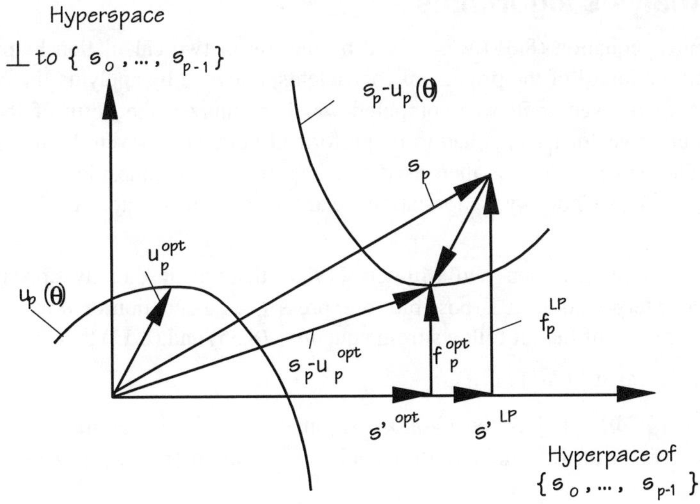

Fig. 8.16. A (very schematic) geometrical interpretation of the GIF problem. Classical linear prediction and GIF results are respectively indicated by ^{LP} and ^{opt} superscripts.

It is important to note that, since the model constrains the LP residual waveform by forcing it to verify $\mathbf{u}_p(\theta)$ and given the fact that the LP filter has minimum phase (for its transfer function has no zero), non-linear phase distortions, which can hardly be avoided in practical speech recording systems, cannot be taken into account by the GIF equations unless phase compensation is applied to the signal before its analysis (Fig. 8.17a).

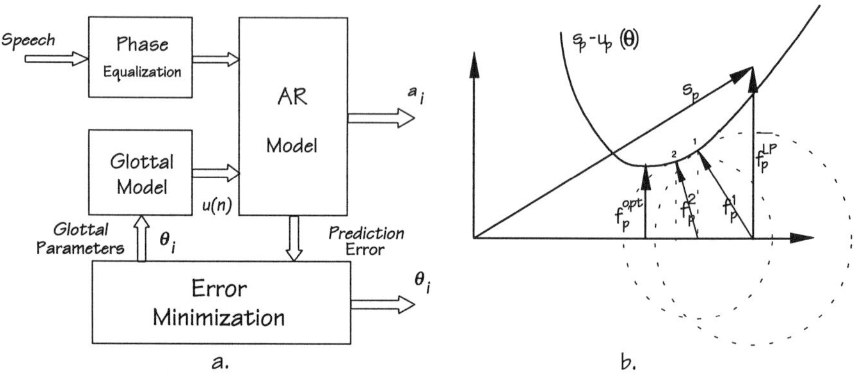

Fig. 8.17. Classical resolution of the GIF equations. Successive prediction vectors are obtained by least-squares estimation; successive glottal vectors θ^i are given by iterative resolution of a nonlinear problem.

8.9.2. Analysis algorithms

Until recently, equation (8.35) was solved by the use of two calculation loops. From a first approximation $\mathbf{a}^1$ of the prediction coefficients, obtained by applying the covariance method on $\mathbf{s}_p$, the vector θ^1 was computed which minimized the norm of the forward prediction error vector $\mathbf{f}_p$. This had to be performed iteratively, given the non-linearities in $\mathbf{u}_p(\theta)$. The value of θ^1 was then used to compute the decomposition of $\mathbf{s}_p$-$\mathbf{u}_p(\theta^1)$ on $\{\mathbf{s}_0,\ldots,\mathbf{s}_{p-1}\}$, with a Cholesky factorization, leading itself to $\mathbf{a}^1$, etc., as shown on Fig. 8.17b.

It was noticed by Isaksson and Millnert (1989) that, as it directly appears in our geometrical interpretation, it is possible to express $\mathbf{f}_p$ as a function of θ and $\{\mathbf{s}_0,\ldots,\mathbf{s}_{p-1}\}$ only.[18] As a matter of fact, it follows from equations (8.35) and (8.11) that

$$\mathbf{f}_p = [\mathbf{I}\text{-}\mathbf{\Phi}\ (\mathbf{\Phi}^T\mathbf{\Phi})^{-1}\ \mathbf{\Phi}^T]\ (\mathbf{s}_p\text{-}\mathbf{u}_p) \quad (8.38)$$

where $[\mathbf{I}\text{-}\mathbf{\Phi}\ (\mathbf{\Phi}^T\mathbf{\Phi})^{-1}\ \mathbf{\Phi}^T]$ is the $(N\text{-}p \times N\text{-}p)$ matrix of the linear transformation that associates a vector of $\Re^{N-p}$ and its orthogonal component on $\{\mathbf{s}_0,\ldots,\mathbf{s}_{p-1}\}$; we shall write it $\mathbf{P}_\perp$.

[18]Such a decomposition between the linear and non-linear parts of a problem is actually a simplified version of a more general framework developed in Cadzow (1990).

The iterative resolution can then be performed through quickly converging algorithms, such as the Gauss-Newton one. Indeed, if we express

$$\mathbf{f_p}(\theta) = \mathbf{P}_\perp (\mathbf{s_p}\text{-}\mathbf{u_p}) = [0,0,...,0]^T , \tag{8.39}$$

to the first order, we find

$$\mathbf{f_p}(\theta^{i+1}) = \mathbf{f_p}(\theta^i) + \mathbf{J_f}(\theta^i)\,(\theta^{i+1}\text{-}\theta^i) \tag{8.40}$$

in which $\mathbf{J_f}(\theta^i)$ is the Jacobian matrix of $\mathbf{f_p}$ with respect to θ, computed at $\theta=\theta^i$:

$$\mathbf{J_f}(\theta^i) = \begin{bmatrix} \left.\frac{\partial f(0)}{\partial \theta_1}\right|_{\theta=\theta^i} & \left.\frac{\partial f(0)}{\partial \theta_2}\right|_{\theta=\theta^i} & \cdots & \left.\frac{\partial f(0)}{\partial \theta_r}\right|_{\theta=\theta^i} \\ \left.\frac{\partial f(1)}{\partial \theta_1}\right|_{\theta=\theta^i} & \left.\frac{\partial f(1)}{\partial \theta_2}\right|_{\theta=\theta^i} & \cdots & \left.\frac{\partial f(1)}{\partial \theta_r}\right|_{\theta=\theta^i} \\ \cdots & \cdots & \cdots & \cdots \\ \left.\frac{\partial f(N-p-1)}{\partial \theta_1}\right|_{\theta=\theta^i} & \left.\frac{\partial f(N-p-1)}{\partial \theta_2}\right|_{\theta=\theta^i} & \cdots & \left.\frac{\partial f(N-p-1)}{\partial \theta_r}\right|_{\theta=\theta^i} \end{bmatrix} \tag{8.41}$$

System (8.40) has again to be solved under least-squares constraints (again with the Cholesky algorithm). For each iteration, a new value of $\mathbf{J_f}(\theta^i)$ must be computed. Given the linear relationship (8.39) between $\mathbf{f_p}(\theta)$ and $(\mathbf{s_p}\text{-}\mathbf{u_p})$, $\mathbf{J_f}(\theta^i)$ can be obtained by

$$\mathbf{J_f}(\theta^i) = \mathbf{P}_\perp \mathbf{J_{s\text{-}u}}(\theta^i) = -\,\mathbf{P}_\perp \mathbf{J_u}(\theta^i) \tag{8.42}$$

where $\mathbf{J_u}(\theta^i)$ is the Jacobian matrix of $\mathbf{u_p}$ with respect to θ, computed at $\theta=\theta^i$. Its computation directly results from partial derivations of $\mathbf{u_p}(\theta)$ with respect to each θ_i parameter. Since $\mathbf{P}_\perp$ depends only on Φ, it can be computed once for all iterations. However, its direct estimation by its expression in (8.38) must unquestionably be avoided, since it requires the inversion of the $(p \times p)$ covariance matrix, and two $(p \times p) \times (p \times N\text{-}p)$ matrix products. What is more, its storage would require to memorize $(N\text{-}p)^2$ real numbers. A more convenient solution consists of computing $\mathbf{P}_\perp$ through the Householder transform.

A more drastic way to simplify the parameter estimation process is to relax the dependence between prediction coefficients and the actual glottal waveform. Analysis is then performed in two steps: prediction coefficients are first estimated with classical LP algorithms, and a glottal waveform is then directly modeled from the prediction residual obtained after inverse filtering the speech signal. Hence, the algorithm depicted in Figure 8.17a is limited to its first iteration (see Wong *et al.*, 1979; Strik *et al.*, 1993, for instance).

8.9.3. Segmental quality : further research issues

The analysis of synthetic GIF speech (speech that had previously been synthesized with a GIF model) with the Isaksson and Millnert approach is found to give accurate values of θ after an average of five iterations, even with very coarse initial values. In contrast, convergence cannot strictly be ensured for real speech. This directly results from the fact that the optimization of a given glottal pulse is highly sensitive to the results obtained for the previous ones. What is more, it confirms that the algorithm is very sensitive to phase distortion (Hedelin, 1989). Hence the use of glottal inverse filtering methods in which the glottal waveform is assumed to be independent of the prediction coefficients. Whatever method used, however, several authors have reported that the introduction of a glottal excitation model in the linear prediction equations sensibly increased the naturalness of the resulting synthetic speech (see Fries, 1994, and Oliveira, 1993, for instance). Further improvements could probably be obtained by using a more realistic glottal model (see Cranen and Schroeter, 1995, for instance)

8.10. Conclusions

In this chapter, we have examined the pros and cons of autoregressive (AR) models, whether glottal (GAR) or not, in the context of high-quality text-to-speech synthesis. We could summarize them as follows:

- AR models exhibit good linear interpolation and smoothing properties, provided the transfer function of the prediction filter is expressed in terms of PARCOR or LSP coefficients.
- They allow very high data compression ratios. A complete diphone database (less than three minutes of speech digitized at 16 kHz) submitted to an LPC analysis of order 18 can be stored in about 100 Kbytes of memory.
- They exhibit straightforward prosody matching capabilities.
- As far as synthesis is concerned, their computational load is not excessive. It does, however, require the use of a fast chip. Regarding analysis complexity, AR models are reasonably simple, which is far from being the case of GAR ones. Yet this is not a major concern for TTS synthesis.
- The main drawback of autoregressive models resides in their poor segmental quality, which is revealed by LPC copy synthesizers. We do believe that GAR models are intrinsically able to remedy this impediment.

References

ALEXANDER, S.T., (1986), *Adaptive Signal Processing: Theory and Applications*, Springer-Verlag, New-York, pp.123-141.

ANANTHAPADMANABHA, T.V., and B. VEGNANARAYANA, (1979), "Epoch Extraction of Voiced Speech", *IEEE Transactions on Acoustics, Speech, and Signal Processing*, vol. ASSP-27, pp. 309-319.

ATAL, B.S., and N. DAVID, (1979), "On Synthesizing Natural-Sounding Speech by Linear Prediction", *Proceedings of the International Conference on Acoustics, Speech, and Signal Processing 79*, pp. 44-47.

ATAL, B.S., and J.R. REMDE, (1982), "A New Model of LPC Excitation for Producing Natural-Sounding Speech at Low Bit Rates", *Proceedings of the International Conference on Acoustics, Speech, and Signal Processing 82*, vol. 1, pp. 614-617.

BOITE, R., and M. KUNT, (1987), *Traitement de la parole*, Presses polytechniques romandes, Lausanne.

CADZOW, J. A., (1990), "Signal Processing via Least Squares Error Modeling", *IEEE ASSP Magazine*, October, pp 12-31.

CASPERS, B., and B.S. ATAL, (1983), "Changing Pitch and Duration in LPC Synthesized Speech Using the Multipulse Excitation", *Journal of the Acoustical Society of America,* vol. 73, S5.

CHENG, Y.M., and D. O'SHAUGNESSY, (1989), "Automatic and Reliable Estimation of Glottal Closure Instant and Period", *IEEE Transactions on Acoustics, Speech, and Signal Processing*, vol. 37, n°12.

COURBON, J.L., and F. EMERARD, (1982), "SPARTE : a Text-to-Speech Machine Using Synthesis by Diphones", *Proceedings of the International Conference on Acoustics, Speech, and Signal Processing 82*, pp. 1597-1600.

CRANEN, B., and J. SHROETER, (1995), "Modeling a Leaky Glottis", *Journal of Phonetics*, 23, pp. 165-177

CROSMER, J.R., and T.P. BARNWELL, (1985), "A Low Bit Rate Segment Vocoder Based on Line Spectrum Pairs", *Proceedings of the International Conference on Acoustics, Speech, and Signal Processing 85*, n° 7.2.

DEL CARPIO, J., (1989), *Réalisation d'un Système de Synthèse de la Parole à partir d'un Texte en Langue Espagnole*, PhD dissertation, Faculté Polytechnique de Mons.

DELLER, J.R., (1982), "Evaluation of Larygeal Dysfunction Based on Features of an Accurate Estimate of the Glottal Waveform", *Proceedings of the International Conference on Acoustics, Speech, and Signal Processing 82*, pp. 759-762.

DELSARTE, F., and Y. GENIN, (1987), "On the Splitting of Classical Algorithms in Linear Prediction Theory", *IEEE Transactions on Acoustics, Speech, and Signal Processing*, vol. ASSP-35, n°5, pp. 645-653.

DEPALLE, P., X. RODET, and G. POIROT, (1990), "Energy and Articulation Rules for Improving Diphone Speech Synthesis", *Proceedings of the First ESCA Workshop on Speech Synthesis,* Autrans, pp. 47-50.

FANT, G., (1960), *Acoustic Theory of Speech Production*, Mouton, The Hague, 1960.

FLANAGAN, J.L., (1972), *Speech Analysis, Synthesis, and Perception*, Springer Verlag, Berlin, p. 77.

FRIES, G., (1994), "Hybrid Time- and Frequency-Domain Speech Synthesis with Extended Glottal Source Generation", *Proceedings of the International Conference on Acoustics, Speech, and Signal Processing 94*, Adelaide, vol. I, pp. 581-584.

FUJISAKI, H., and M. LJUNGQVIST, (1986), "Proposal and Evaluation of Models for the Glottal Source Waveform", *Proceedings of the International Conference on Acoustics, Speech, and Signal Processing 86*, Tokyo, 31.2, pp. 1605-1607.

FUJISAKI, H., and M. LJUNGQVIST, (1987), "Estimation of Voice Source and Vocal Tract Parameters Based on ARMA Analysis and a Model for the Glottal Source Waveform", *Proceedings of the International Conference on Acoustics, Speech, and Signal Processing 87*, 15.4, pp. 637-640.

FULDSETH, A., E. HARBORG, F.T. JOHANSEN, and J.E. KNUDSEN, (1991), "A Real-Time Implementable 7 kHz Speech Coder at 16 kbits/s", *Proceedings of Eurospeech 91,* pp. 897-900.

GOLUB, G.H., and C.F. VAN LOAN, (1989), *Matrix Computations*, Johns Hopkins University Press, London, p. 243.

HEDELIN, P., (1984), "A Glottal LPC Vocoder", *Proceedings of the International Conference on Acoustics, Speech, and Signal Processing 84*, pp. 1.6.1-1.6.4.

HEDELIN, P., (1988), "Phase Compensation in All-Pole Speech Analysis", *Proceedings of the International Conference on Acoustics, Speech, and Signal Processing 88*, 8.10, pp. 339-342.

HEDELIN, P., (1989), "High-Quality LPC Vocoding", *Proceedings of the International Conference on Acoustics, Speech, and Signal Processing 89*, 9.9, pp. 465-468.

HESS, W., (1992), "Pitch and Voicing Determination", in *Advances in Speech Signal Processing,* S. Furui, M. Sondhi, eds., Dekker, New York, pp. 3-48.

HUNT, M.J., J.S. BRIDLE, and J.N. HOLMES, "Interactive Digital Inverse Filtering and its Relation to Linear Prediction Methods", *Proceedings of the International Conference on Acoustics, Speech, and Signal Processing 78*, pp. 15-19.

HUNT, M.J., D.A. ZWIERZYNSKI, and R.C. CARR, (1989), "Issues in High-Quality LPC Analysis ans Synthesis", *Proceedings of Eurospeech 89*, vol.2, pp. 348-351.

ISAKSSON, A., and M. MILLNERT, (1989), "Inverse Glottal Filtering Using a Parameterized Input Model", *Signal Processing*, n°18, pp. 435-445.

ITAKURA, F., and S. SAITO, (1969), "Speech Analysis-Synthesis System based on the Partial Autocorrelation Coefficients", *Proceedings of the Acoustical Society of Japan Meeting*.

ITAKURA, F., (1975), "Line Spectrum Representation of Linear Predictive Coefficients of Speech Signals", *Journal of the Acoustical Society of America*, vol. 57, S35(A), 1975.

KABAL, P., and P. RAMACHANDRAN, (1986), "The Computation of Line Spectral Frequencies using Tchebyshev Polynomials", *IEEE Transactions on Acoustics, Speech, and Signal Processing*, vol. ASSP-34, n°6.

KANG, G.S., and S.S. EVERETT, (1985), "Improvement of the Excitation Source in the Narrow-Band Linear Prediction Vocoder", *IEEE Transactions on Acoustics, Speech, and Signal Processing*, Vol. ASSP-33, n°2, pp. 377-386.

KAY, S. M., (1988), *Modern Spectral Estimation*, Prentice Hall Signal Processing Series, p. 222.

KROON, P., (1985), *Time-Domain Coding of (Near) Toll Quality Speech at Rates below 16 kB/s*, Ph.D. dissertation, Techniche Hogeschool, Delft.

LOBO, A.P., and W.A. AINSWORTH, (1989), "Evaluation of a Glottal ARMA Modeling Scheme", *Proceedings of Eurospeech 89*, vol. 2,pp. 027-030.

MAC AULAY, R.J., and T.F. QUATIERI, (1986), "Speech Analysis/Synthesis based on a Sinusoidal Representation", *IEEE Transactions on Acoustics, Speech, and Signal Processing*, vol. 34, pp. 744-754.

MAKHOUL, J., R. VISWANATHAN, R. SHARTZ, and A.W.F. HIGGINS, (1978), "A Mixed-Source Model for Compression and Synthesis", *Proceedings of the International Conference on Acoustics, Speech, and Signal Processing 78*, pp. 163-166.

MALLAT, S. G., (1989), "Multiresolution Approximations and Wavelet Orthonormal Bases of $L^2(R)$", *Transactions of the American Mathematical Society*, vol. 315, n°1, pp. 69-87.

MARKEL, J.D., and A.H. GRAY Jr, (1976), *Linear Prediction of Speech*, Springer Verlag, New York, pp. 10-42.

MILEKOVIC, P., (1986), "Glottal Inverse Filtering by Joint Estimation of an AR System with a Linear Input Model", *IEEE Transactions on Acoustics, Speech, and Signal Processing*, vol. 34, n°1.

MOULINES, E., and F. CHARPENTIER, (1988), "Diphone Synthesis Using a Multipulse LPC Technique", *Proceedings of the seventh FASE International Conference*, Edinburgh, pp. 47-53.

OLIVEIRA, L.C., (1993), "Estimation of Source Parameters by Frequency Analysis", *Proceedings of Eurospeech 93*, Berlin, vol. 1, pp. 99-102.

PAPAMICHALIS, P.E., (1987), *Pratical approaches to speech coding*, Prentice Hall.

ROY, G., and P. KABAL, (1991), "Wideband Speech Coding at 16 kbits/sec", *Proceedings of the International Conference on Acoustics, Speech, and Signal Processing 91*, vol. 1, pp. 17-20.

STRIK, H., B. CRANEN, and L. BOVES, (1993), "Fitting an LF-model to Inverse Filter Signals", *Proceedings of Eurospeech 93*, Berlin, vol. 1, pp. 103-106.

STRUBE, H.W., (1974), "Determination of the Instant of Glottal Closure from the Speech Wave", *Journal of the Acoustical Society of America*, vol. 56, pp. 1625-1629.

SUGAMURA, N., and F. ITAKURA, (1986), "Speech Analysis and Synthesis Methods Developed at ECL in NTT - From LPC to LSP", *Speech Communication*, June, pp. 199-215.

TREMAIN, T.E., (1982), "The Government Standard Linear Predictive Coding Algorithm : LPC-10", *Speech Technology*, vol. 1, n°2, April, pp. 40-49.

VAN COILE, B.W., and J.P. MARTENS, (1989), "Dutch Text-to-Speech Aids for the Vocally Handicapped", *Proceedings of Eurospeech 89*, vol.1, pp. 590-593.

VARGA, A., and F. FALLSIDE, (1987), "A Technique for Using Multipulse Linear Predictive Speech Synthesis in Text-to-Speech Type Systems", *IEEE Transactions on Acoustics, Speech, and Signal Processing*, vol. 35, pp. 386-387.

WHITE, S., (1990), "Codeur CELP à Débit Variable : Application au Codage de Diphones", *Proc. 19èmes Journées d'Etudes sur la Parole*, Montréal.

WONG, D.Y., J.D. MARKEL, and A.H. GRAY Jr, (1979), "Least Squares Glottal Inverse Filtering of the Acoustic Speech Waveform", *IEEE Transactions on Acoustics, Speech, and Signal Processing*, vol. ASSP-27, n°4, pp. 350-353

Chapter Nine

HYBRID HARMONIC / STOCHASTIC SYNTHESIS

Where the noise came from, she couldn't make out: the air seemed full of it.

Lewis Carroll, *Through the Looking Glass*

Among the many signal analysis and synthesis techniques, spectral techniques take first place. This is widely due to the fact that imaginary exponentials are eigenfunctions of the convolution operation, which completely describes linear time invariant systems in the time domain, and to the existence of Fast Fourier transforms (FFT) which make it possible to swap between time and frequency spaces in a fraction of the time needed to perform any other orthogonal decomposition. These techniques have been widely applied to speech processing. As a matter of fact, one can reasonably assume that speech sounds are produced by a linear system. Frequency domain approaches therefore lead to a very important class of phenomenological models—that is, *spectral models*.

The point is that sounds evolve over a period of time, which addresses the more general (and complex) issue of time-varying linear systems spectral analysis (Almeida & Tribolet 83). Hopefully, speech presents the essential feature that, if ever one finds a way to determine its instantaneous transfer function[1], it appears to remain approximately[2] constant over a period of time that covers several times its impulse response (about 30 ms). This interesting property, mostly due to physiological constraints, is often denoted in stochastic terms as *quasi-stationarity*. It results in considerable simplifications in the analysis algorithms and allows spectral models to exhibit important data compression capabilities, as well as simple and efficient interpolation properties. Both these features are very appreciated in TTS systems.

When applied to voiced signals, spectral models naturally give birth to harmonic ones, in which spectral measures are interpreted as the result of the analysis of harmonically related sine waves:

---

[1]Which could easily be computed if the vocal and nasal tracts were artificially frozen.

[2]The approximation being very rough for plosives and semiconsonants.

$$\tilde{s}(t) = \sum_i a_i(t)\cos(\phi_i(t)) \quad \text{, with} \qquad \phi_i(t) = \phi_i(0) + \int_0^t i\omega_O(\tau)\,d\tau \tag{9.1}$$

or alternatively, in terms of discrete signals: [3]

$$\tilde{s}(n) = \sum_i a_i(n)\cos(\phi_i(n)) \quad \text{, with} \qquad \phi_i(n) = \phi_i(0) + \sum_{j=1}^{i} \left[i\,\omega_O(j)\right] T \tag{9.2}$$

where i denotes the index of harmonics, and $a_i(t)$ and $i\ \omega_O(t)$ are their amplitudes and instantaneous frequencies, and T is the sampling time interval.

Such descriptions can be extended to unvoiced sounds in many ways. Sinusoidal modeling techniques, which relax the harmonicity constraint in unvoiced frames, were first simultaneously proposed by MacAulay and Quatieri (1984) and by Almeida and Silva (1984), and further refined in MacAulay and Quatieri (1986) and later on in Marques and Almeida (1989), who first proposed a frequency-varying sinusoidal model. Their overall segmental quality is reported as being higher than with classical LPC synthesizers, but some buzziness remains, especially when synthesizing voiced fricatives, due to the remaining voiced/unvoiced decision at the frame level, as for classical LPC models.[4]

Alternatively, a series of advanced hybrid harmonic/stochastic models have recently appeared, which transfer V/UV decisions to frequency bands or even transform them into more flexible frequency dependent V/UV ratios. The resulting additional degrees of freedom allow a better simulation of mixed sounds, for which fricative noise and periodic vibration of the vocal folds are not mutually exclusive, as shown on Fig. 9.1. These models are now reviewed, and their performances are analyzed in the context of TTS synthesis.

9.1. Hybrid models

The idea of associating noise in time-varying proportions to periodical signals, so as to produce natural sounding speech sounds, first appeared in (Fujimura, 1968), far before the development of sinusoidal models. It was not, however, until the appearing of the multiband excited (MBE) model (Griffin, 1987) that high quality speech was synthesized on the basis of mixed sinusoidal/stochastic signals. A parallel hybrid approach was proposed by Abrantes *et al.* (1991). Both models are essentially similar; the main difference between them lies in the analysis algorithms provided to best compute their parameters. Similarly, a hybrid model of the *LPC error* signal has been proposed by Wéry *et al.* (1989), the analysis stage of which has lately been turned into the powerful MBE analysis method (Deketelaere *et al.*, 1992). More recently, d'Alessandro *et al.* (1995) have proposed a powerful periodic/aperiodic decomposition algorithm for the

---

[3]For the sake of clarity, we shall keep expressing mathematical ideas in terms of continuous-time expressions when possible.

[4]It is striking, however, to see how efficiently sinusoidal models are able to produce purely unvoiced sounds, with a restricted number of sinusoids.

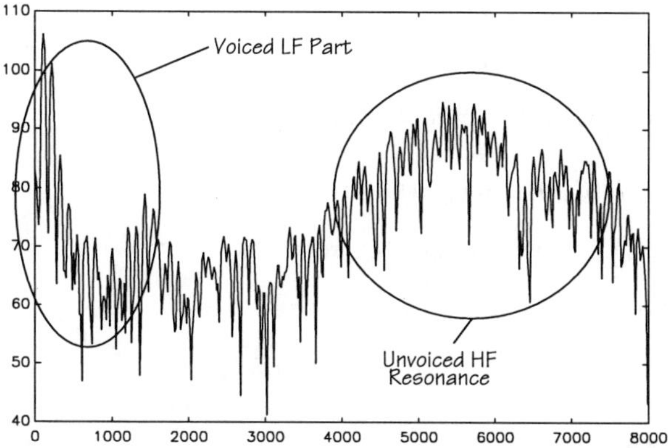

Fig. 9.1. Amplitude spectrum of a realization of the vowel [z]. Stochastic components are mainly encountered in high frequencies, above 4 kHz. As a consequence, flexible V/UV models are still of a higher importance in the context of high-quality TTS systems than for low-bit rate vocoders, for which they were originally developed.

LPC error signal and studied its performances (Darsinos *et al.*, 1995). Finally, hybrid expressions have also recently appeared in the context of music synthesis (Serra and Smith, 1990). The related model is somewhat peculiar in the sense that it also releases the harmonicity constraint, so that it should rather be classified as *hybrid sinusoidal/stochastic model.*

Hybrid models basically add a stochastic component $\tilde{s}_r(t)$ to equation (9.1):

$$\tilde{s}(t) = \tilde{s}_p(t) + \tilde{s}_r(t) = \sum_i a_i(t)\cos(\phi_i(t)) + \tilde{s}_r(t) \tag{9.3}$$

in which, for the sake of simplicity, $\tilde{s}_r(t)$ is often assumed to be Gaussian in theoretical developments while uniform distributions are generally used in synthesis algorithms. In the Gaussian case, it is completely determined by its power density spectrum $\tilde{S}_{rr}(\omega)$. As for sinusoids amplitudes and frequencies, $\tilde{S}_{rr}(\omega)$ is supposed to vary slowly with time, according to the quasi-stationarity hypothesis.

The signal described by equation (9.3) can alternatively be understood as the output of the filter model of Fig. 9.2 (Abrantes *et al.*, 1991), in which $u(t)$ is a white Gaussian random signal with unitary variance, and $e_p(t)$ is a periodic excitation given by

$$e_p(t) = \sum_i \cos(\theta_i(t)) \qquad \text{with} \qquad \theta_i(t) = \int_0^t i\omega_O(\tau)\,d\tau \tag{9.4}$$

Clearly, all the parameters of the model, except one (ω_O), have been included in the spectral shaping filters $h_p(t,\tau)$ and $h_r(t,\tau)$, in which t denotes a slow time dependency of the filters parameters. These are therefore estimated and stored once every 10 ms, and temporal linear interpolation on the resulting spectra is assumed to describe speech in a realistic way. If, in a given short time interval centered around time t^i, the transfer functions of the filters are given by

$$H_p(t^i,\omega) = \left|H_p(t^i,\omega)\right| \exp(j\xi_p(t^i,\omega)) \tag{9.5}$$

$$H_r(t^i,\omega) = \left|H_r(t^i,\omega)\right| \exp(j\xi_r(t^i,\omega)) \tag{9.6}$$

then the parameters of the model can be expressed as

$$\omega_0{}^i = \omega_0(t^i) \tag{9.7}$$

$$a_j{}^i = \left|H_p(t^i, j\omega_0{}^i)\right| \quad , \quad \phi_j^i = \xi_p(t^i, j\omega_0{}^i) \tag{9.8}$$

$$\tilde{S}_{rr}^i(\omega) = \left|H_r(t^i,\omega)\right|^2 \tag{9.9}$$

where upper indices refer to frames while lower ones refer to harmonics.[5]

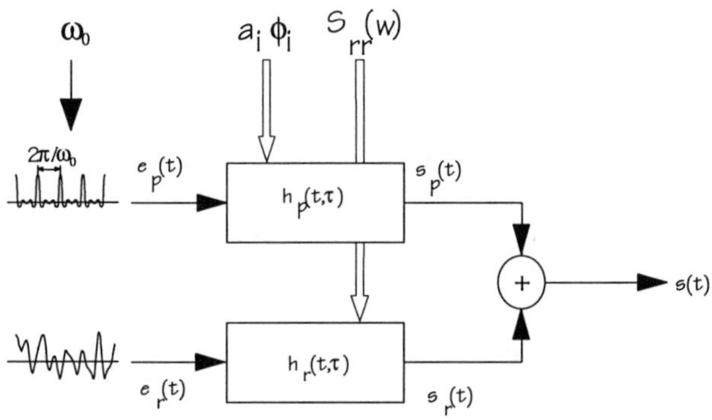

Fig. 9.2. The hybrid harmonic/stochastic model.

$H_r(t^i,\omega)$ itself can be parametrized in many ways, in the form of autocorrelation or PARCOR coefficients, for example. In genuine hybrid H/S models, however, it is roughly approximated as the superposition of N_{NBN} band limited noises with contiguous flat spectra, the central frequencies and bandwidths of which are either linearly distributed on the frequency axis (Abrantes *et al.*, 1991; Griffin, 1987) or adjusted on the critical bands of the human ear, so as to maintain a constant psycho-acoustical low distortion on the frequency range (Holger and Kolpatzik, 1991). As we shall see in Section 9.6, related to hybrid synthesis, such descriptions better suit synthesis algorithms in respect to computational load.

---

[5] Notice that such a harmonic sampling holds only under stationarity assumptions. This is why we stated that « t denotes a slow time dependency ». Additionally, it is assumed that the source components do not change too abruptly.

9.2. Hybrid analysis

9.2.1. Spectral analysis of speech

The basic tool for spectral analysis of quasi-stationary signals is the *short-term Fourier transform* (STFT) (see Allen, 1977, for details). For analog signals, it is given by

$$x(t) \leftrightarrow X(t,\omega) = \int_{-\infty}^{\infty} x(\tau)\, w(t-\tau)\, e^{-j\omega\tau}\, d\tau \tag{9.10}$$

in which $w(t)$ is a finite duration weighting window.

In practice, it is given by a discrete (DFT) form, on a finite set of values of $\omega = \{0, \omega_N, 2\omega_N, \ldots, (N-1)\,\omega_N\}$:

$$x(n) \leftrightarrow X(n,k) = \sum_{i=0}^{N-1} x(i)\, w(n-i)\, e^{-jk\omega_N i} \tag{9.11}$$

which in turn is easily computed with an FFT algorithm, provided N is restricted to 2^m and $\omega_N = 2\pi/N$. For speech signals, a real FFT algorithm can be used, which provides only the $N/2+1$ independent values of $X(n,k)$.

Spectral analysis consists in repeating STFT computations for successive values of n and interpreting the results. Its ability to effectively describe the original signal, as well as to allow fruitful interpretations, obviously depends on the window length L (in seconds), the window shift S (in seconds), the window shape, and the number of frequency bins N. Their respective influence are now briefly summarized.

STFTs are traditionally given two interpretations. The *filter-bank* interpretation consists of considering $X(t, \omega_C)$ for fixed ω_C. From equation (9.10), the result is clearly the output of a filter, the impulse response of which is $w(t)$, to the excitation $x(t)e^{-j\omega_c t}$. Conversely, the *spectral convolution* interpretation examines $X(t_C, \omega)$ for fixed t_C. Equation (9.10) then describes $X(t_C, \omega)$ as the Fourier transform of the product of $x(t)$ and $w(t_C-t)$, which results in a spectral convolution of their respective spectra.

From the filter-bank interpretation, the maximum frequency of $X(t, \omega_C)$ is limited by the filter bandwidth B, so that Shannon's theorem imposes a sampling frequency $1/S$ greater or equal to $2B$. Since this restriction is independent of ω_C, it theoretically conditions the window shift. In the particular case of speech, however, this constraint is often relaxed, given the strong redundancy speech signals exhibit.

The spectral convolution interpretation is of a much higher importance in H/S analysis. In the case of a single sinusoid with frequency ω_0 and amplitude A, it produces, instead of the expected spectral line at $\omega=\omega_0$, the complete window spectrum at $\omega=\pm\omega_0$:

$$X(0,\omega)\big|_{x(t)=A\cos(\omega_0 t+\varphi)} = \frac{A}{2}\left[W(\omega-\omega_0)\, e^{j\varphi} + W(\omega+\omega_0)\, e^{-j\varphi}\right] \tag{9.12}$$

Its maximum amplitude is obtained for $\omega \approx \omega_0$:

$$\max\left(\left|X(0,\omega)\right|\right) \approx \frac{A}{2}\int_{-\infty}^{\infty} w(\tau)\, d\tau \tag{9.13}$$

If a *periodic signal* is to be inspected, it makes sense to arrange the analysis so as to maintain some evidence of its spectral line composition in the STFT. The effect of the analysis window shape and length on its Fourier Transform are extensively discussed in (Harris, 1978). It is shown that, for a given value of the window length L, the choice of a window shape results from a tradeoff between the window spectral main lobe width (which is precisely equal to twice the bandwidth in the filter bank interpretation) and its level with respect to sidelobes (often denoted as *spectral leakage*). A hamming window is often used, for which $B \approx 2/L$, and spectral leakage is maintained more than 40 dB below the main lobe level. It is clear from Fig. 9.3, in which STFTs of a purely periodic sound are plotted as a function of L, that harmonic analysis benefits from window length greater than two times the signal period, while extending the window length over four times the signal period does not further decrease the noise-like effect due to secondary lobes.

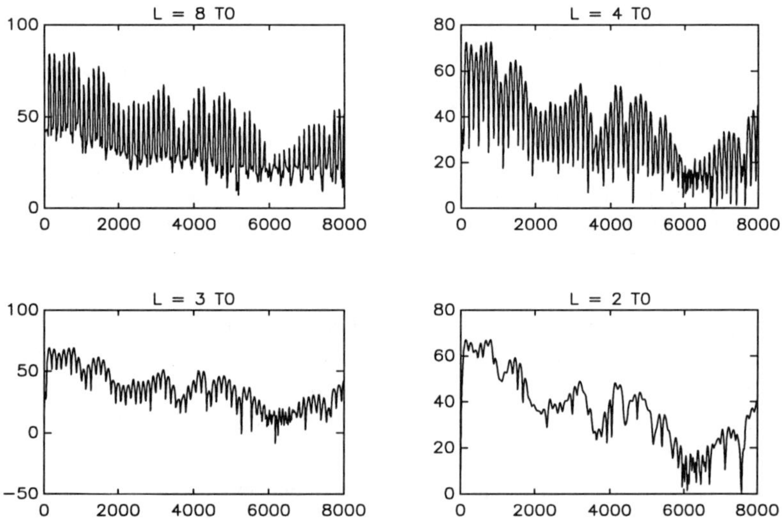

Fig. 9.3. The effect of the analysis window length on the STFT of a periodic signal (vowel [a], F_0=132 Hz, hamming window).

In practice, fundamental frequency seldom decreases below 80 Hz, so that a window length of about 30ms is a reasonable value. A window shift of about 10ms is often adopted, which slightly infringes on the theoretical maximum value of $L/4$.

STFTs also take part in the spectral estimation of *stochastic signals*, mainly in the form of the well-known *periodogram*:

$$P_{XX}(t,\omega) = \frac{|X(t,\omega)|^2}{\int_{-\infty}^{\infty} w^2(\tau)\,d\tau} \tag{9.14}$$

and its ensemble average, which practically reduces to a time-average, termed as the *averaged periodogram* $AP_{xx}(\omega,t)$, assuming the ergodicity of $P_{xx}(\omega,t)$ with respect to *t*:

$$AP_{XX}(t,\omega) = E\big(P_{XX}(t,\omega)\big) = \lim_{N\to\infty} \frac{\sum_{i=-N}^{N} P_{XX}(t - i\,\Delta t,\omega)}{N} \tag{9.15}$$

Provided Δt is chosen so as to avoid an excessive correlation between successive periodogram estimates, it is well known (Kay, 1988) that, for finite values of N, the first and second moments of the averaged periodogram computed on zero-mean white Gaussian noise are given by

$$E\big[AP_{XX}(t,\omega)\big] = S_{XX}(\omega) \qquad E\Big[\big(AP_{XX}(t,\omega) - S_{XX}(\omega)\big)^2\Big] = \frac{S_{XX}(\omega)}{N} \tag{9.16}$$

In a word, the interpretation of STFT results strongly depends on *a priori* assumptions on the analysis signal. What now in the case of hybrid H/S signals?

9.2.2. Approximation criteria

For a given speech frame, hybrid H/S analysis consists, as shown in Fig. 9.2, of simultaneously optimizing the fundamental frequency ω_0 and performing the approximation of both the H_p and H_r filters[6], so as to minimize the error between the original frame and the one that would be synthesized from its model.

When no stochastic component is encountered, the problem of estimating ω_0 together with the harmonic amplitudes and phases a_i and ϕ_i, assuming they are constant in the analysis frame, has typically been addressed by harmonic coding techniques (Marques *et al.*, 1990). Its solution is based on the least-squares minimization of the cost function $e_p(t)$ for a given value of *t*:

$$e_p(t) = \int_{-\infty}^{\infty} w^2(t-\tau)\left|s_p(\tau) - \tilde{s}_p(\tau)\right|^2 d\tau = \frac{1}{2\pi}\int_{-\infty}^{\infty}\left|S_p(t,\omega) - \tilde{S}_p(t,\omega)\right|^2 d\omega \tag{9.17}$$

in which *S(.,.)* denotes a short-time Fourier transform. In case of sinusoids in white Gaussian noise, it is shown in Kay (1988) that criterion (9.17) corresponds to a Maximum Likelihood Estimation of the sinusoid parameters[7].

Sometimes, the simpler (9.18) formulation is used, with $n=2$, in which phase information is no longer taken into account, so that only harmonic amplitudes are provided by the (real-valued) least squares decomposition.

---

[6]In order to increase readability, all references to frame numbers or to frame reference instants will be removed when only the current frame is considered.

[7]Sinusoids estimation criteria based on the computation of eigenvectors of the signal covariance matrix, such as in Pisarenko harmonic decomposition or in the MUSIC algorithm (Kay, 1988), are not considered here, given their high computational complexity (the order of the covariance matrix must be greater than the number of sinusoids).

$$e_p(t) = \frac{1}{2\pi} \int_{-\infty}^{\infty} \left(\left| S_p(t,\omega) \right|^n - \left| \tilde{S}_p(t,\omega) \right|^n \right)^2 d\omega \tag{9.18}$$

Similarly, if one assumes that no harmonic component exists in $\tilde{s}(t)$, the optimization of H_r can be done in the least mean squares sense—that is, by minimizing the expected value of the squared difference between the analysis and synthesis samples:

$$e_r(t) = E\left(\int_{-\infty}^{\infty} w^2(t-\tau) \left| s_r(\tau) - \tilde{s}_r(\tau) \right|^2 d\tau \right) = \frac{1}{2\pi} E\left(\int_{-\infty}^{\infty} \left| S_r(t,\omega) - \tilde{S}_r(t,\omega) \right|^2 d\omega \right) \tag{9.19}$$

However, the more tractable equation (9.20) is generally preferred to (9.19), since the actual noise waveform is of much lesser perceptual importance than its amplitude spectrum:

$$e_r(t) = \frac{1}{2\pi} \int_{-\infty}^{\infty} E\left(\left| S_r(t,\omega) \right|^2 \right) - E\left(\left| \tilde{S}_r(t,\omega) \right|^2 \right) d\omega \tag{9.20}$$

$$= \frac{1}{2\pi} \int_{-\infty}^{\infty} \left(E\left(\left| S_r(t,\omega) \right|^2 \right) - \left| H_r(t,\omega) \right|^2 \right) d\omega \tag{9.21}$$

The point is that both components coexist in $s(t)$ and there is no *a priori* means of separating them, so that the optimization of $e_p(t) + e_r(t)$ with equations (9.17) + (9.20) is impossible.

A more pragmatic two steps approach is therefore adopted by (Griffin, 1987) in the context of the MBE analysis algorithm.[8] The parameters of the periodic part are first deduced from equation (9.17),[9] and $s_r(t)$ is suboptimally introduced in (9.20) as

$$s_r(t) \approx s(t) - \tilde{s}_p(t) \tag{9.22}$$

Equation (9.17) obviously is nonlinear in ω_0, but can be made linear in $\{a_i, \phi_i\}$ for a given value of ω_0, provided equation (9.1) is changed into its complex form:

$$\tilde{s}_p(t) = \sum_{i=-N}^{N} \underline{a}_i \, e^{j i \omega_0 t} \qquad \text{, with} \qquad \underline{a}_{-i} = \underline{a}_i^* \tag{9.23}$$

in which N is the number of harmonics accounted for and $\underline{a}_i = a_i \, e^{j\phi_{i0}}$

[8] It should be emphasized that the speech model underlying MBE speech coding is more restrictive than what we term here as *hybrid H/S* model, in that it uses no harmonic phases and assigns frequency-dependent voiced/unvoiced *decisions* instead of *ratios,* yet decisions are associated to wide frequency bands rather than to individual harmonics. When we refer to the MBE model in this book, we address only the first steps of its analysis algorithm—(9.17), (9.22), and (9.20).

[9] Things happen to be a little bit more complex in practice, since criterion (10) does not intrinsically prevent computed pitch values from presenting abrupt variations, known as frequency doubling or halving. As in most other efficient pitch extraction methods, the MBE analysis algorithm makes use of dynamic programming to ensure a continuous pitch evolution when possible.

Since finite-duration weighted exponential functions are not, in all generality, orthogonal to one another, least-squares approximation (9.18) is practically expressed by a Yule-Walker system of linear equations

$$\mathbf{R}\underline{\mathbf{a}} = \mathbf{r} \tag{9.24}$$

in which **R** is a classical Toeplitz matrix given by the scalar products of basis imaginary functions[10] and **r** contains the scalar products of the analysis signal with the basis functions. In practice, **R** is approximately made orthogonal by adequately choosing the shape and length of the weighting window.

Equation (9.24) denotes, in the time or frequency domain, the orthogonality of the error function with the basis exponential functions. It is solved on a grid of frequency values, and the set of $\{\omega_0, a_i, \phi_i\}$ parameters with the lowest error is retained. It is shown in Griffin (1987) that the pitch period estimation accuracy achievable with this method is about one eighth of a sample, which gives a precision of 8 Hz for the last harmonic, whatever the sampling frequency, for f_0=100 Hz. The analysis algorithm requires the computation of only one STFT per analysis frame: $S(t,\omega)$.

Once the values of $\{\omega_0, a_i, \phi_i\}$ have been found, $\tilde{s}_r(t)$ is computed from (9.22) in the frequency domain by subtracting from $S(t,\omega)$ the complex values of the resulting $\tilde{S}_p(t,\omega)$. The remaining spectrum is interpreted as the STFT of some noise. Only its amplitude spectrum is therefore examined. As introduced in Section 9.1, it is approximated as the superposition of N_{NBN} band limited noise signals, the RMS value of which is estimated by a frequency-domain averaged periodogram, as the mean spectral amplitude in the corresponding noise band:

$$\sigma_i^2(t) = \frac{1}{B_i} \int_{\omega_i - \frac{B_i}{2}}^{\omega_i + \frac{B_i}{2}} P_{XX}(\omega,t)\, d\omega \tag{9.25}$$

in which ω_i is the center frequency of band i, B_i is its bandwidth, and $P_{XX}(\omega,t)$ is the averaged periodogram at time t.

9.3. Database compression

Harmonic and hybrid coders have originally been proposed for narrow band speech coding (F_s=8 kHz), as a means to bridge the gap between waveform coders operating down to 12 kbits/s, yielding toll-quality output, and LPC vocoders, producing unnatural voices. Recently, most coding research teams involved in sinusoidal, harmonic, or hybrid synthesis have struggled to tally the parameters of their coders into the 4800 bits/s required by the new federal standard of the U.S. Department of Defense (see Hardwick and Lim, 1988; Marques *et al.*, 1990; MacAulay and Quatieri, 1991, Abrantes and

[10] In practice, integrals are replaced by sums. Therefore, when computations are performed in the frequency domain, STFTs are first estimated on a number of frequency bins that is at least equal to the number of harmonics to be estimated. In practice, a better selectivity is obtained when the number of frequency bins is chosen as several times the number of harmonics. For 16 kHz, we used 1024 bins.

Marques, 1992). Deketelaere *et al.* (1992) have further reduced the bit rate to 2300 bits/s, and more recently to 900 bits/s thanks to vector quantization techniques.

Some general trends can be outlined. Regarding sinusoid amplitudes, important data reduction can be achieved by storing the envelope spectrum in the form of LPC parameters (a residual error variance σ and a set of LSP coefficients) (Rowe *et al.*, 1991). Amplitudes are simply reconstructed by spectral sampling of the LPC envelope. The idea is readily extended to stochastic components variances, with the difference that a reduced prediction order can be used, as they do not have the same perceptual importance as voiced amplitudes. Abrantes and Marques (1992) have chosen 10 and 6 for the voiced and unvoiced prediction orders, respectively.

Phase coding has proven to be a more difficult task, since coarse quantization results in systems that are reverberant. Marques *et al.* (1990) have proposed to predict harmonic phases from frame to frame and encode the prediction error. Their prediction model is a simple quadratic one that assumes a linear temporal evolution of harmonic frequencies inside frames, from which phases are computed by integration. In contrast, an analytical model of phase as a function of frequency has been developed by MacAulay and Quatieri (1991). Voiced phases are depicted as the superposition of a linear component, which depends on the particular position of the glottal pulse within the analysis period and a minimum phase contribution due to the vocal tract.

To our knowledge, there has been no attempt to optimize the compression performances of hybrid models in wide-band conditions. One can therefore only put forward the coarse approximation of 10 kbits/s.[11]

9.4. Prosody matching with the H/S model

The H/S model lies halfway between production and phenomenological models, in that it is not based on strong hypotheses on the signal to analyze. As such, it benefits from the (quite) simple and efficient prosody matching capabilities of the former, while keeping the powerful descriptive potential of the latter.

Modifying durations is performed in the form of a dynamic change in the frames reference instants t^i by a simple look-up at the time matching function $t'(t)$ computed as in Section 7.2.2. Phases have to be modified accordingly, in order to preserve some sort of phase coherency among neighboring frames. This can be achieved by propagating phase variations from the first frame to the last one, as expressed by

$$\Delta t^i = t'(t^i) - t^i$$

$$\Delta\phi_j^i = \Delta\phi_j^{i-1} + j\,\frac{(\omega_0{}^i + \omega_0{}^{i-1})}{2}\,\Delta t^i \tag{9.26}$$

for $i = 2 \ldots N_F$ and $j = 1 \ldots \min(N_H^{i-1}, N_H^i)$, and assuming $\Delta\phi_j^1 = 0$, for all j.

where N_F is the number of frames of the current segment, and N_H^i is the number of harmonics of its ith frame.

[11] For the French diphone database we have used in Dutoit (1993, 1994a), this figure results in memory requirements below 250 kbytes.

Notice that, unlike N_{NBN} (the number of band-limited noise functions chosen to describe the unvoiced part of each frame, which is fixed, and therefore does not refer to a particular frame), N_H^i depends on ω_0^i. It may thus happen that $N_H^{i-1} < N_H^i$ or $N_H^{i-1} > N_H^i$, respectively termed as *births* and *deaths* cases by MacAulay and Quatieri (1986). In (9.26), j is limited by the number of harmonics of the previous frame, since phase coherency is essentially meaningless when harmonic births are encountered.

Changing the pitch of a segment is not straightforward, even though f_0 explicitly appears in the parameters of the model. Noise-band variances should obviously be left untouched, while harmonic amplitudes have to be adapted to their new frequencies, so as to maintain the original spectral envelope information. In practice, a simple piecewise-linear interpolation of the original amplitudes gives a good approximation of the new ones. As a matter of fact, one can reasonably assume that, provided the original pitch frequency is not too high, most of the significant spectral details are maintained. In all generality, a constant amplitude factor should also be applied on harmonics in order to preserve the initial signal power. Practical experiments, however, revealed that naturalness was preserved even when no such correction was applied (Dutoit, 1993). Finally, harmonic phases have to be adjusted, in the same way as in (9.26):

$$\Delta\omega_0{}^i = \omega_0{}'(t^i) - \omega_0{}^i$$

$$\Delta\phi_j^i = \Delta\phi_j^{i-1} + j\,\frac{(\Delta\omega_0{}^i + \Delta\omega_0{}^{i-1})}{2}(t^i - t^{i-1}) \qquad (9.27)$$

for $i = 2 \ldots N_F$ and $j = 1 \ldots \min(N_H^{i-1}, N_H^i)$, and assuming $\Delta\phi_j^1 = 0$, for all j.

9.5. Segment concatenation

Hybrid H/S models are inherently able to smooth spectral amplitude and phase mismatches,[12] which naturally occur at segments borders, in a highly efficient way.

Phase discontinuities arise only when voiced sounds are produced. They partly originate in the fact that the position of the first analysis frame in the right segment has been chosen independently of the position of the last analysis frame of the left segment when the parametric segment database has been created. In this case, however, they can be completely eliminated by propagating the phase discontinuity on all the frames of the the right segment. Using the same notations and frame indices as in (7.1) and (7.2)

$$\phi_{R}{}_j^i = \phi_{R}{}_j^i + (\phi_{L}{}_j^0 - \phi_{R}{}_j^0) \qquad (9.28)$$

for $i = 1 \ldots N_{FR}$ and $j = 1 \ldots \min(N_{HR}^{0}, N_{HL}^{0}, N_{HL}^{i})$

It will be hypothesized in Section 9.7 that phase mismatches also arise due to coarticulation, so that simply propagating phase differences only partially solves the problem.

---

[12]Pitch mismatches cannot be encountered here since the segments to be chained have been previously submitted to a powerful prosody matching algorithm, with similar values of synthesis ω_0 at the left and right of the concatenation point.

As far as spectral amplitude mismatches are concerned, the simple linear smoothing laws introduced in (7.1) and (7.2) are theoretically less powerful when applied on harmonic amplitudes a_i and narrow band noises variances σ_i^2 than on PARCOR coefficients. Indeed, given the linear dependence between spectral and temporal domains, linearly interpolating two spectral envelopes is equivalent to simultaneously fading out and in the corresponding sounds rather than progressively adapting their formant frequencies and bandwidths (Fig. 9.4). Dutoit (1993) found, however, that provided initial sounds are acoustically related, as is the case in segment concatenation synthesis, the transitions obtained with the aforementioned spectral linear laws are perceived as natural. Nevertheless, a number of more complex methods have been proposed to allow more realistic formant movements. In (Banga *et al.*, 1993), for instance, a_i and σ_i are casted into an approximated autoregressive representation, in the form of PARCOR coefficients. A further step toward a more realistic concatenation scheme was taken by Grau *et al.* (1993) who modeled the spectral envelope of speech sounds as a sum of formants whose parameters (amplitude, center frequency, bandwidth) were interpolated between diphones.

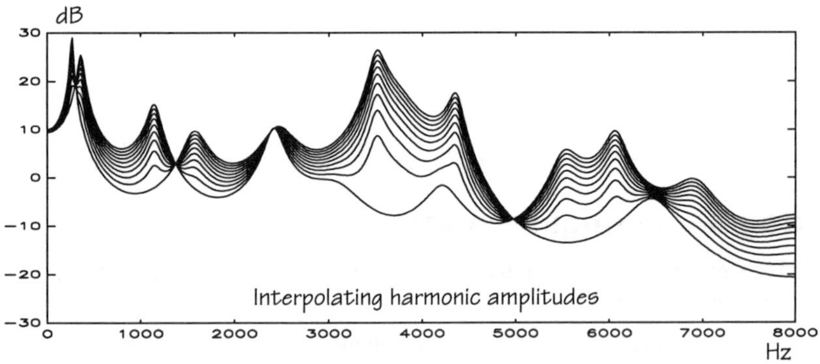

Fig. 9.4. Linear smoothing of harmonic amplitudes with the hybrid H/S model. Smoothing is performed between [l] phones, respectively encountered in the same diphones [ɑl] and [lo] as in Fig. 8.10.

In contrast, the accuracy of the concatenation process depends on the values of M_L and M_R (respectively, the number of left and right frames to participate in the smoothing). As a matter of fact, imposing constant values, independently of the segments to be concatenated, typically leads to overcoarticulated (underarticulated) speech, especially for highly transient sounds such as plosives or semi-vowels. Voice-doubling effects, which are even worse than the initial spectral discontinuities, are also encountered when important mismatches are spread over a time interval that is too long. In order to avoid this problem, Dutoit (1994b) has proposed a segment-dependent concatenation strategy, based on some observed inaccuracies of hybrid H/S analysis algorithms. Indeed, as will be further investigated in Section 9.7.2, some non-periodic sounds are typically misinterpreted by hybrid H/S analysis algorithms, in the sense that important stochastic components are erroneously found. They are:

- Semivowels, quickly pronounced labials, as well as the explosion of plosives, all of which hardly respect the quasi-stationarity hypothesis. Stochastic components are then found to be distributed uniformly on the frequency axis.
- Vowels with rapidly changing pitch, which prevent them from being properly accounted for by the analysis algorithms. Errors are mostly encountered in higher-order harmonics, because their frequencies are submitted to larger variations.

It is possible to take advantage of this feature to roughly classify each analysis frame into steady states and transients, on the simple basis of both the ratio R_{VUV} and the sum S_{VUV} of their voiced and unvoiced low-frequency instantaneous powers P_V and P_{UV}, respectively computed as:

$$P_V^{LF} = \sum_{j=1}^{N_H^{LF}} \frac{a_j^{\;2}}{2} \quad , \quad P_{UV}^{LF} = \frac{\sum_{j=1}^{N_{NBN}^{LF}} B_j \sigma_j^2}{\sum_{j=1}^{N_{NBN}^{LF}} B_j} \quad , \quad R_{VUV} = \frac{P_V}{P_{UV}} \quad , \quad S_{VUV} = P_V + P_{UV} \tag{9.29}$$

in which N_H^{LF} and N_{NBN}^{LF} respectively refer to the number of harmonics and narrow band noises in the low-frequency part of the spectrum (typically up to 1000 Hz). Clearly, the aim of this restriction is to make parameters approximately insensitive to pitch variation effects. Classification can then be achieved with the following criteria:[13]

$$\begin{array}{ll} \text{Steady states:} & (R_{VUV} \geq \alpha_1) \text{ or } (S_{VUV} < \alpha_2) \\ \text{Transients:} & (R_{VUV} < \alpha_1) \text{ and } (S_{VUV} \geq \alpha_2) \end{array} \tag{9.30}$$

in which the thresholds have to be adjusted experimentally.

Frames are then examined around the concatenation point and interpolation is constrained to the steady-state area around it. Furthermore, the number of frames to be interpolated in both segments is limited to a maximum N_I, in order to avoid voice-doubling effects. The interpolation zone is finally imposed to be symmetric with respect to the concatenation point, that is M_L and M_R are changed into $min(M_L, M_R)$.

Practical experiments have shown that the segmental quality obtained with the aforementioned algorithm is such that few concatenation points, if any, can be detected.

9.6. Hybrid synthesis

Many algorithms have been proposed for synthesizing speech from its hybrid H/S description. They mainly differ in the way they handle the implicitly time-varying nature of the H/S parameters.

---

[13] In Dutoit (1994b), these results have been further developed so as to obtain a voiced/unvoiced decision from R_{VUV} and S_{VUV}.

In the *parameter interpolation* approach, harmonics amplitudes and frequencies as well as narrow band noise variances are linearly interpolated on a frame-by-frame basis, so as to provide smooth temporal evolutions:

$$\Psi(t)=\Psi^{k-1}+(\Psi^{k}-\Psi^{k-1})\frac{t-t^{k-1}}{t^{k}-t^{k-1}} \quad \text{, for } t^{k-1}\le t<t^{k} \tag{9.31}$$

in which Ψ successively stands for a_i, and σ_j^2, for $i=1\ldots\min(N_H^{k-1},N_H^k)$ and $j=1\ldots N_{NBN}$. In order to account for births and deaths of harmonics, equations (9.31) are simply added to the set

$$a_i(t)=a_i^k\frac{t-t^{k-1}}{t^{k}-t^{k-1}} \quad \text{, for } t^{k-1}\le t<t^{k} \text{ and } i=N_H^{k-1}\ldots N_H^{k} \text{ in case of births} \tag{9.32}$$

$$a_i(t)=a_i^{k-1}\frac{t^{k}-t}{t^{k}-t^{k-1}} \quad \text{, for } t^{k-1}\le t<t^{k} \text{ and } i=N_H^{k}\ldots N_H^{k-1} \text{ in case of deaths} \tag{9.33}$$

The hybrid synthesis equations are then simply an extended version of equation 9.1:

$$\tilde{s}(t)=\sum_i a_i(t)\cos(\phi_i(t))+\sum_{j=1}^{N_{NBN}}\sigma_j(t)\eta_j(t) \tag{9.34}$$

$$\text{with } \phi_i(t)=\phi_i(0)+\int_0^t i\omega_O(\tau)\,d\tau \tag{9.35}$$

where $\eta_j(t)$ are narrow-band basis functions with contiguous bandwidths B_j, central frequencies ω_j, and unitary variance. In practice, they are obtained by band-pass filtering a looped white noise signal, the duration of which is chosen so that looping is not audible (0.5 sec is a minimum value). In case all bandwidths have been chosen to be identical, equation (9.36) can be given the amplitude modulated cosine form

$$\tilde{s}(t)=\sum_i a_i(t)\cos(\phi_i(t))+\sum_{j=1}^{N_{NBN}}2\sigma_j(t)\eta_0(t-t_j)\cos(\varphi_j(t)) \tag{9.36}$$

$$\text{with } \varphi_j(t)=\varphi_j(0)+\omega_j t \tag{9.37}$$

where $\eta_0(t)$ is a low-pass filtered noise with unitary variance, $\varphi_j(0)$ are random initial phases, and t_j are random time offsets that are required to avoid synchronicity of stochastic components. This formulation strongly reduces the amount of memory needed to store the stochastic basis functions. What is more, if ω_j are imposed to be submultiples of the sampling frequency, then $\cos(\varphi_j(t))$ can easily be obtained from tabulated values of $\cos(\varphi_1(t))$.

As can be expected, problems arise when straightforwardly trying to use equation (9.34), even if $\omega_O(t)$ is also given a piecewise linear temporal evolution. Indeed, phases are overdetermined, since continuously integrating harmonic frequencies over time from initial phase values at t^0 is generally not compatible with the subsequent phase values resulting from hybrid analyses at t^1, t^2, ...

The most straightforward way to tackle this problem is simply to ignore it, that is not to take phases into account. Equations (9.34) and (9.35) then hold, with initial values of phases chosen randomly in [0,2π]. It has been reported by MacAulay and Quatieri (1986) in the context of sinusoidal synthesis and verified by many others that such a strategy importantly decreases segmental quality by overharmonizing vocalic sounds, especially breathy ones such as voiced fricatives. An elegant solution has been given by MacAulay and Quatieri (1986) who have shown that, provided $\theta_j(t)$ is described by a third-order polynomial between t^{k-1} and t^k, as in

$$\theta_j(t) = \zeta_j + \gamma_j t + \alpha_j t^2 + \beta_j t^3 \quad \text{for } t^{k-1} \leq t < t^k \text{ and } i = 1 \ldots \min(N_H^{k-1}, N_H^k) \tag{9.38}$$

one can easily compute the values of ζ_j, γ_j, α_j, β_j, and an additional parameter M_j so that $\theta_j(t^{k-1}) = \theta_j^{k-1}$, $\theta_j(t^k) = \theta_j^k + 2\pi M_j$, $\dot{\theta}_j(t^{k-1}) = j\omega_0^{k-1}$, $\dot{\theta}_j(t^k) = j\omega_0^k$, and $\dot{\theta}_j(t)$ is maximally smooth from t^{k-1} to t^k:

$$\zeta_j = \theta_j^{k-1} \quad , \quad \gamma_j = \omega_0^{k-1} \tag{9.39}$$

$$M_j = \frac{1}{2\pi}\left[\left(\theta_j^{k-1} + j\omega_0^{k-1} - \theta_j^k\right) + j\left(\omega_0^k - \omega_0^{k-1}\right)\frac{t^k - t^{k-1}}{2}\right] \tag{9.40}$$

$$\begin{bmatrix} \alpha_j \\ \beta_j \end{bmatrix} = \begin{bmatrix} \dfrac{3}{(t^k - t^{k-1})^2} & \dfrac{-1}{(t^k - t^{k-1})} \\ \dfrac{-2}{(t^k - t^{k-1})^3} & \dfrac{1}{(t^k - t^{k-1})^2} \end{bmatrix} \begin{bmatrix} \theta_j^k - \theta_j^{k-1} - j\omega_0^{k-1}\left(t^k - t^{k-1}\right) + 2\pi M \\ j\left(\omega_0^k - \omega_0^{k-1}\right) \end{bmatrix} \tag{9.41}$$

Abrantes *et al.* (1991) have implemented this method in a hybrid H/S copy synthesizer with a sampling frequency of 8kHz, and reported almost non-perceptible differences between original and synthetic speech, although a small background noise could be heard. As far as computational complexity is concerned, it is easily found that equations (9.35), (9.36), and (9.38) imply about ($10\ \hat{N}_H + 4\ N_{NBN}$) additions and multiplications and $\hat{N}_H$ cosine computation per synthesis sample,[14] where $\hat{N}_H$ denotes the average number of harmonics.

In the *overlap-add* approach, synthesis frames are produced independently of one another with fixed values of the synthesis parameters, on overlapping time intervals:

[14]The relative importance of cosine computations and elementary operations actually depends on the target processor. Cosines are hardware operations for Intel 387 and 486 processors, while they have to be evaluated as series or tabulated on DSPs.

$$\tilde{s}^k(t) = \sum_i a_i^k \cos\left(i\omega_0^k (t - t^k) + \phi_i^k\right) + \sum_{j=1}^{N_{NBN}} \sigma_j^k \eta_j(t) = \tilde{s}_p^k(t) + \tilde{s}_r^k(t) \quad (9.42)$$

$$\text{for } t \in \left[t^k - OlaFact \frac{(t^{k+1} - t^k)}{2}, t^{k+1} + OlaFact \frac{(t^{k+1} - t^k)}{2} \right]$$

where *OlaFact* is the overlap factor (typically = 1), defined as the ratio of the synthesis frame duration on its time coverage $(t^{k+1} - t^k)$, minus one. Each frame is sent to an overlap-adder with a shift value of $(t^{k+1} - t^{k-1}) / 2$ (see Fig. 9.5).

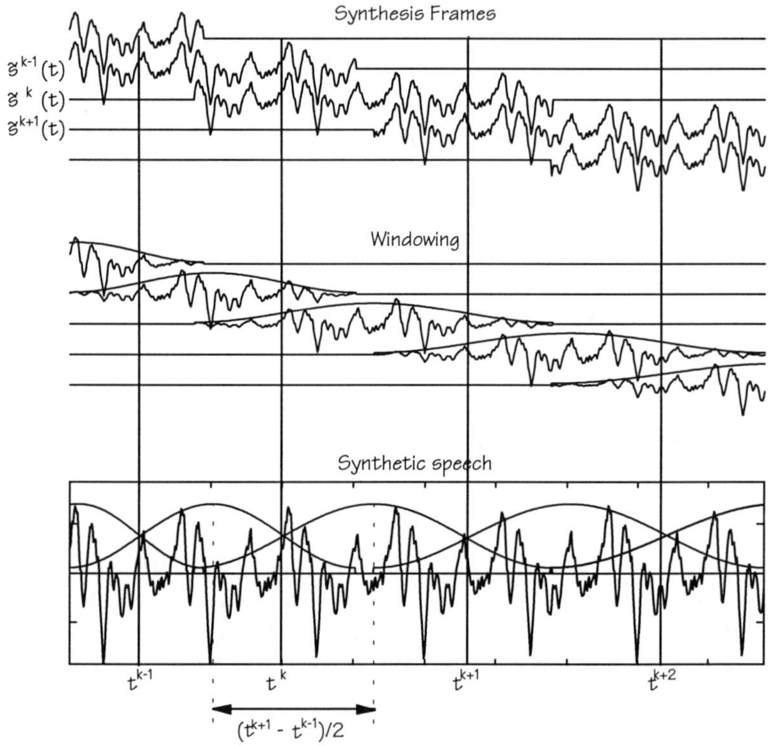

Fig. 9.5. Overlapping synthesis frames in the OLA approach.

The main advantage of this approach is that equation (9.42) can be efficiently computed on a set of discrete instants with an IFFT algorithm (Dutoit and Gosselin, 1996). Indeed, let us first consider the FFT of a sequence of $N=2^m$ complex exponentials with zero initial phase and fundamental frequency ω:

$$\left\{1, e^{j\omega T_s}, \ldots, e^{j(N-1)\omega T_s}\right\} \xleftrightarrow{\text{FFT}} \left\{X_0, X_1, \ldots, X_{N_{FFT}-1}\right\} \quad (9.43)$$

It is fairly easy to show that

$$X_l = D(\omega - l\omega_N) = \exp\left(j\frac{(N-1)(\omega - l\omega_N)}{2}\right)\frac{\sin\left(N\frac{(\omega - l\omega_N)}{2}\right)}{\sin\left(\frac{\omega - l\omega_N}{2}\right)} \tag{9.44}$$

(with $\omega_N = \frac{2\pi F_s}{N}$) in which we recognize the Dirichelet kernel D(ω). Now consider the FFT of the following set of N samples of $\tilde{s}_p^{\,k}(t)$:

$$\left\{x(0), x(T_s), x(2T_s), \ldots, x\left((t^{k+1} - t^k)\ OlaFact\right), x(N_s T_s), \ldots, x((N-1)T_s)\right\} \xleftrightarrow{\text{FFT}} \{X_0, X_1, \ldots, X_{N-1}\} \tag{9.45}$$

in which $x(t)$ is defined as

$$x(t) = \tilde{s}_p^{\,k}\left(t - \left(t^k - OlaFact\frac{(t^{k+1} - t^k)}{2}\right)\right) \tag{9.46}$$

One can reasonably assume that the contributions of harmonics are limited to their first Dirichelet lobes, which moreover do not significantly overlap provided N is greater than four times the synthesis period (expressed in samples). It follows that

$$\arg(X_l) = \phi_i^k - i\omega_0^k\ OlaFact\frac{(t^{k+1} - t^k)}{2} + \arg\left(D(i\omega_0^k - l\omega_N)\right) \tag{9.47}$$

$$|X_l| = \frac{a_i^k}{2}\left|D(i\omega_0^k - l\omega_N)\right| \tag{9.48}$$

$$\text{for}\quad l\omega_N \ \text{ in } \left[(i-\frac{1}{2})\omega_0^k, (i+\frac{1}{2})\omega_0^k\right[$$

Similarly, the FFT of N samples of the stochastic term $\tilde{s}_r^{\,k}(t)$ are realistically given by

$$|X_l| = \sqrt{N}\,\sigma_i^k \quad \text{and} \quad \arg(|X_l|) = \text{random}(2\pi) \tag{9.49}$$

$$\text{for}\quad l\omega_N \ \text{ in } \left[\omega_i - \frac{B_i}{2}, \omega_i + \frac{B_i}{2}\right[$$

By summing both contributions and computing their IFFT, one obtains N samples of $x(t)$, from which only the first N_s samples are retained.

Dutoit and Gosselin (1996) implemented this method in a hybrid H/S copy synthesizer at 16 kHz as well as in the corresponding concatenative synthesizer and noticed few audible differences between original and synthesized signals,[15] if each OLA frame duration was kept small enough (typically less than 10 ms). A similar effect has been reported by MacAulay and Quatieri (1986): larger frames, especially the ones with rapidly varying pitch, were subject to buzziness, due to the overlapping of signals with very different

[15]They are detailed in the next section.

fundamental frequencies. In the context of text-to-speech synthesis, however, the duration of synthesis frames is imposed by the prosody matching block, so that the aforementioned condition cannot be automatically satisfied. Dutoit and Gosselin thus modified the synthesis algorithm by splitting long frames into subframes. Harmonics frequencies, amplitudes and noise variances at the beginning of each sub-frame were linearly interpolated from their values at t^k and t^{k+1}. Harmonics phases were simply deduced from their values at t^k, as the integral of the related frequencies. On the basis of the corresponding limitation of N_s to 160, they tried to minimize the value of N and found that 512 FFT points were necessary when synthesizing male voices with possible pitch frequency excursion below 100 Hz.[16] As a matter of fact, a value of 256 points invalidates approximations (9.47) and (9.49) in that case, as harmonics contributions to the overall spectrum severely overlap.

Regarding computational efficiency, real IFFT's require about $2N\ log_2(N)$ operations (Sorensen *et al.*, 1987) and are computed every 10 ms, which makes about sixty operations per sample in our case. Computing one-half of the 512 points FFT with equations (9.47), (9.48), and (9.49)—that is 256 complex values—requires twelve operations for the modulus and the argument of each FFT bin, assuming the Dirichelet Kernel is tabulated. Deriving real and imaginary parts can itself be done with about ten operations per bin, assuming again that a *cos* function can be found in tables. This makes a total of about thirty-five additional operations per sample for the computation of each complex FFT. Since OLA itself only implies between two and five times *(OLAFactor+1)* operations per sample, depending on whether weighting windows have to be computed or not, the complete OLA/IFFT approach approximately corresponds to 100 operations per sample: at least ten times less than with the parameters interpolation one and approximately 50 percent more than the computational load of an LPC synthesizer of order 18.

9.7. Segmental quality

We have mentioned in the previous section that the segmental quality obtained with H/S copy synthesis is such that one can hardly distinguish the original sound from the synthetic one, at least when the original phases are taken into account by the synthesis algorithm. More particularly, Dutoit and Gosselin (1996) reported that transients were slightly degraded (which is not really astonishing given the quasi-stationarity assumption underlying the model) and that, more surprisingly, some extra breath appears in sonorants. They also reported that the segmental quality obtained in a concatenation-based synthesis framework was quite different from the one obtained in copy synthesis experiments, in that the former sounds more buzzy than the latter. What is more, even though these problems were already encountered in low bit rate voice coding systems (with F_S = 8 kHz), for which hybrid models were originally developed (Hardwick and Lim, 1988), their importance increases when higher sampling frequencies are used.

[16]This, by the way, simply means that female voices are economically interesting in the context of hybrid H/S synthesis.

In order to examine the possible origins or these problems, Dutoit and Gosselin successively investigated: the importance of the analysis criterion, the analysis errors due to pitch and spectral amplitude variations, and the effect of coarticulation on phase values.

First, they browsed an enlarged set of analysis criteria, starting from the ones proposed in (Griffin, 1987) and (Abrantes *et al.*, 1991) and found that the latter approach reduces analysis errors, in the sense that smaller sinusoids are erroneously found in white noise. Experiments on real speech, however, showed that Griffin's criterion should be preferred, at least in wide-band conditions. To understand why, they investigated the influence of time-varying frequency and amplitude on the accuracy of the hybrid H/S analysis of sinusoids and on the related analysis error (which is interpreted as noise in the hybrid model) and showed that even moderate pitch changes can cause significant analysis errors with Griffin's criterion, mostly in high frequency, while amplitude variations equally affect the whole spectrum. These effects can be understood as the origin of the extra breath encountered when synthesizing sonorants with Griffin's criterion. The effect of pitch and spectral amplitude variations is much worse with on Abrantes *et al.*'s criterion: synthetic speech is unquestionably breathy, at least in the HF part of the spectrum.

Dutoit and Gosselin also found that the overall segmental quality of a hybrid H/S concatenation-based synthesizer is highly sensitive to phase modifications. When producing speech by segment concatenation, a simple propagation of phase mismatches at the concatenation point does not seem to be sufficient, as it gives birth to some hollow effect in voiced fricatives, while these effects were not encountered in hybrid H/S copy synthesis. Concatenation seems to have an important effect on the temporal evolution of measured initial phase vectors. Further work is needed to design efficient phase smoothing algorithms in that context.

It should be noted, however, that despite these inaccuracies, speech produced by a hybrid H/S TTS system is unquestionably superior to its LPC counterpart (see Dutoit, 1994a, for comparative tests). It has been recently put forward by Boeffard and Violaro (1994) that the H/S model can improve the robustness of TTS synthesizers for large prosodic variations, which makes it a candidate for high-quality speech synthesis. It compares favorably with the increasingly used time-domain techniques (see Chapter Ten) regarding its ability to dynamically modify a speaker's voice, so as to produce different speaking styles (Laroche *et al.*, 1993; Stylianou *et al.*, 1995a). Recent refinements in the model (such as the use of a time-domain envelope for the synthesis of the stochastic part of voiced fricatives) and advances in the analysis algorithms (mainly, the use of pitch-synchronous analysis techniques) have even been reported by Stylianou *et al.* (1995b), who have obtained high-quality pitch- and time-scale modifications of continuous speech. All these remarks make the H/S model a very good candidate for high quality, wide-band TTS synthesis.

9.8. Summary

In this chapter, we have studied the performance of the hybrid harmonic/stochastic model in the context of text-to-speech concatenative synthesis. We could summarize them as follows:

- Regarding computational and storage efficiency:

  A fast OLA/IFFT synthesis algorithm has been presented as an alternative to the computationally expensive parameters interpolation approach. The resulting computational load reduces to about one and a half times the number of operations required by an LPC synthesizer of order 18.

  Database storage bit rates of about 10kbits/s can reasonably be achieved, although to our knowledge this has not been tested.

- Regarding segmental quality:

  The segmental quality obtained with H/S copy synthesis is such that one can hardly distinguish the original file from the synthetic one, at least when the original phases are taken into account by the synthesis algorithm. Some extra breath, however, appears in sonorants, but this effect can be attenuated by an adequate choice of the analysis criteria.

  In the context of concatenation-based synthesis, segment concatenation is efficiently described by simple spectral linear smoothing laws, as far as harmonics amplitudes and noise bands variances are concerned. The overall segmental quality of a hybrid H/S concatenation-based synthesizer, however, is highly sensitive to phase modifications. When producing speech by segment concatenation, a simple propagation of phase mismatches at the concatenation point does not seem to be sufficient, as it gives birth to some hollow characteristic in voiced fricatives.

  Despite these inaccuracies, however, speech produced by a hybrid H/S TTS system is unquestionably superior to its LPC counterpart. Its quality could still be enhanced, by studying the aforementioned phase concatenation problems.

References

ABRANTES, A.J. and J.S. MARQUES, (1992), "Hybrid Harmonic Coding of Speech", *Proceedings of Eusipco 92*, Brussels, pp. 487-491.

ABRANTES, A.J., J.S. MARQUES, and I.M. TRANSCOSO, (1991), "Hybrid Sinusoidal Modeling of Speech without Voicing Decision", *Proceedings of Eurospeech 91*, Paris, pp. 231-234.

ALLEN, J.B., (1977), "Short-Term Spectral Analysis, Synthesis, and Modification by the Discrete Fourier Transform", *IEEE Transactions on Acoustics, Speech, and Signal Processing*, vol. 25, n°3, pp. 235-238.

ALMEIDA, L.B., and F.M. SILVA, (1984), "Variable-Frequency Synthesis: an Improved Harmonic Coding Scheme", *Proceedings of the International Conference on Acoustics, Speech, and Signal Processing 84*, pp. 27.5.1-4

ALMEIDA, L.B., and J.M. TRIBOLET, (1983), "Nonstationary Modeling of Voiced Speech", *IEEE Transactions on Acoustics, Speech, and Signal Processing*, vol. 31, pp. 664-678.

BANGA, E.R., E. LOPEZ-GONZALO, and C. GARCIA-MATEO, (1993), "A Text-to-Speech System for Spanish with a Frequency Domain Based Prosodic Modification Algorithm", *Proceedings of the International Conference on Acoustics, Speech, and Signal Processing 93*, vol. 2, pp. 183-186.

BOEFFARD, O., and F. VIOLARO, (1994), "Improving the Robustness of Text-to-Speech Synthesizers for Large Prosodic Variations", *Proceedings of the second ESCA/IEEE Workshop on Speech Synthesis,* New-Paltz, NY, pp. 111-114.

D'ALESSANDRO, C., B. YEGNANARAYANA, and V. DARSINOS, (1995), "Decomposition of Speech into Deterministic and Stochastic Components", *Proceedings of the International Conference on Acoustics, Speech, and Signal Processing 95,* Detroit, pp. 760-763.

DARSINOS, V., C. D'ALESSANDRO, and B. YEGNANARAYANA, (1995), "Evaluation of a Periodic/Aperiodic Decomposition Algorithm", *Proc. of Eurospeech 95,* Madrid, pp. 393-396.

DEKETELAERE, S., H. LEICH, B. WERY, I. DEMAN, M. DOTHEY, (1992), "A New Quantization Technique for LSP Parameters and its Application to Low Bit Rate Multi-Band Excited Vocoders", *Proceedings of Eusipco 92,* Brussels, pp. 475-478.

DUTOIT, T., (1993), *High Quality Text-to-Speech Synthesis of the French Language,* Ph.D. dissertation, Faculté Polytechnique de Mons.

DUTOIT, T., (1994), "Text-to-Speech Synthesis: a Comparison of Four Candidate Algorithms", *Proceedings of the International Conference on Acoustics, Speech, and Signal Processing 94,* Adelaide, vol. 1, pp. 565-568.

DUTOIT, T., (1994), "On the Ability of Various Speech Models to Smooth Segment Discontinuities in the Context of Text-to-Speech Synthesis by Concatenation", *Proceedings of Eusipco 94,* vol. 1, pp. 8-12.

DUTOIT, T., and B. GOSSELIN, (1996), "On the Use of a Hybrid Harmonic/Stochastic Model for TTS Synthesis-by-Concatenation", *Speech Communication,* 19, pp. 119-143.

FUJIMURA, O., (1968), "An Approximation to Voice Aperiodicity", *IEEE Transactions on Audio and Electroacoustics,* March, pp. 68-72.

GRADHSTEYN, I.S. and I.M. RYZHIK, (1965), *Table of Integral Series and Products,* Academic Press, New York, p. 485.

GRAU, S., C. D'ALESSANDRO, and G. RICHARD, (1993), "A Speech Formant Synthesizer Based on Harmonic+Random Formant-Waveforms Representations", *Proceedings of Eurospeech 93,* Berlin, pp. 1697-1700.

GRIFFIN, D.W., (1987), *Multi-Band Excitation Vocoder,* Ph.D. dissertation, MIT, Cambridge.

HARDWICK, J.C., and J.S. LIM, (1988), "A 4.8 Kbps Multi-Band Excitation Speech Coder", *Proceedings of the International Conference on Acoustics, Speech, and Signal Processing 88,* pp. 374-377.

HARRIS, F.J., (1978), "On the Use of Windows for Harmonic Analysis with the Discrete Fourier Transform", *Proceedings of the IEEE,* vol. 66, n°1, pp. 51-83.

HOLGER, C., and B. KOLPATZIK, (1991), "Speech coding using nonstationary sinusoidal modeling and narrow-band basis functions", *Proceedings of the International Conference on Acoustics, Speech, and Signal Processing 91,* pp. 581-584.

KAY, S. M., (1988), *Modern Spectral Estimation,* Prentice Hall Signal Processing Series, Englewood Cliffs, NJ.

LAROCHE, J., Y. STILYANOU, and E. MOULINES, (1993), "HNS: Speech Modification Based on a Harmonic + Noise Model", *Proceedings of the International Conference on Acoustics, Speech, and Signal Processing 93,* Minneapolis, vol. II, pp. 550-553.

MAC AULAY, R.J., and T.F. QUATIERI, (1984), "Magnitude Only Reconstruction Using a Sinusoidal Speech Model", *Proceedings of the International Conference on Acoustics, Speech, and Signal Processing 84,* pp. 27.6.1-27.6.4.

MAC AULAY, R.J., and T.F. QUATIERI, (1986), "Speech Analysis/Synthesis Based on a Sinusoidal Representation", *IEEE Trans. on Acoustics, Speech, and Signal Processing,* vol. 34, pp. 744-754.

MAC AULAY, R.J., and T.F. QUATIERI, (1991), "Sine-Wave Phase Coding at Low Data Rates", *Proceedings of the International Conference on Acoustics, Speech, and Signal Processing 91,* S9.1, pp. 577-580.

MARQUES, J., and L. ALMEIDA, (1989), "Frequency-Varying Sinusoidal Modeling of Speech", *IEEE Transactions on Acoustics, Speech, and Signal Processing,* vol. 37, n° 5, pp. 763-765.

MARQUES, J., L. ALMEIDA, and J. TRIBOLET, (1990), "Harmonic Coding at 4.8 Kbps", *Proceedings of the International Conference on Acoustics, Speech, and Signal Processing 90,* vol. 1, pp. 17-20.

ROWE, D., W. COWLEY, and A. PERKIS, (1991), "A Multi-Band Excitation Linear Predictive Speech coder", *Proc. Eurospeech 91,* Genova, pp. 239-243.

SERRA, X., and J. SMITH III, (1990), "Spectral Modeling Synthesis: A Sound Analysis/Synthesis System Based on a Deterministic plus Stochastic Decomposition.", *Computer Music Journal,* vol. 14, n° 4, pp. 12-24.

SORENSEN, H.V., D.L. JONES, M.T. HEIDEMAN, and C.S. BURRUS, (1987), "Real-Valued Fast Fourier Transform Algorithms", *IEEE Transactions on Acoustics, Speech, and Signal Processing,* vol. 35, n° 6, pp. 849-863.

STYLIANOU, Y., O. CAPPE, and E. MOULINES, (1995), "Statistical Methods for Vice Quality Transformation", *Proceedings of Eurospeech 95,* Madrid, pp. 447-450.

STYLIANOU, Y., J. LAROCHE, and E. MOULINES, (1995), "High-Quality Speech Modification based on a Harmonic + Noise Model", *Proceedings of Eurospeech 95,* Madrid, pp. 451-454.

WERY, B., A. LEROUX, H.P. DELBROUCK, and J. LECLERC, (1989), "Un Nouveau Procédé de Codage Paramétrique dans le Domaine Fréquentiel", *Traitement du Signal,* vol. 6, n°4, pp. 277-289.

Chapter Ten

TIME-DOMAIN ALGORITHMS

The Queen jumped up and bawled out, "He's murdering the time! Off with his head!"

Lewis Carroll, *Alice's Adventures in Wonderland*

Among the recently developed text-to-speech synthesis techniques, the pitch-synchronous overlap-add (PSOLA) algorithms (Charpentier and Stella, 1986; Moulines and Charpentier, 1990; see also Lent, 1989) have drawn considerable attention, given their exceptional segmental and suprasegmental efficiency, associated with a virtually unequaled simplicity in the case of their time-domain version: TD-PSOLA[1] (Hamon *et al.*, 1989). They extend the early time domain harmonic scaling (TDHS) of Malah (1979) and the synchronized overlap-add (SOLA) technique of Roucos and Wilgus (1985), which highlighted the efficiency of simple block operations for time scaling of speech. The distinctive feature of the TD-PSOLA method, already introduced in Lukaszewicz and Karjalainen (1987), is the idea that it is also possible to perform the required pitch modifications directly on continuous waveforms, without the use of any parametric model, thereby increasing the segmental quality provided by TTS synthesizers.

If we restrict ourselves to a purely periodic signal $s(n)$, it is possible to get a pitch-modified version $\tilde{s}(n)$ of $s(n)$ by summing *OLA frames* $s_i(n)$, extracted pitch-synchronously from $s(n)$ with a weighting window $w(n)$, and changing the time-shift between frames from the original pitch period T_0 to the desired period T:

$$s_i(n) = s(n)\, w(n - i\, T_0) \tag{10.1}$$

$$\tilde{s}(n) = \sum_{i=-\infty}^{\infty} s_i(n - i(T - T_0)) \tag{10.2}$$

[1] PSOLA/TD® is a registered trademark of France Telecom.

If $T \neq T_0$, the operation results, according to the Poisson formula[2] (Papoulis, 1962), in a reharmonization of the spectrum of $s_i(n)$ (which, if we assume perfect periodicity, does not depend on i when the original voiced sound is purely periodic) with the fundamental frequency $1/T$:

$$\text{if } s_i(n) \xleftrightarrow{\Im} S_i(\omega) \text{ then } \tilde{s}(n) \xleftrightarrow{\Im} \frac{2\pi}{T} \sum_{i=-\infty}^{\infty} S_i\left(i\frac{2\pi}{T}\right) \tag{10.3}$$

Consequently, provided $w(n)$ can be chosen so that the spectrum of $s_i(n)$ closely matches the spectral envelope of $s(n)$, equation (10.2) provides a simple and efficient way to change the pitch of a periodic signal (Fig. 10.1).

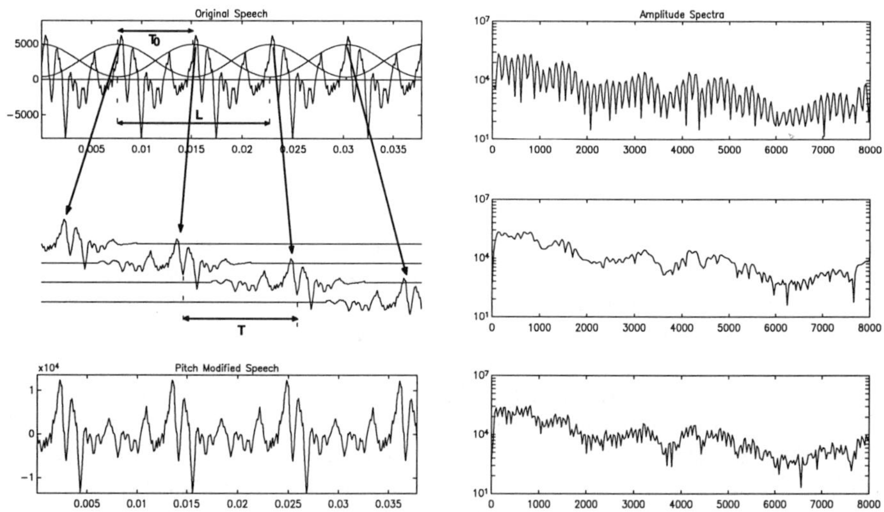

Fig. 10.1. The TD-PSOLA reharmonization process. The pitch-modified waveform (bottom plot) has the same spectral envelope as the original waveform (top plot).

TD-PSOLA is now widely used for concatenative synthesis, because of its high synthesis quality and extraordinarily low computational load (real-time synthesis can be achieved with an Intel 386 processor). Synthesized speech, however, is not perfect. Simplicity, as we shall see, is achieved at the expense of the ease of spectral matching between adjacent segments, a problem which is more easily solved with parametric synthesizers.

---

[2]According to the Poisson formula, summing an infinite number of shifted versions of a given signal $f(t)$ results in sampling its Fourier transform with a sampling period equal to the inverse of the time shift:

$$\text{if } f(t) \xleftrightarrow{\Im} F(\omega)\,, \text{ then } \sum_{n=-\infty}^{+\infty} f(t-nT_0) \xleftrightarrow{\Im} \frac{2\pi}{T_0} \sum_{n=-\infty}^{+\infty} F(\frac{n}{T_0})\,\delta(w-\frac{n}{T_0})$$

10.1. The TD-PSOLA "model"

It is quite strange to think of TD-PSOLA as a model. Indeed, when applied in the context of a copy synthesizer—that is, with $T=T_0$—equation (10.2) simply results in a reconstructed signal that is approximately proportional to the original one:

$$\tilde{s}(n) = \sum_{i=-\infty}^{\infty} s(n)w(n-iT) = s(n)\hat{w}(n) \approx K\, s(n) \tag{10.4}$$

provided the spectrum of the analysis window is narrow enough in comparison with $2\pi/T_0$ (its size is large enough with respect to T_0) for the Poisson formula to produce a Dirac pulse spectrum for $\hat{w}(n)$. In the particular case of a triangular window, the size of which is chosen as twice the pitch period, approximation (10.4) even reduces to a true equality, with K=1.

We shall nevertheless use the terminology of models since, like with any other concatenative TTS systems, PSOLA-based systems conform to the general block diagram of Fig. 7.2, and therefore require their segment database to be stored in the form of lists of parameters (one list for each segment). In the particular case of TD-PSOLA, parameter lists reduce to sequences of pitch markers η^i indicating the center of OLA frames. These are positioned pitch-synchronously on the voiced parts of segments with the help of a pitch extraction algorithm, and regularly spaced on unvoiced sounds. Segment waveforms are stored as such.

In practice, the length of the weighting window *w(n)* is implicitly adapted to the local pitch period, so that the samples of $s_i(n)$ only differ from 0 on an interval which depends on the *overlap factor* F_R, defined as the ratio of the size L of the weighting window *w(n)* by the analysis pitch period, minus one ($F_R = L / T_0 - 1$). Hence, TD-PSOLA can be seen as an intermediate between two extreme situations, neither of which offers satisfactory synthesis results: if F_R is too large, spectral lines appear in the spectrum of $s_i(n)$, which prevent *s(n)* from being reharmonized; if F_R is too small, a coarse harmonization is produced. Moreover, approximation (10.4) no longer holds. The fact is that, instead of suffering from both problems, the intermediate case gives a very good quality for reasonable values of F_R: if $F_R \approx 1$ (and if the original signal is complex enough!), the spectrum of $s_i(n)$ approximates the envelope of the spectrum of *s(n)* and the re-harmonization operation changes the pitch without affecting the formant frequencies and bandwidths (Fig. 10.2).

10.2. Database compression

Since the TD-PSOLA "model" does not require any parameter estimation stage (except pitch marking), it is not linked to any specific data reduction algorithm. Or, in other words, it can *a priori* be associated with any existing speech coding compression technique. In practice, the range of useful coding algorithms is drastically constrained by the intrinsically high quality and simplicity of PSOLA. Indeed, the difficult but necessary tradeoff between the storage reduction achieved by a given compression technique and the distortion it introduces in synthetic speech is compounded by the fact that the

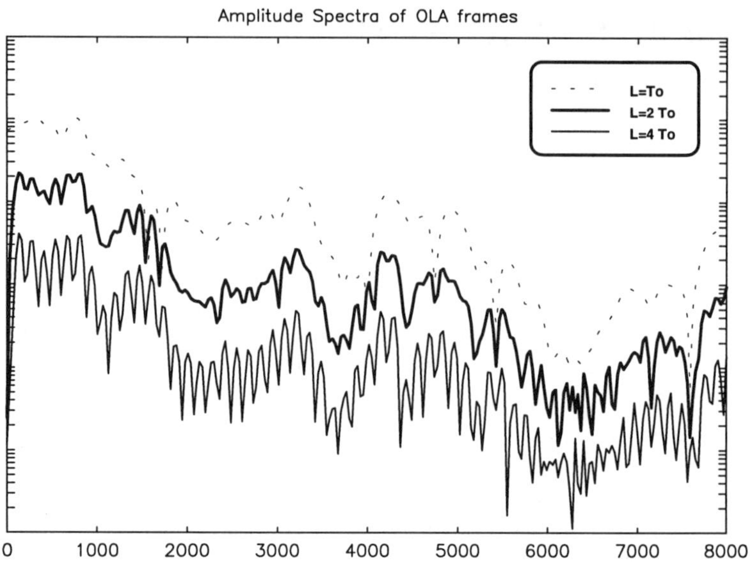

Fig. 10.2. Amplitude spectra of OLA frames extracted from the French vowel [a], for several values of F_R.(F_S: 16 kHz; Window: Hamming).

computational costs of most coding algorithms deprive PSOLA of its initial straightforwardness.

Waveform coding techniques typically require less computational power, but offer lower data reduction ratios. A zero-tap DPCM coder seems to be best suited to the low computational requirements of TD-PSOLA, since it results in only one extra addition per sample at synthesis time.[3] A transmission rate of 40 kb/s has been reported by (Le Faucheur *et al.*, 1991) for a sampling frequency of 8 kHz. A little bit less than 80 kb/s seems to be a reasonable value for 16 kHz, given the additional redundancy of speech in higher frequencies. This, for instance, reduces the size of a complete French diphone database recorded at 16 kHz from more than 5 Mbytes down to approximately 1.7 Mbytes.

10.3. Prosody matching

Modifying both the duration and the pitch of a given segment can be performed in a straightforward way. Let us first assume that, prior to any processing, we associate with each analysis pitch mark η^i the value of its local pitch period T_0, so that a pair (η^i, T_0^i) suffices to define the corresponding analysis OLA frame $s_i(n)$. It is easy to derive the position of synthesis pitch marks η'^j: the shift between two successive synthesis pitch

---

[3] *N*-taps differential coders, either adaptive or not, involve $(2N+1)$ elementary operations per sample, which prevent them from being implemented on common low-cost processors.

marks η'^j and η'^{j+1} is adjusted to the value of the synthesis pitch period at time η'^j, given by the pitch function $T_0(t')$ (see Section 7.2.2) in which t' refers to the position of the last synthesis pitch mark η'^j. By applying the time alignment function $t'(t)$ on each initial pitch mark position η^i, we also obtain the corresponding *analysis pitch mark on the synthesis time axis*, denoted as $v^i=t'(\eta^i)$. It is then easy to associate an analysis pitch mark η^i to each synthetic pitch mark η'^j by looking for the closest analysis pitch mark on the synthesis time axis (Fig. 10.3). Given the time distortion introduced by $t'(t)$, some analaysis pitch marks are not be associated with synthetic pitch marks, and others are duplicating. A list of synthesis parameters is finally created, defined by triplets (η'^j, η^i, $T_0{}^i$) in which η'^j gives the synthesis pitch marks positions and η^i, $T_0{}^i$ refers to the analysis OLA frame associated to the current synthesis pitch mark. These triplets are indicated as $\eta'^j{}_i$ in Figure 10.3.

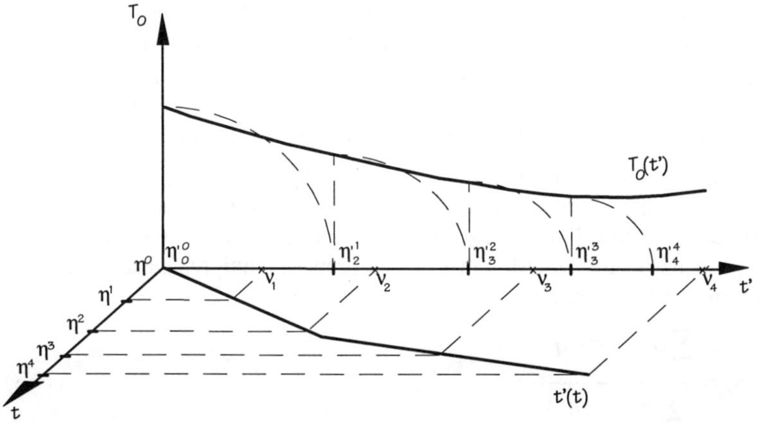

Fig. 10.3. Pitch and timing modifications with TD-PSOLA. Starting from five analysis OLA frames on the analysis time axis (t), four synthesis pitch marks are positioned on the synthesis time axis (t'), and respectively associated to an analysis frame. In the example, frame 1 has been eliminated, while frame 2 has been duplicated.

The acoustical distortions involved in such a prosodic modification scheme are surprisingly low. Pitch factors from 0.5 to 2 are typical. Stretching factors typically range from 0.25 to 2. Lower values of the duration parameter prevent normal speech from being understood (which happens to be the case for human speech, too), while audible buzziness appears in unvoiced sound when larger values are applied, due to the regular repetition of identical OLA frames. In order to prevent the algorithm from introducing such an artificial short-term autocorrelation in the synthesis signal, Charpentier and Moulines (1989) have proposed reversing the time axis of every repeated version of an unvoiced OLA frame. With such an artifice, we found that speech can reasonably be slowed down by a factor of four, even though some tonal effect is still encountered in voiced fricatives, which typically combine voiced and unvoiced frequency regions and therefore cannot be time reversed.

10.4. Speech synthesis

Synthesis can be performed as in equations (10.1) and (10.2), the size of the weighting window being adjusted from frame to frame:

$$w_j(n) = w_1\left(n\frac{T_e}{T_0{}^j}\frac{1}{F_R}\right) \tag{10.5}$$

$$s_j(n) = s(n)\, w_j(n-\eta^j) \tag{10.6}$$

$$\tilde{s}(n) = \sum_{j=-\infty}^{\infty} s_i(n-\eta'^j) \tag{10.7}$$

in which $w_1(n)$ is a weighting window that takes its values in the range [0..1]. OLA frames are extracted from segment waveforms at the position indicated by their pitch marker η^j and sent to an overlap-adder, similar to the one of Section 9.6, with a shift value equal to $(\eta'^j-\eta'^{j-1})$. With the standard value of $F_R = 1$, the infinite summation above is limited to a maximum of four terms, when the *pitch factor*, defined as the ratio of the local synthesis pitch period to the original one ($F_P=T/T_0$), is varied in the range [0.5...2].

It should be noted, however, that the accuracy of approximation (10.4) depends on the value of the synthesis pitch period rather than on the analysis period. When $F_P >1$, it obviously tends to fail. Even in case $F_P <1$, the value of K is highly dependent on the pitch factor. Charpentier and Moulines (1989) propose multiplying each synthesis sample by two normalization factors:

$$\tilde{s}(n) = \frac{\sum_i \alpha_i\, s_i(n-\eta'^i)}{\sum_i w_i(n-\eta'^i)} \tag{10.8}$$

where α_ι is introduced to compensate for the dependence of K on F_P, and the denominator plays the role of a dynamic compensation factor, which counterbalances the variations of K with n. Dutoit (1993) noted no significant quality degradation when the denominator of (10.8) was not used and thus eliminated it. Additionally, getting rid of α_ι did not result in unnatural speech, though it frequently produced arithmetic overflows for high values of F_P. The problem was solved by imposing $\alpha_\iota=1/F_P$.

Regarding computational complexity, one could hardly imagine a simpler algorithm.[4] Since the weighting window length depends on the analysis shift, weighting values have to be computed for every OLA frame. Assuming they are tabulated for a reference length, it is easy to show that ($4F_R$+3) elementary operations per sample are required,[5] which makes seven in the intermediate case of F_R=1.

---

[4]Though it is shown to be possible in Section 10.6.5.

[5]This value actually depends on the pitch factor. It is given for $F_P = 1$.

10.5. Segmental quality with TD-PSOLA

As introduced in Section 10.2, the segmental quality provided by TD-PSOLA in the context of copy synthesis is close to perfection. Furthermore, important prosodic modification can be achieved with almost no effect on segmental quality, as long as they are performed on continuous signals.

Some segmental problems arise when the algorithm is used to concatenate speech segments. Specifically, the OLA operation, when performed inside a quasi-stationary segment, approximates the theoretical case described in (10.3) (where the added OLA frames are all identical). Things may not be the same at the boundary of two segments extracted from different words, due to three incoherence events, respectively related to phase, pitch, and overall spectral envelope mismatches (Dutoit and Leich, 1993), and illustrated in Figs. 10.4, 10.5, and 10.6 for the concatenation of vowels [a] sampled at 16 kHz. For each, the top, middle, and bottom plots respectively show the last frames of the first segment, the first frames of the second one, and the result after OLA. OLA windows are positioned as they would be in the TTS segment database.

10.5.1 Phase mismatch

Even if the wave shapes to be concatenated are identical, the OLA operation may result in an unexpected discontinuity if the respective windows are not positioned at the same place within the period (Fig. 10.4).

The reader may argue that the algorithm responsible for this positioning should take this effect into consideration and try to choose the glottis closure instant (the only reference instant available in voiced speech) as the center of each window. This is the solution normally adopted (Di Francesco and Moulines, 1989).[6] It has two major drawbacks: it requires expert assistance and it is not very precise if performed automatically (see Cheng and O'Shaughnessy, 1989, for a review), and time consuming if performed visually. Even so, it cannot completely avoid phase mismatches (Macchi *et al.*, 1993), given the fact that once the first window has been positioned on the first voiced segment of the database, all the others are imposed, through period by period propagation. As a consequence, phase coherency is rather hard to maintain, although it is possible.

Dutoit (1993), for instance, developed a computer-aided pitch marking system (with the help of interactive visualization tools) and observed a strict methodology. All [aX] diphones were first marked, where X stands for any sonorant. Pitch marks were always positioned on the first negative peak of the first period in the initial [a], which was found to be reliable. Subsequent pitch marks were propagated from period to period on the basis of a preliminary pitch analysis with the highly accurate pitch determination algorithm implemented in the MBE analyzer of Chapter Nine (precision : one-eighth of a sample). Each period was visually checked and possibly corrected, at each propagation

[6]Synchronization on the glottal closure instant actually presents another advantage: it has been reported by Charpentier and Moulines (1989) that, even when simply processing continuous speech, the re-harmonization process described by equation (10.3) was better achieved when pitch marks were chosen in the vicinity of the main excitation within the pitch period.

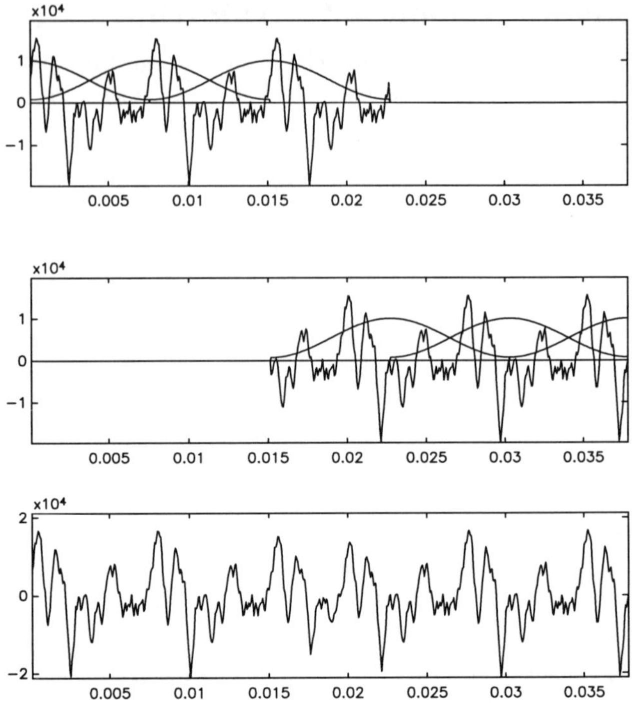

Fig. 10.4. Phase mismatch: Waveforms are identical but the OLA windows are not centered on the same relative positions within periods.

step. Dutoit then obtained a reference position of pitch marks in all sonorants, which was found to be somewhat different from the expected first negative peak. Although this simple positioning rule was approximately respected for most vowels, it was not in nasal consonants, in the steady state of voiced plosives, nor in voiced fricatives (see Fig. 10.5). This effect was later confirmed on other diphones. In the case of nasals and plosives, it can be simply understood as the influence of the low-pass filtering of the glottal pulses by the nasal tract, which results in long fall and rise times. As for fricatives, higher-frequency formants are responsible for their highly noisy waveform and therefore prevent glottal pulses from being reliably detected by simple visual inspection.

A possible solution to this problem would be to measure glottal activity directly, with the help of a electro-laryngograph (Hess and Indefrey, 1987), instead of trying to recompute it from speech data. To our knowledge, this has never been published in the context of PSOLA synthesis, although it has been investigated (Conkie, 1996).

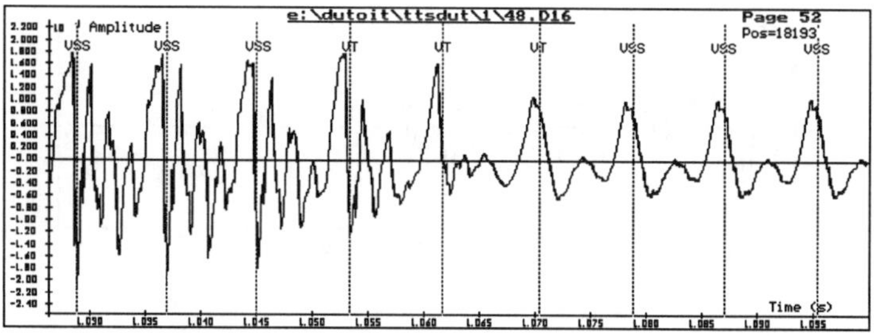

Fig. 10.5. Pitch marks in diphone [an] are supposed to fall on the first negative peak of each period. It happens not to be the case in practice. In this particular case, it would have been wiser to position pitch marks on the first positive peak, which is found to be more reliable.

10.5.2 Pitch mismatch

Signals having identical spectral envelopes and windowed on similar relative positions cannot be properly overlap-added if their periods are very different. Again, this situation cannot easily be avoided in a database of several thousand segments (Fig. 10.6). It is nevertheless possible to minimize this factor by recruiting professional speakers to record the segments database and training them to read the corpus with the most constant pitch possible.

10.5.3 Spectral envelope mismatch

Being a time-domain synthesis technique, TD-PSOLA has no straightforward means of adapting the spectral envelopes of concatenated segments to one another (Fig. 10.7). At best, spectral envelope mismatches can be minimized, by using context-dependent segment boundaries (Conkie and Isard, 1994). Spectral envelope mismatch, however, is well known as one of the most important drawbacks of concatenative synthesis. It is mostly due to speaker's phonemic variability and to co-articulation effects intrinsic to any language. Parametric synthesizers, such as the LPC and hybrid harmonic/stochastic ones examined in Chapters Eight and Nine respectively, can naturally cope with this spectral envelope mismatch. So does FD-PSOLA, the computationally expensive frequency-domain equivalent of TD-PSOLA (Charpentier and Moulines, 1988). The standard implementation of TD-PSOLA cannot. As a result, synthesizing highly fluent speech with TD-PSOLA often requires a painstaking trial and error process in which the most probable diphone sequences are systematically tested, and diphones are rejected when an important spectral envelope mismatch is encountered. Other versions of the rejected diphones are then extracted from words with a different phonetic context than the initial one, tests are run again, and so on.

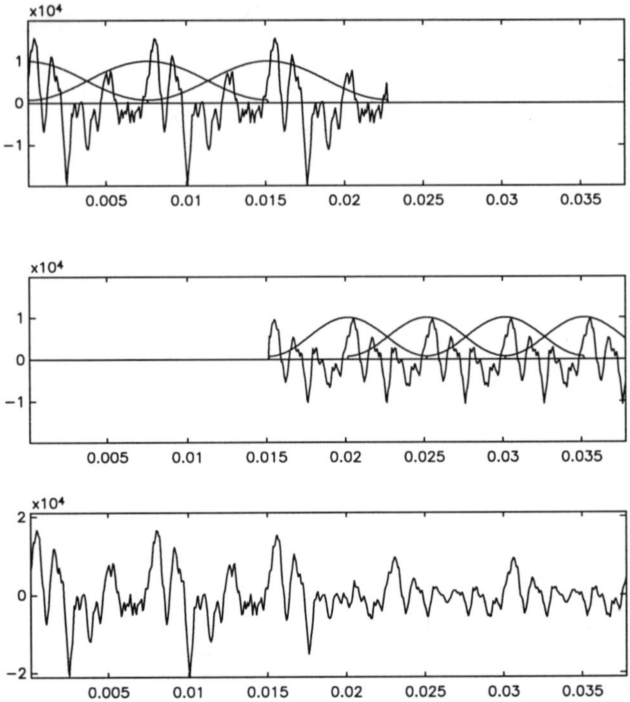

Fig. 10.6. Pitch mismatch: Both segments have exactly the same spectral envelope but were pronounced with different pitches. OLA windows are positioned coherently. The synthetic signal is obtained after changing the interval between the frames of the right segment, in order to impose a constant pitch.

As we shall see in the next sections, this can be avoided in two ways. A first solution consists of *resynthesizing the PSOLA segment database* (the original speech segments, since PSOLA involves no parametric analysis) so as to provide segments with interesting properties and to turn them into account at synthesis time. Resynthesis is performed once on the database, off-line. This is the case of the so-called multiband resynthesis PSOLA (MBR-PSOLA) technique (Dutoit and Leich, 1993) and of the resynthesis performed in the PHRITTS TTS system (Meyer *et al.*, 1993), which happen to provide an efficient solution to pitch and phase mismatches as well. It is also possible to *associate TD-PSOLA with a flexible parametric speech model* and to perform spectral smoothing in the parametric domain. One can truly associate a parametric synthesizer with TD-PSOLA, as with the so-called linear prediction PSOLA (LP-PSOLA) approach, in which prosodic modifications are performed with the TD-PSOLA algorithm on the excitation of the all pole synthesis filter rather than on direct speech (Moulines and Charpentier, 1990).

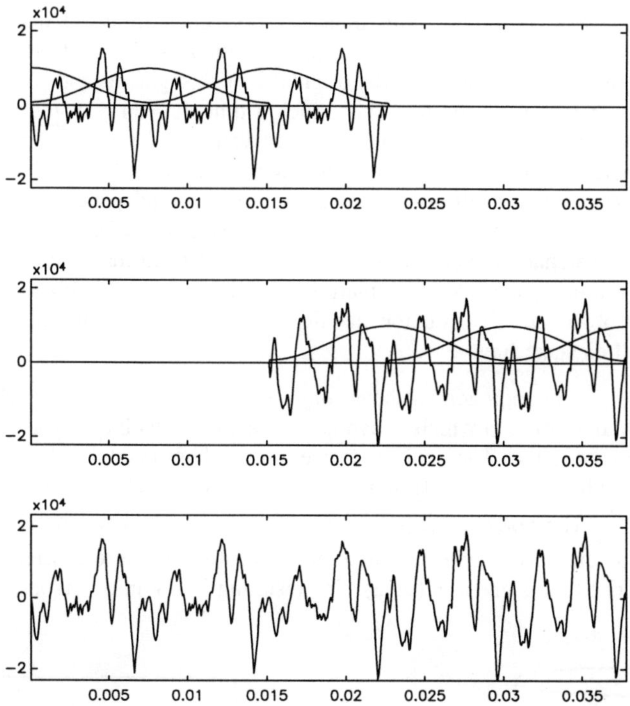

Fig. 10.7. Spectral envelope mismatch: The left segment is the diphone [ma], the right one is [am]. Pitch is constant, and windows are again positioned coherently. The spectral discontinuity is concentrated in one period.

10.6. Resynthesizing the segment database

10.6.1. The resynthesis process

In order to approach the optimum case of identical OLA windows, as required for equation (10.3) to be truly applicable, the speech database should exhibit the following characteristics:

1. All the words should be pronounced with a constant pitch;
2. A coherent positioning of OLA windows should be possible;
3. Spectral interpolation between concatenated windows should be easy to perform, without an excessive increase of the computational load.

A first step in this direction consists of eliminating pitch mismatches thanks to a pitch-constant resynthesis of all the pitch periods of the segment database. In order to maintain the naturalness of synthetic speech, however, this operation should only minimally affect the quality of TD-PSOLA segments.

Meyer *et al.* (1993) proposed, with their PIOLA (pitch inflected overlap and add) method, to perform such a transparent resynthesis with TD-PSOLA itself, by positioning OLA frames pitch synchronously on the original speech segments and simply adding the resulting OLA frames with constant shift. Such a time-domain resynthesis does not get rid of phase mismatches: the combination of resynthesized segments into speech still requires a coherent positoning of OLA frames at synthesis time (although it is useless for the resynthesis operation itself, as segments are processed independently of each other). In order to cope with it, PIOLA performs, at synthesis time, a phase optimization on the segments to concatenate. Given the position of the last pitch mark on the left segment, the optimal position of the first pitch mark on the right segment is obtained (by finding the maximum of the cross-correlation function, for instance), and the signal of the right segment is shifted accordingly.

A somewhat different approach has been followed by Dutoit and Leich (1993) with their MBR-PSOLA algorithm, in which re-synthesis is achieved with the hybrid H/S model of Chapter Nine (Fig. 10.8). The analysis frame rate is chosen equal to the synthesis pitch period, which is itself computed from the average fundamental frequency throughout the database: ω_0^{SYN}. Harmonic amplitudes, phases, and narrow band noise variances are computed with the MBE algorithm, and submitted to a frame classification algorithm similar to the one that is given in (9.32). The resulting V/UV information is used to drive

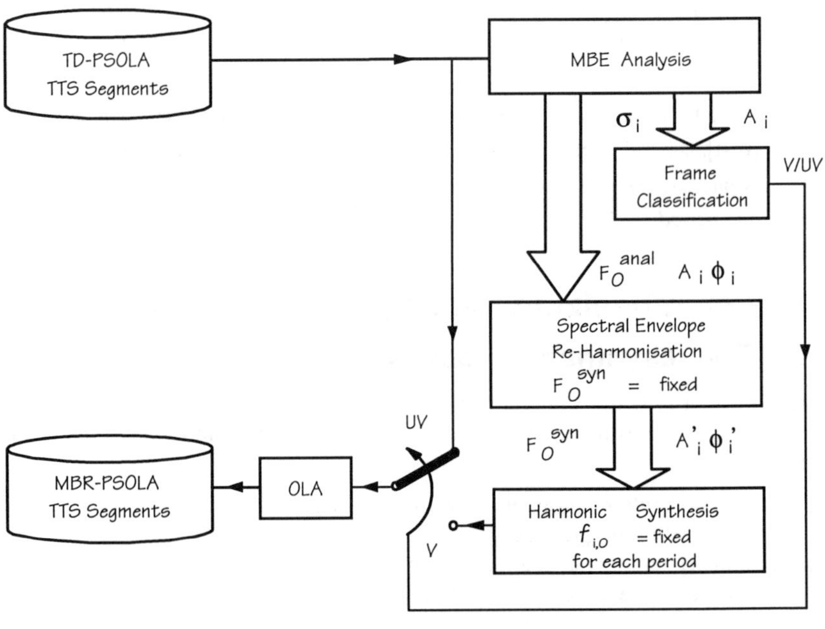

Fig 10.8. The MBE resynthesis operation of MBR-PSOLA.

analysis frames either directly to the OLA in the case of unvoiced frames, which are left untouched, or through the complete resynthesis process. In this case, phases are set to fixed value and amplitudes are submitted to the prosody matching algorithm given in Section 9.4, with time and frequency alignment functions set to $t'(t)=t$ (no time axis distortion) and $\omega_0(t')=\omega_0^{SYN}$. Variances are no longer used. Purely harmonic pitch periods that fully satisfy condition 1 are finally produced and sent to the overlap-adder, with an OLA factor equal to 1 (i.e. synthesis frames are two pitch periods long). The advantage of resetting harmonic phases to well chosen values at the beginning of each pitch period (which can only be achieved with a parametric resynthesis operation), is that phase mismatches are automatically eliminated. Pitch marking turns out to be automatic. As a matter of fact, given the constant pitch period and initial harmonic phases imposed during the resynthesis operation, pitch markers can be given any relative position within resynthesized periods (starting from the reset instant), provided it is kept constant throughout the voiced part of the database. Since the reset instant is totally controlled, pitch markers are implicitly imposed.

What is more, a very important by-product of the resynthesis process is that the concatenation of two segments can now be achieved by a simple linear interpolation in the time domain. As a matter of fact, in the particular case of two signals with the same pitch and provided their harmonic phases are identical (which is imposed by MBR-PSOLA and only approximately verified with the approach of Meyer *et al.*, 1993), it is easily understood that spectral envelope interpolation is equivalent to direct temporal interpolation.[7, 8]. This enables the TTS synthesizer to fulfill condition 3 by performing the desired spectral matching directly on neighboring segment waveforms, prior to the OLA itself.[9] Smoothing is again performed with equations (7.1) and (7.2), in which p_L^0 and p_R^0 respectively denote the last and first OLA windows of the left and right concatenated segments (Fig. 10.9). Spectral envelope interpolation is achieved through a very simple *time-domain linear smoothing* of the voiced pitch periods.

In the case of MBR-PSOLA, the quality of the linear smoothing achieved by this algorithm is therefore exactly equivalent to that obtained with hybrid H/S parameters (see Fig. 9.4), except that only voiced parts are smoothed here. In most cases the smoothing algorithm succeeds in its task: few concatenation points, if any, can still be detected (see Fig. 10.10). These are mostly concentrated at the border of unvoiced segments, or when voiced transients are encountered at segments borders. The use of higher-level segments, such as triphones, may help in this case, even though we believe that it is not really worth trying to further refine concatenation with this algorithm.

---

[7]Temporal interpolation of signals is always equivalent to their *complex* spectral interpolation.

[8]This can be seen as a (much simpler) temporal equivalent to the spectral linear interpolation proposed in Charpentier and Moulines (1988).

[9]Such a smoothing strategy, however, only holds for the concatenation of voiced segments. Unvoiced spectral amplitude mismatches are left unsolved.

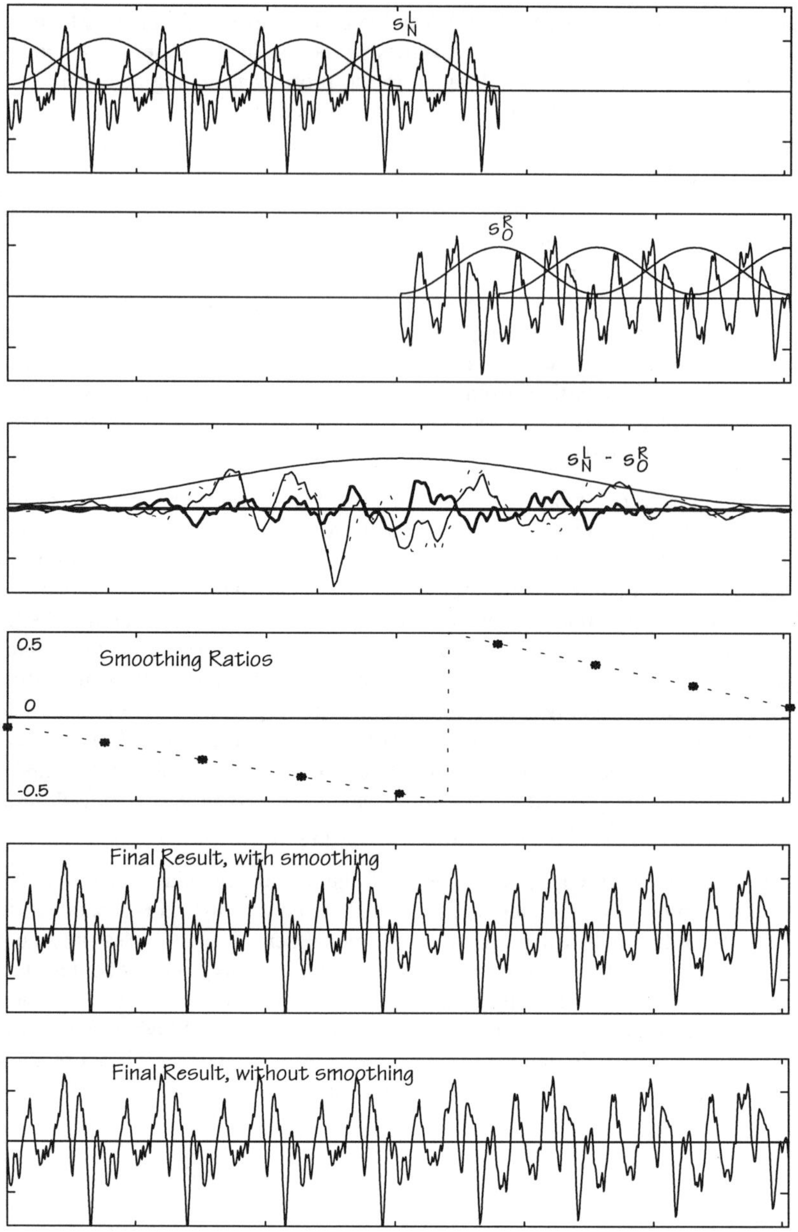

Fig. 10.9. The time-domain linear smoothing process. $M_L = 5$; $M_R = 4$. Clearly, waveform transitions are more regular when smoothing is applied. This results in more fluent speech.

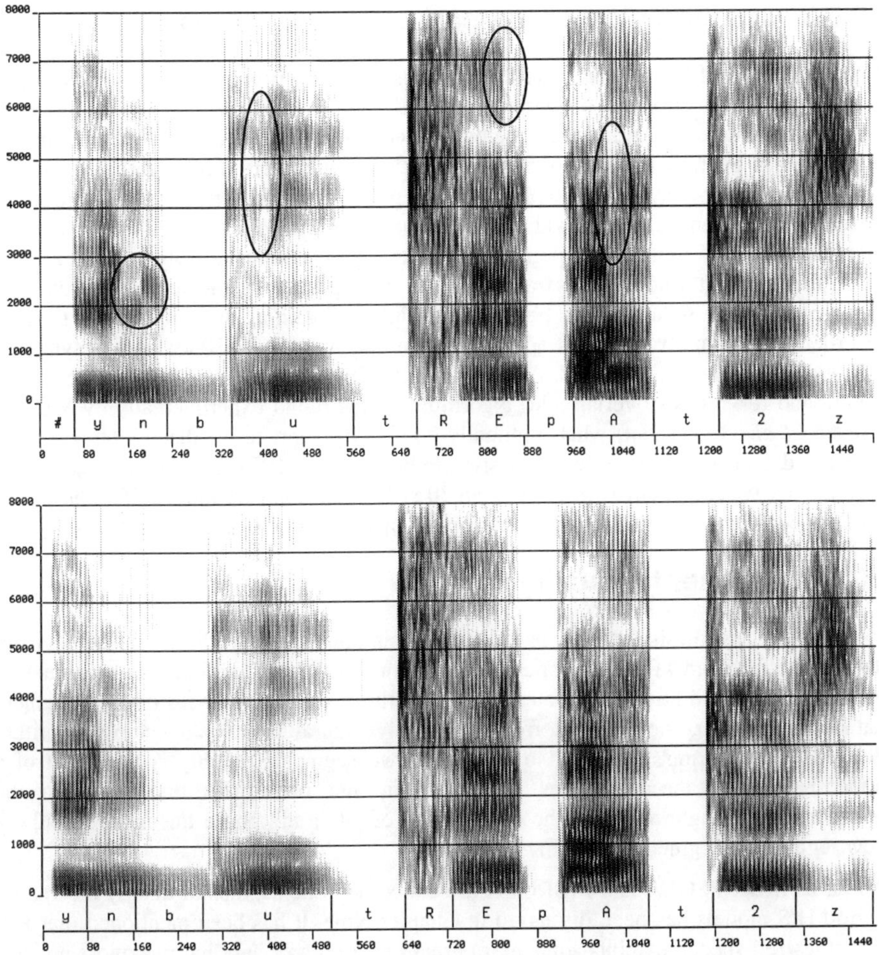

Fig. 10.10. A spectrograph showing the effect of the time-domain linear smoothing process. Synthesis of the French phrase *une boue très pâteuse*. Top: without smoothing (spectral discontinuities are spotted); bottom: smoothed.

10.6.2. Modified PSOLA synthesis

Synthesizing speech with the new OLA frames is essentially the same as with TD-PSOLA. Better still, computational load is reduced to the very minimal value of $(3F_R+2)$ operations per sample, for one and the same weighting window is used for all OLA frames. Taking temporal interpolation into account, this figure goes back to $(5F_R+2)$: exactly as with TD-PSOLA for $F_R=1$.

10.6.3. Compression of the new segment database

Another benefit of the constant pitch resynthesis of voiced frames is that it is possible to enhance data reduction by developing tailored segment compression algorithms with limited computational cost. Indeed, the best suited but rather inefficient DPCM technique mentioned in Section 10.2 can now be applied on a pitch period basis rather than on a sample one. In this way, each sample in a given pitch period gets stored in the form of the difference between its current value and the value of the corresponding sample in the previous period[10]. Since pitch-period waveforms evolve rather slowly with time, and given the fact that voiced OLA frames fill up to 75 percent of the segment database[11], significant storage reduction can be expected while maintaining the simplicity of DPCM. In order to maximally take this advantage into account, the MBR-PSOLA resynthesis operation was recently modified, thereby turning MBR-PSOLA into the new MBROLA (multiband resynthesis overlap-add) algorithm. It was found experimentally by Van der Vreken *et al.* (1997) that, when optimally combining short term (that is, on a sample basis) and long term (on a period basis) predictions on MBROLA diphone databases, high quality was maintained with less than 30 kb/s (to be compared to the 256 kb/s of the initial speech data at 16 kHz), with very limited increase in computational cost.

10.6.4. Segmental quality

PSOLA TTS synthesis systems based on a pitch-constant resynthesis of the segment database differ from TD-PSOLA-based systems in that they exhibit limited concatenation mismatches and additional concatenative capabilities due to their customized segment database. As such, they benefit from the highly natural prosody matching algorithm derived from Poisson's formula, so that their own segmental quality is the result of a tradeoff between the distortions introduced by the resynthesis operation and the enhancement brought about by the cancellation of pitch and phase mismatches and the new segment-to-segment smoothing strategy.

As far as MBR-PSOLA and MBROLA are concerned, the segmental quality offered by hybrid H/S models has been discussed in Chapter Nine. It has been mentioned that H/S resynthesized speech exhibits some extra breath in sonorants and that buzziness appears in concatenating voiced fricatives, even when the original phases are maintained. In order to avoid the first error, the resynthesis operation used in MBR-PSOLA and MBROLA is constrained to voiced frames, in which only harmonic components are taken into account. As a result, voiced frames are resynthesized without their stochastic component, thereby avoiding extra breath. In contrast, given the coarse phase reset additionally required by MBR-PSOLA, some buzziness is inevitably introduced in the segment database. In practice, it is found to be highly dependent on the phase reset values. It is well known that constant or linearly distributed harmonic phases lead to

[10]Except of course the first pitch period of any segment, which can merely be coded with a classical sample-to-sample DPCM technique.

[11]For the French diphone database used in Dutoit (1993), periods were distributed as follows: VSS: 11680, VT: 2943, UVSS: 3020, UVT: 1438 for a total of 19081 OLA Frames. Such a number of voiced frames (76 percent) is not surprising: vowels are less numerous but much more lengthy than consonants.

somewhat metallic sounds. In Dutoit and Leich (1993), phases were given random (but fixed) initial values, resulting in a higher quality synthetic voice. It remains, however, that fluency in MBR-PSOLA is achieved at the expense of some extra buzziness.

To a lesser extent, the TD-PSOLA resynthesis approach of Meyer *et al.* (1993) also slightly affects the final segmental quality of synthetic speech. TD-PSOLA does not strictly maintains the spectral envelope of the original speech signal (see Fig. 10.1 again). The resulting degradation is reinforced by the fact that it is applied twice (once off-line and once at synthesis time). What is more, linear smoothing in the time domain occasionally introduces phase distortions, since the harmonic phases of the left segment cannot be guaranteed to be identical to the corresponding phases in the right segment.

10.7. Combining PSOLA with a parametric synthesizer

We have seen in Section 10.2 that the TD-PSOLA algorithm does not easily lend itself to high data compression ratios. Besides, being a time-domain synthesis technique, it does not provide efficient means of smoothing spectral envelope mismatches (except if the base speech corpus has been recorded by a speaker capable of maintaining a constant pitch and constant harmonic phases throughout his (or her) reading, as mentioned in Section 10.5.3). The *linear prediction PSOLA* (LP-PSOLA) approach (Moulines and Charpentier, 1988) was introduced as a means of combining the high compression and smoothing capabilities of the AR model with the natural prosodic modifications provided by TD-PSOLA. LP-PSOLA techniques thus associate a parametric LP synthesizer with a TD-PSOLA algorithm. Prosodic modifications are performed with the TD-PSOLA algorithm on the excitation signal of the all pole synthesis filter rather than on direct speech. The output of the overlap-adder is simply connected to an LPC synthesis filter, on the coefficients of which linear smoothing rules are applied (typically on LARs). Several LP-PSOLA techniques can be derived from this principle, depending the type of excitation signal used.

In the *multipulse linear prediction coding PSOLA* algorithm (MPLPC/PSOLA) of Moulines and Charpentier (1988), for instance, TD-PSOLA is applied on the multipulse excitation signal of each synthesis segment. In order to avoid the need for a careful marking of the excitation waveform (hence, to prevent further phase mismatches at the concatenation of segments: pulse positions are not coherent from one segment to another, because they originate in MPLPC analysis frames centered on different positions within periods) they first apply a phase equalization stage of the MPLPC excitation by computing FFTs and zeroing phases in the low-frequency range, and they obtain the phase equalized signal with overlapped IFFTs. This results in excitation periods with imposed harmonic phases, like in MBR-PSOLA except that all computations are performed at synthesis time while they are applied once for all (i.e. off line) with MBR-PSOLA. With this method, Moulines and Charpentier have stored a complete French diphone database recorded at 16 kHz in 700 kbytes of memory—about four times the size of the corresponding LPC diphone database, and one seventh of its uncoded size (see also Le Faucheur *et al.*, 1991). The synthetic speech quality was judged equivalent to that obtained with TD-PSOLA (Moulines and Charpentier, 1990), but they could not avoid using a DSP (TMS320C25) for the real-time implementation. A (wide-band) variable rate CELP coding technique has also been proposed as an alternative to MPLPC for segments database compression (White, 1990). An average bit rate of 18 kbits/s has been reported, with almost no effect on segmental quality, which corresponds to a database size of about 400 kbytes. There again, a DSP was needed.

Recently, a *residue excited PSOLA* (RELP-PSOLA) technique has been tested as a candidate algorithm for high quality speech synthesis (Macchi *et al.*, 1993). No particular speech coding is used, and TD-PSOLA is directly applied the LP residual signal. The resulting algorithm thus looses its data compression capabilities, but maintains the ability of the AR model to smooth out segment mismatches. It was tested on continuous speech and found to be amenable to high-quality prosodic modifications, although a more simple RELP technique based on truncating the pitch period residual or expanding it with zeros when needed performed equally well. As any other LP-PSOLA technique, RELP-PSOLA is about ten times more computationally intensive then TD-PSOLA alone.

10.8 Conclusions

In this tenth and penultimate chapter, we have introduced some increasingly used time-domain synthesis technique and browsed related synthesis algorithms.

TD-PSOLA offers high-quality synthetic speech with very limited computational cost: ten times less than with the corresponding LPC synthesizer. On the other hand, it provides virtually no efficient coding technique and is subject to pitch, phase, and spectral envelope mismatch.

One way of eliminating these problems consists of resynthesizing, once and for all, the voiced parts of all segments with constant synthesis pitch and possibly with fixed initial phases for each period, as performed by the PIOLA and MBR-PSOLA algorithms. Concatenation is then also performed by linear interpolation in the time domain. This time, spectral envelope mismatches can be attenuated, at the expense of some additional buzziness introduced by the resynthesis operation (at least for MBR-PSOLA). The MBR-PSOLA has recently been turned into the more efficient MBROLA one, which exhibits very good database coding performances.

LP-PSOLA techniques associate a parametric LP synthesizer with a TD-PSOLA algorithm. Prosodic modifications are performed with the TD-PSOLA algorithm on linear prediction residual waveforms rather than on direct speech. The output of the overlap-adder is simply connected to an LPC synthesis filter, on the coefficients of which linear smoothing rules are applied. LP-PSOLA techniques, however, suffer from the computational load imposed by linear prediction. They are about ten times more computationally intensive then TD-PSOLA alone.

References

CHARPENTIER, F., and E. MOULINES, (1988), "TTS Algorithms Based on FFT Synthesis", *Proceedings of the International Conference on Acoustics, Speech, and Signal Processing 88*, pp. 667-670.

CHARPENTIER, F., and E. MOULINES, (1989), "Pitch-Synchronous Waveform Processing Techniques for Text-to-Speech Synthesis Using Diphones", *Proceedings of Eurospeech 89*, Paris, vol.2, pp. 13-19.

CHARPENTIER, F., and M.G. STELLA, (1986), "Diphone Synthesis Using an Overlap-Add Technique for Speech Waveforms Concatenation", *Proceedings of the International Conference on Acoustics, Speech, and Signal Processing 86*, pp. 2015-2018.

CHENG, Y.M., and D. O'SHAUGNESSY, (1986), "Automatic and Reliable Estimation of Glottal Closure Instant and Period", *IEEE Transactions on Acoustics, Speech, and Signal Processing*, vol. 37, n°12.

CONKIE, A., (1996), Private Communication.

DI FRANCESCO, R., and E. MOULINES, (1989) "Detection of the Glottal Closure by Jumps in the Statistical Properties of the Signal", *Proceedings of Eurospeech 89*, Paris, vol. 2, pp. 39-41.

DUTOIT, T., (1993), *High-Quality Text-to-Speech Synthesis of the French Language*, Ph.D. dissertation, Faculté Polytechnique de Mons.

DUTOIT, T., and H. LEICH, (1993), "MBR-PSOLA: Text-to-Speech Synthesis Based on an MBE Resynthesis of the Segments Database", *Speech Communication*, n°13, pp. 435-440.

DUTOIT, T., (1994), "On the Ability of Various Speech Models to Smooth Segment Discontinuities in the Context of Text-to-Speech Synthesis by Concatenation", *Proceedings of Eusipco 94*, vol. 1, pp. 8-12.

HAMON, C., E. MOULINES, and F. CHARPENTIER, (1989), " A Diphone System Based on Time-Domain Prosodic Modifications of Speech", *Proceedings of the International Conference on Acoustics, Speech, and Signal Processing 89*, S5.7, pp. 238-241.

HESS, W., and H. INDEFREY, (1987), "Accurate Time-Domain Pitch Determination of Speech Signals by Means of a Laryngograph", *Speech Communication*, n°6, pp. 55-68.

LE FAUCHEUR, L., O. BOËFFARD, B. CHERBONNEL, AND S. WHITE, (1991), "Un Algorithme de Synthèse de Parole de Haute Qualité", *Proceedings of the Séminaire SFA/GCP*, pp. 104-107.

LENT, K., (1989), "An Efficient Method for Pitch-Shifting Digitally Sampled Sounds", *Computer Music Journal*, vol. 13, n°4, pp. 65-71.

LUKASZEWICZ, K., and M. KARJALAINEN, (1987), "Microphonemic Method of Speech Synthesis", *Proceedings of the International Conference on Acoustics, Speech, and Signal Processing 89*, Dallas, vol. 3, pp. 1426-1429.

MACCHI, M., M.J. ALTOM, D. KAHN, S. SINGHAL, and M. SPIEGEL, (1993), "Intelligibility as a Function of Speech Coding Method for Template-Based Speech Synthesis", *Proceedings of Eurospeech 93*, Berlin, pp. 893-896.

MALAH, D., (1979), "Time-Domain Algorithms for Harmonic Bandwidth Reduction and Time-Scaling of Pitch Signals", *IEEE Transactions on Acoustics, Speech, and Signal Processing*, vol. 27, n° 2, pp. 121-133.

MEYER, P., H.W. RÜHL, R. KRÜGER, M. KUGLER, L.L.M. VOGTEN, A. DIRKSEN, AND K. BELHOULA, (1993), "PHRITTS - A Text-to-Speech Synthesizer for the German Language", *Proceedings of Eurospeech 93*, Berlin, pp. 877-890.

MOULINES, E., and F. CHARPENTIER, (1988), "Diphone Synthesis Using a Multipulse LPC Technique", *Proceedings of the FASE International Conference*, Edinburgh, pp. 47-51.

MOULINES, E., and F. CHARPENTIER, (1990), "Pitch Synchronous Waveform Processing Techniques for Text-to-Speech Synthesis Using Diphones", *Speech Communication*, vol. 9, n°5-6.

PAPOULIS, A., (1962), *The Fourier Integral and its Applications*, McGraw Hill, New-York, p 47.

ROUCOS, S., and A. WILGUS, (1985), "High-Quality Time Scale Modification of Speech", *Proceedings of the International Conference on Acoustics, Speech, and Signal Processing 85*, pp. 236-239.

VAN DER VREKEN, O., N. PIERRET, T. DUTOIT, V. PAGEL, and F. BATAILLE, (1997), "A Simple and Efficient Algorithm for the Compression of MBROLA Segment Databases", to appear in *Proceedings of Eurospeech 97*, Patras.

WHITE, S., (1990), "Codeur CELP à Débit Variable: Application au Codage de Diphones", *Proceedings of the 19è Journées d'Etudes sur la Parole*, Montréal, pp. 167-170.

Chapter Eleven

CONCLUSIONS AND PERSPECTIVES

The confusion got worse every moment, and Alice was very glad to get out of the wood into an open place.

Lewis Carroll, *Through the Looking Glass*

The practical goal of any good textbook is to present matters in such a way that readers are less confused *after* reading it than *before*. Or, to put it in information-theoretic terms, it is to decrease the entropy of one's knowledge in a specified area (TTS synthesis in this case) by organizing it into a maximally coherent view: a sort of a book-sized classification tree! It should be clear that, although we have examined the structure of this tree, by pointing out hypotheses and choices (branches) when they arise and by detailing the resulting text analysis and speech synthesis techniques (leaves), a comparative recapitulation is necessary to achieve our goal. This chapter presents such a synopsis, as well as a list of research topics that are known or believed to yield significant challenges for tomorrow's TTS synthesis.

11.1. Synopsis

When developing a TTS system, one has to evaluate many alternatives, regarding both NLP and DSP techniques to be implemented so as to obtain the best speech quality within the limits imposed by economical criteria. As a result of these constraints, which are conditioned by the potential market of the application, it is impossible to give absolute preference for particular techniques. In this section, we shall therefore quote the analysis and synthesis methods presented throughout this textbook and sketch their pros and cons for each of the operations performed by a TTS system.

11.1.1. Natural language processing and TTS synthesis

Table 11.1 summarizes the use and interest of the most widely used NLP methods in the context of TTS synthesis: expert systems on one hand, with regular or augmented context free rules, and corpus-based methods on the other hand, with classification and regression trees, multilayer perceptrons, and hidden Markov models. Analysis scores are mentioned when available, although numerical results should be considered with great care, as they often originate from analyses performed on different text corpora. Since Chapters Four to Six have successively presented each application and browsed the related NLP methods, we shall examine Table 11.1 on a column basis.[1] It reads as follows:

1. Regular rewrite rules appear to be useful at any stage of the analysis process. This should not be surprising: they have almost always been tested and adopted first, given their intrinsic simplicity and their relatively low computational cost for parsing (not to be confused with their inference, which can be time consuming for even expert linguists, as happens to be the case for phonetization). What is more, regular rules are easy to read. Their use has been made popular by the SPE formalism. Their limitations, however, originate in the relative complexity of their inference and in the constraints they impose on the language examined: no hierarchical organization can be highlighted or, *a fortiori*, put to profit.

2. This drawback (of no hierarchical organization) is eliminated with augmented context-free rules (e.g., definite-clause grammars), at the expense of a much higher computational cost (again, for parsing). The Prolog inference engine constitutes the most straightforward implementation of a DCG parser but, being based on a top-down depth-first search strategy, it is not efficient. Bottom-up, head-first search chart parsers should be preferred in most cases. The potential power of DCGs for syntactic-prosodic parsing has not been fully exploited yet, as no satisfactory syntactic-prosodic mapping is available.

3. Although they have quite a long history, classification and regression trees (and related techniques) have only recently been used for TTS synthesis, given the availability of very large labeled-text corpora. They have proved very useful for tasks in which a decision *a priori* depends on many heterogenous factors in such a way that hand-derived rules are hard to write even for human experts. Such tasks include the detection of prosodic phrase boundaries, the disambiguation of homographs, or automatic phonetization. When compared to other corpus-based approaches, they have the advantages of readability, and of low memory requirements. Being based on a greedy inference principle, however, they may fail to detect analysis shortcuts.

4. Multilayer perceptrons used as classifiers can also be used to take contextual decisions. After the network has been trained, which is a highly computationally expensive operation, all its knowledge is concentrated in its synaptic weights. Hence the amount of memory needed to store it depends on the network size. It is typically higher than with CARTs but lower than with corresponding *n*-grams. MLPs, however, should not be seen as the solution to all problems that they have often been taken for, as revealed by their low phonetization score.

---

[1]Notice that neither Table 11.1 nor Table 11.2 mentions automatic prosody generation from syntactic-prosodic descriptions (the bottom subblock of Fig. 3.6). This task, indeed, lies at the border between symbolic and numerical processing.

		EXPERT SYSTEMS		CORPUS - BASED METHODS		
		Regular rewrite rules	**Augmented CF rules**	**CARTs and related techniques**	**MLPs**	**HMMs**
T	**Preprocessing**	Can do most of the job	-	For sentence end detection	For sentence end detection	For sentence end detection
A	**Morphological analysis**	Can do most of the job	Interesting for highly compound languages	-	-	-
S	**Contextual analysis**	Constraint grammars (95-97 %)	-	Decision lists (97% for homograph disambiguation)	*N*-word contextual window (95%)	Bi- and trigrams (95%)
K	**Syntactic-prosodic parsing**	Hand-derived heuristics (limited but very simple)	DCGs +syntactic-prosodic mapping	Decision trees (90% on phrase boundaries)	-	-
S	**Phonetization**	Efficient LTS or morphophonemic rules (95-100%)	-	Tries (84% on words) also for phonetic postprocessing	*N*-character contextual window (50% on words)	-
	Readability	Good	Very good	Fair	None	Low
C	**Computational cost (parsing)**	Low	High	Low	Low, fixed	Low, fixed
O S T	**Human/ computational cost (inference)**	High (except for heuristics)	High	High, but automatic	Very high, but automatic	Very high, but automatic
S	**Memory at synthesis time**	Lots of rules, but low memory	Very low	Low	Synaptic weights: high	Emission and transition matrices: can be very high

Table 11.1. A summary of the use and interest of some NLP techniques in the context of TTS synthesis. Note: percentages should be compared with care, as they have generally not been obtained under similar test conditions.

5. Last but not least, HMMs have made a noticeable impact in TTS synthesis in the form of bi- and trigrams. They are now being implemented in most state-of-the-art systems, given the high contextual disambiguation they provide combined with a low computational cost at synthesis time.

Notice, though, that data preparation is essential for the inference stage of almost all corpus-based methods. Even complex inference algorithms, do not do well with raw information. CARTs should be supplied with the best contextual factors available, MLP inputs and outputs should be coded so as to emphasize similarities in their values when they exist, and *n*-grams require a careful selection of the set of possible parts-of-speech for a language. Hence, automatic inference involves a lot of trial and error, which is a function of the researcher's own experience: after all, the really intelligent part of the process is still left to humans, as is strictly the case for expert systems.

11.1.2. Digital signal processing and TTS synthesis

Table 11.2 summarizes the features of most of the speech models mentioned in Part Two, regarding analysis, coding, prosody matching, concatenation, and synthesis capabilities, as well as intelligibility and naturalness, in the context of TTS synthesis by concatenation. This time, we shall browse it line by line, since each model has already been presented separately. Table 11.2 reads as follows:

1. TD- and LP-PSOLA are somewhat inferior to other algorithms regarding the availability of automatic analysis procedures, a key point for rapidly developing multilingual TTS systems.[2] Among models for which automatic analysis algorithms exist, the glottal autoregressive model quickly leads to the most complex analysis, as a function of the closeness of glottal waveform models to genuine glottal volume velocity. It is followed by hybrid H/S and MBR-PSOLA (or MBROLA), and by AR and LPC-PSOLA. Analysis complexity, however, is of little importance here, as it is performed once and for all, off-line.

2. Prosody matching gives good results with all parametric models. Yet none of them accounts for the pitch-to-formant dependency encountered in natural speech, although they theoretically could. Even MBR-PSOLA and TD-PSOLA merely provide fair prosody matching performances. Indeed, since the spectrum of OLA frames only approximates the spectral envelope of the sound they have been extracted from, varying pitch with the PSOLA algorithm inevitably introduces a small spectral distortion. This has been verified experimentally by Macchi *et al.* (1993), on the basis of intelligibility tests. Moreover, it has been reported by Boeffard and Violaro (1994) that pitch modification of continuous speech with a hybrid H/S model is preferred over its PSOLA counterpart under severe pitch modifications.

---

[2]Remember, however, that automatic pitch marking can be accepted, provided phase equalization is performed at synthesis time when concatenating segments (as in the PHRITTS system), which highly increases computational load.

	AR	Glottal AR	hybrid H/S	TD-PSOLA	MBR-PSOLA	LP-PSOLA
Analysis	Automatic, easy	Automatic, but possibly complex	Automatic; careful design needed	Semi-automatic (pitch marking)	Automatic; careful design needed	semi-automatic (pitch marking)
Coding (bit rate + database size for 5 min of speech at 16 kHz)	≈4 kb/s ≈100 kb	Depends on the glottal model typ: ≈10 kb/s ≈200 kb	≈10 kb/s ≈200 kb	≈80 kb/s ≈1.7 Mb	Virtually unequaled with MBROLA ≈30 kb/s ≈600Kb	depends on the residual model RELP ≈80 kb/s ≈1.7 Mb
Prosody matching	Good	Very good if it accounts for glottis/vocal tract coupling	Good	Fair	Fair	Fair
Segment concatenation	Linear smoothing of LARs or LSPs ≈ natural formant transitions	Linear smoothing of GAR parameters cf. AR model	Linear smoothing ≡ fade in / fade out	Poor: pitch, phase, spectral amplitude mismatches	cf. hybrid H/S for voiced sounds and TD-PSOLA for unvoiced ones	cf. AR model
Computational load at synthesis time	70 operations per sample	Not significantly different from AR synthesis	≈100 operations per sample (OLA / IFFT method)	Average: 7 operations per sample	Average: 5 operations per sample (7 including linear smoothing)	cf. AR model
Modeling quality (as revealed by copy synthesis)	Poor	Potentially very good	Very good	Almost perfect (no model)	Very good, cf. hybrid H/S	Almost perfect with RELP
intelligibility	Low	-	High	Almost perfect	Very high	Almost perfect
fluidity	High	-	High	Fair	High	High
naturalness	Low	-	High	High	High	Very high

Table 11.2. A comparison of synthesizers by concatenation based on the autoregressive, glottal autoregressive, hybrid harmonic/stochastic, time domain PSOLA, multiband resynthesis, and linear prediction PSOLA models.

3. As far as segment concatenation capabilities are concerned, which is an important feature of TTS synthesis, AR models (whether glottal or not, and including LP-PSOLA) are theoretically slightly superior to hybrid H/S, which is approximately equivalent to MBR-PSOLA (the difference being that MBR-PSOLA reduces to a simple OLA when chaining unvoiced sounds, for which less concatenation problems are encountered anyway). TD-PSOLA exhibits virtually no segment concatenation capabilities. This is in agreement with the quality assessment tests reported in Dutoit (1994), in which listeners were asked to rate the perceived *fluidity*[3] of speech synthesized with four TTS systems based on the same diphone database and provided with identical inputs, except that it was found that the important intrinsic analysis errors encountered with the AR model happen to conceal its good smoothing properties.

Switching to more economics-based criteria, one notices that

4. An efficient segment database compression algorithm exists for LPC and hybrid H/S synthesizers. MBR-PSOLA has recently been turned into the new MBROLA synthesis technique, which exhibits virtually unequaled data compression capabilities (as compared with other time-domain synthesis techniques) with limited additional computational cost.
5. As a result of the computational complexity of their respective synthesizers, the AR, GAR, hybrid H/S, and LP-PSOLA approaches clearly require a fast chip. In contrast, both TD and MBR-PSOLA run in real time on an Intel386 processor.

Finally, regarding the quality and intelligibility of these models[4] in the context of concatenation-based synthesis, it appears that:

6. Dutoit (1994) has also compared the CVC intelligibility of the above-mentioned four TTS systems and found that TD-PSOLA and MBR-PSOLA have the highest CVC-scores, with a slight advantage for TD-PSOLA. As expected, the hybrid H/S synthesizer is much more intelligible than the AR synthesizer.
7. Regarding naturalness, LP-PSOLA (more specifically, RELP-PSOLA) prevails *de facto*, since it does not imply strong hypotheses on speech signals, thereby avoiding the corresponding intrinsic modeling errors, while it provides efficient segment smoothing. Although TD-PSOLA benefits from similar advantages (it does not make use of any speech model), it cannot avoid pitch, phase, and more important spectral amplitude mismatches. In contrast, MBR-PSOLA introduces some off-line distortion of speech segments, but benefits from its enhanced fluidity. In the tests reported in Dutoit (1994), it was found to be as natural as TD-PSOLA. The hybrid H/S model follows, well ahead the AR model.[5]

---

[3]Fluidity was subjectively defined as "related to the amount of concatenation points that can still be perceived in synthetic utterances."

[4]Except for the glottal AR model which, to our knowledge, has never been applied to concatenative synthesis.

[5] Readers who wish to get a comparative taste of the quality provided by LPC, H/S, TD-PSOLA and MBROLA concatenative synthesizers should consult the URL address :

`http ://tcts.fpms.ac.be/synthesis/modelcmp.html`

11.2. Prospects

Speech synthesis is not a solved problem. If TTS systems now achieve fair *intelligibility* scores, their *quality* remains questionable (at least with respect to the high standards imposed by end-users) for their widespread integration into viable consumer electronics products. Some important issues need to be further addressed, which have only been briefly mentioned in this book:

- Although the latest speech processing techniques summarized in Section 11.1.2 have recently allowed researchers to significantly enhance the segmental quality of state-of-the-art speech synthesizers, it seems that it is still possible to refine the underlying speech models and to fine-tune the related analysis and synthesis algorithms. A good example has been recently given by Stylianou *et al.* (1995), where a pitch-synchronous analysis with a hybrid harmonic/stochastic model is used to obtain very high-quality pitch- and time-scale modifications of continuous speech.
- How to best account for coarticulatory phenomena? In the context of concatenation-based synthesis, this question mostly reduces to the question of how to derive *optimized sets of segments* from speech data (see Nakajima 94 or Wang *et al.* 93, for instance).
- Concatenative synthesis, however, is probably not the best synthesis strategy, at least in its current form. Although time-domain concatenative synthesizers clearly outperform rule-based synthesizers, they make little use of phonetic knowledge. In the long run, speech synthesis research might well take a new direction by truly investigating *corpus-based models of speech segments* (as opposed to corpus-based *instances* of segments) that would learn to provide acoustic data for a given segment in a given context, thereby unifying in some way concatenation- and rule-based synthesis (see Tuerk and Robinson, 1993; Donovan and Woodland, 1995; see also Grau *et al.*, 1993, who provide a speech model suitable for both concatenation- and rule-based synthesis).
- How to best formalize the relationship between syntax, semantics, pragmatics and prosody, and how to derive natural sounding intonation and duration from abstract prosodic patterns? Deriving high-level linguistic data from any text remains an open problem. However, text-to-speech synthesis increasingly involves applications in speech generation, where phrasing and focus information is available. Even "psychological" data can be exploited, thereby allowing TTS systems to be provided with several speaking styles. These issues have only recently been addressed (see Black & Taylor, 1994, for instance).
- A fundamental feature of speech has seldom been taken into consideration by TTS systems: *variability*. Phonemes and prosodic patterns, although abstract units, have a conscious basis, while articulatory gestures, formant tracks, pitch curves, durations and intensity, the only measurable quantities, are not under total control. Yet both aspects of speech are submitted to a particular kind of variability which cannot be confused with randomness in that variations maintain some hidden coherency with each other. Hence, even slightly randomizing abstract choices or adding small random

components to segmental and supra-segmental data merely results in a decrease of the overall naturalness. The high naturalness of the unit selection algorithm proposed by Black and Campbell (1995), and Hunt and Black (1996), for example, is partly due to the fact that they precisely allow pitch and duration values to vary to a certain extent, so as to minimize a unit concatenation cost, but that this variation is constrained by their database.

- How to account for speaker and speaking style effects? As opposed to rule-based synthesis, concatenative TTS synthesizers do not easily lend themselves to changes in voice quality or identity. Typically, they incorporate as many segment databases as speaker voices. A new area of speech processing called *voice conversion* could be an alternative (see the special issue of *Speech Communication*, vol. 16, 1995).
- In many cases, multilingual TTS systems are collections of monolingual ones. This is particularly true for their most language-dependent component, the NLP module. Algorithms and formalisms should increasingly be developed with the goal of more clearly separating language-specific data from the actual tools used to handle them.
- Speech assessment and evaluation is closely related to progress in speech technology. The availability of comparative test data is a key point for developers, customers, and funding authorities. As A.I.C. Monhagan (1994) recently pointed out, however: "We all want someone else to solve the evaluation problem!" TTS systems indeed are so deeply involved in natural language and digital signal processing that the simple definition of a test paradigm is itself an open problem (see van Santen, 1993; Pols and Jekosch, 1994).

To have a chance of successfully tackling these problems, one needs to gain access to many information sources. Designing a high-quality TTS system, indeed, is an increasingly complex task, which typically requires the cooperation of various specialists. As a result of this multidisciplinarity, however, it has become more and more difficult to achieve it solely with public funding. Hence this paradox, denoted by Cole *et al.* (1995): "As for industry funding, the results are not generally in the public domain, and consequently speech research has suffered. In contrast to a decade ago, it is difficult to gain access to a state-of-the-art synthesis system that will allow full control of the parameters necessary for conducting speech research."

Let us try to change this! [6,7]

---

[6] A first step in this direction has recently been taken by Dutoit *et al.* (1996), in the form of the MBROLA Project. The aim of this internet project, initiated by TCTS Lab, Faculté Polytechnique de Mons (Belgium), is to obtain a set of MBROLA speech synthesizers for as many languages and voices as possible, free of use for noncommercial applications. The ultimate goal is to boost up academic research on speech synthesis, and particularly on prosody generation, known as one of the biggest challenges taken up by text-to-speech systems for the years to come. See the MBROLA WWW home page: `http://tcts.fpms.ac.be/synthesis/mbrola.html`

[7] In the same spirit, the Center for Speech Technology Research (CSTR), Edinburgh, has recently made a major contribution to the development of freely available, high quality TTS synthesis for non-commercial purposes. Alan W Black and Paul Taylor's free Festival Speech Synthesis system, indeed, is nothing less than a generic, portable and extensible TTS system which lays the foundations of truly global TTS research. It will undoubtedly be a de facto standard in the field. See the Festival WWW home page:
`http://www.cstr.ed.ac.uk/projects/festival.html`

References

BLACK, A.W., and P.A. TAYLOR, (1994), "Synthesizing Conversational Intonation from a Linguistically Rich Input", *Proceedings of the Second ESCA/IEEE Workshop on Speech Synthesis,* New-Paltz, NY, pp. 175-178.

BLACK, A.W., and N. CAMPBELL (1995), "Optimizing Selection of Units from Speech Databases for Concatenative Synthesis", *Proceedings of Eurospeech 95*, Madrid , pp. 581-584.

BOEFFARD, O., and F. VIOLARO, (1994), "Improving the Robustness of Text-to-Speech Synthesizers for Large Prosodic Variations", *Proceedings of the Second ESCA/IEEE Workshop on Speech Synthesis,* New-Paltz, NY, pp. 111-114.

COLE, R., L. HIRSCHMAN, L. ATLAS, M. BECKMAN, A. BIERMANN, M. BUSH, M. CLEMENTS, J. COHEN, O. GARCIA, B. HANSON, H. HERMENSKY, S. LEVINSON, K. McKEOWN, N. MORGAN, D.G. NOVICK. M. OSTENDORF, S. OVIATT, P. PRICE, H. SILVERMAN, J. SPITZ, A. WAIBEL, C. WEINSTEIN, S. ZAHORIAN, V. ZUE, (1995), "The Challenge of Spoken Language Systems: Research Directions for the Nineties", *IEEE Transactions on Speech and Audio Processing*, vol. 3, n°1, pp. 1-21.

DONOVAN, R.E., and P.C. WOODLAND, (1995), "Automatic Speech Synthesizer Parameter Estimation Using HMMs", *Proceedings of the International Conference on Acoustics, Speech, and Signal Processing 95*, Detroit, pp. 640-643.

DUTOIT , T., (1994), "Text-to-Speech Synthesis: a Comparison of Four Candidate Algorithms", *Proceedings of the International Conference on Acoustics, Speech, and Signal Processing 94*, Adelaïde, vol. I, pp. 565-568.

DUTOIT, T., V. PAGEL, N. PIERRET, O. VAN DER VREKEN, and F. BATAILLE, (1996), "The MBROLA Project: Towards a Set of High-Quality Speech Synthesizers Free of Use for Non-Commercial Purposes", *Proceedings of the International Conference on Spoken Language Processing 96*, Philadelphia.

GRAU, S., C. D'ALESSANDRO, and G. RICHARD, (1993), "A Speech Formant Synthesizer Based on Harmonic+Random Formant-Waveforms Representations", *Proceedings of Eurospeech 93*, Berlin, pp. 1697-1700.

HUNT, A.J., and A.W. BLACK, "Unit Selection in a Concatenative Speech Synthesis System Using a Large Speech Database", *Proceedings of the International Conference on Acoustics, Speech, and Signal Processing 96*, Atlanta.

MACCHI, M., M.J. ALTOM, D. KAHN, S. SINGHAL, and M. SPIEGEL, (1993), "Intelligibility as a Function of Speech Coding Method for Template-Based Speech Synthesis", *Proceedings of Eurospeech 93*, Berlin, pp. 893-896.

NAKAJIMA, S., (1994), "Automatic Synthesis Unit Generation for English Speech Synthesis Based on Multi-Layered Context Oriented Clustering", *Speech Communication*, n°14, pp. 313-324.

POLS, L.C.W., and U. JEKOSCH, (1994), "A Structured Way of Looking at the Performance of Text-to-Speech Systems", *Proceedings of the Second ESCA/IEEE Workshop on Speech Synthesis,* New-Paltz, NY, pp. 203-206.

STYLIANOU, Y., J. LAROCHE, and E. MOULINES, (1995), "High-Quality Speech Modification based on a Harmonic + Noise Model", *Proceedings of Eurospeech 95,* Madrid, pp. 451-454.

TUERK, C., and T. ROBINSON, (1993), "Speech Synthesis Using Artificial Neural Networks Trained on Cepstral Coefficients", *Proceedings of Eurospeech 93*, Berlin, pp. 1713-1716.

VAN SANTEN, J.P.H., (1993), "Perceptual Experiments for Diagnostic Testing of Text-to-Speech Systems", *Computer, Speech and Language*, n°7, pp. 49-100.

WANG, W.J., W.N. CAMPBELL, N. IWAHASHI, and Y. SAGISAKA, (1993), "Tree-Based Unit Sselection for English Speech Synthesis", *Proceedings of the International Conference on Acoustics, Speech, and Signal Processing 93*, vol. II, pp.191-194.

Index

Affix, 79

Allophone, 9

Aphasias, 20

Apraxias, 23

Aprosodias, 22

Ariculation

place of, 7

manner of, 7

ARPA-bet, 107

Assessment, 195-197, 278

Assimilation, 107, 187

ATN, 51, 65, 153

Autocorrelation method, 206, 210

Autoregressive model, 201-202

Bigram, *see* n-gram

Bleeding, 114

Bottom-up, 50

Breadth first search, 47, 72

BSU, 73

B-tree, 84

CART, 157-159, 66, 96, 122, 125, 165

CELP, *see* Codebook

Chart parsing, 50, 66, 153

CF-PSG, *see* Grammars - context free

Coarticulation, 9, 124, 179, 181, 239, 247, 277

COC, *see* Context oriented clustering

Codebook, 220, 268

Consonant cluster simplification, 107

Context oriented clustering, 188, 190

Covariance method, 205

CPA, 107

Cross-validation, 159

Data structures

linear, 59

efficient, 84-85

DCG, *see* Grammars - definite clause

Decision lists, 96

Demisyllable, 188

Depth first search, 47, 72

Diphone, 180, 187, 189

DPCM, 256, 266

Duration, 132-133

generating, 163-165

intrinsic, 164

Elasticity principle, 163

Entropy, 157

Equalization, 183

Error-driven learning, 96

Feature structure, 52, 60, 83

Feeding, 114

Focus, 132, 147-148

Formant, 5

synthesis, 179

Fourier transform

short term, 2, 233-235, 237

fast, 244-246, 267

Poisson formula, 252-253

FS, *see* Feature structure

FSA, 40, 51, 76, 92, 142-144, 150

FSU, 74

FST, 43, 116

FSTN, 40

Function words, 77
Fundamental frequency 2, 132
F0, *see* Fundamental frequency

GAR, 221
GIF, *see* Glottal -inverse filtering
Glissando threshold, 138
Glottal
inverse filtering, 220-225
flow, 5, 202
closure, 221, 257
Graphemic syntax, 117, 119-120
Grammar
constraint, 83, 96-100, 152
context free, 47-52, 66
definite clause, 52, 60, 66-69, 83, 153
local, 95-100
phrase structure, 38
regular, 40-47
unification, 52-54, 153

Harmonic
coding, 220, 233
model, 230
birth, death, 242
re-synthesis, 263
Hash table, 84
Heterophonic homographs, 88, 108-109, 161
Hidden Markov Model, 41,
tagger, 92
training, 45
phonetization, 121
speech segmentation, 192
HMM, *see* Hidden Markov Model

Inference, 38
regular, 45
context free, 49
Intelligibility, 14, 195-197
Intensity, 2, 25, 132
Intermediate phrases, 149
Intonation, 129
acceptably neutral, 148
languages, 145
phrases, 149
IPA, 6, 10, 107, 140
Isochrony, 163

KIMMO, *see* Two-level morphology

LAR, 213-216, 267
Lattice, 211-212, 217
Letter to sound , 107, 115
Lexeme, 78
Lexicon, 52
Logatom, 190
LPC, 203, 231, 238
LP-PSOLA, 267-268
LSP, 213-217
LTS, *see* Letter to sound

MBE, 231, 236-237
MBR-PSOLA, 262-265
MBROLA, 266, 278
Microprosody, 132
MLDS, *see* Multilevel data structure
MLRR, *see* Multilevel rewrite rule
MLP, *see* Multilayer Perceptron
Modeling errors

extrinsic, intrinsic, 179, 192
Morph, *see* Morpheme
Morpheme, 78, 111
Morphology, 78,162
two-level, 83
Morphophonology; 109, 113
two-level, 114
MP-LPC, *see* Multipulse
Multigram, 92
Multilayer Perceptron, 43
tagger, 93
training, 46
phonetization, 123
Multilevel data structure, 60-62, 65
Multipulse, 220
Mutual information, 158

Naturalness, 14, 195-197
N-gram, 90, 96-97
Non-determinism, 39

Open lexicon strategy, 86, 88
Orthogonal, 204, 208, 224, 237
Overlap-add, 244-246, 251

Parallel distributed processing, 18
PARCOR coefficients, 209, 213-217, 232, 240
Parsing, 38
regular, 47
Part-of-speech, 78, 89
PATR, 52
Perceptual center, 164
Periodogram, 235, 237
Phase, 203, 224, 239, 243, 247
mismatch, 257-258
reset, 263, 266
Phenomenological model, 193
Phonation type, 7
Phone, 8
Phoneme, 8
Phonetic
classes, 7
liaisons, 108
post-processing, 125
Phonetization, 14, 67
dictionary-based, 109-110, 126
expert systems, 118-119
rule-based, 109-110, 125
trained systems, 120-122
two-level, 119
see also LTS
PIOLA, 262
Pitch, 131
asynchronous, 203
contour, 140-141
factor, 256
mismatch, 259
synchronous, 253
Polygram, 92
Prediction residual, 203-204
Probability
emission, 41, 93
transition, 41, 93
Production model, 193
Production rule, *see* Rewrite rule
Proper names, 109, 125-126, 162
Prosodemes, 139
Prosody, 129
acoustic stylization, 136-137
generation, 165-169

levels of representation, 130
matching, 184, 194,
LPC, 214
H/S, 238-239
PSOLA, 256-257
meanings of, 132
models, 133-145
perceptual stylization, 137-139
PSG, *see* Grammars - phrase structure
PSOLA, 220, 251-262
Psychoacoustics, 25, 130
Psycholinguistics, 26

Quasi-stationarity, 229, 231, 240, 246, 257
Recording
system, 2, 203
speech corpus, 178, 182
Recurent neural network, 167
RELP, 268
Rewrite rule, 38
multilevel, , 64-66, 69, 83, 114, 119, 157, 161
Root, 79, 109
RTN, 47

SAMPA, 107
Sampling theorem, 2
Schwas, 108
Segmental quality, 14, 179, 187-195
LP model, 218-219
GAR model, 226
TD-PSOLA, 257-261
MBR-PSOLA, 266
Segment concatenation, 186, 195
LPC, 214-216
H/S, 239-241, 246-247
TD-PSOLA, 259-261
MBR-PSOLA, 263-265
SPE, 114
Speaking style, 277
Spectral enveloppe, 5
Spectrogram, 5
Stem, 79
STFT, *see* Fourier transform - short term
Stress, 131, 160-163
fixed, free, 145
Syllabic length, 131
Synthesis
articulatory, 193
by-concatenation, 180-195
by-rule, 178-180
copy, 134, 192, 246

Tagging, 89
Target theory, 24
ToBI, 143
Toeplitz, 210-211
Tonal segment, 137
Tone 142
language, 145
sequence, 142-144
Tonetics, 140
Top-down, 50
Transduction, 38
two-level, 83
Trie, 84, 122
Trigram, *see* N-gram
Triphone, 188

Variability, 277
Vocal cords, 5

Vocoder, 194, 237
Vowel harmony, 107
V/UV ratio, 231

Text, Speech and Language Technology

1. H. Bunt and M. Tomita (eds.): *Recent Advances in Parsing Technology*. 1996 ISBN 0-7923-4152-X
2. S. Young and G. Bloothooft (eds.): *Corpus-Based Methods in Language anc Speech Processing*. 1997 ISBN 0-7923-4463-4
3. T. Dutoit: *An Introduction to Text-to-Speech Synthesis*. 1997 ISBN 0-7923-4498-7

KLUWER ACADEMIC PUBLISHERS – DORDRECHT / BOSTON / LONDON